Parents' Guide to
Common Core Arithmetic:
How to Help Your Child, 3rd ed.

Herbert S. Gaskill, Ph.D.
and
Catherine M. Gaskill, B.A.

©2014

Contents

iv

This book is dedicated to the children who need to learn arithmetic and to the parents willing to help.

Prefaces

The central expectation of every parent when their child graduates from high school is:

My child is ready to take on the world.

What we mean by this is: my child is ready to succeed in college; or, my child is ready to succeed in the job market.

For large numbers of children in North America, this expectation is not fulfilled because the level of educational achievement, particularly in respect to basic arithmetic, is inadequate (see §1.3).[1]

Based on 40 years of experience teaching post-secondary mathematics, the first author concludes that the essential educational failure to teach our children what they need to succeed in the future occurs during the primary and elementary grades. This conclusion is supported by data available at the National Center for Educational Statistics and by internal test data collected by some Canadian provinces showing large numbers of students fail to meet minimal competency requirements (see §1.5).

This book was written to help parents support and encourage their children in the critical study of arithmetic.

Acknowledgements

The authors particularly thank Dr. John Baldwin for pointing them in the direction of the CCSS-M. We appreciate the support of Dr. Chris Radford, the Head of the Department of Mathematics and Statistics at Memorial. Various folks made incredibly helpful comments during the preparation of the manuscript. We thank Richard Ellis, Matthew Gaskill, Jane Mckay and Heather Gaskill.

Herbert Gaskill
Memorial University of Newfoundland
gaskillmath@gmail.com

---

[1] Most of the data discussed is associated with the US. However, as reported on the Huffington Post Canada website, some 27% of Canadian university graduates are functionally illiterate. *Shocking Number Of Canadian University Grads Don't Hit Basic Literacy Benchmark*, **The Huffington Post Canada, Posted: 04/29/2014**

As a person with little training in math, my job here has been to comment on how understandable the text is. In the process I have come to appreciate how understandable arithmetic is.

I know how busy and fast moving parents are but here is a chance to slow down with a clean sheet of paper and pencil for a completely different kind of adventure. This material demands thoughtful reading but the very precision of it is what's fun. There is a clear reason for everything, no ambiguities: you can figure it out. Not only that but you can also share this adventure with your child for his/her certain benefit.

Enjoy!

Catherine Gaskill
St John's, Newfoundland

I cannot put out a second edition of this book with out mentioning *The Public School Advantage* by C.A. and S.T. Lubienski (available from Amazon). This book provides a careful examination of the key testing data that has been used fallaciously to support the contention that *private is better*. **This contention is simply wrong** and parents contemplating spending big bucks on a private school should be aware of the facts before going down that road.

Parents should also be aware of the www.parcconline.org/parcc-assessment website. It has practice tests that provide additional insight into the CCSS-M.

Herbert Gaskill
Memorial University of Newfoundland
July, 2014

This past August research documenting the importance of rote learning of math was published that should effect how we teach math to every child in North America. The information can be found at two websites:

http://www.medicaldaily.com/math-skills-childhood-can-permanently-affect-brain-formation-later-life-298516
and
http://www.nature.com/neuro/journal/vaop/ncurrent/full/nn.3788.html
This work with be discussed at several places in the book.

Herbert Gaskill
Memorial University of Newfoundland
October, 2014

Chapter 1

Common Core Arithmetic and Your Child

The new **Common Core State Standards** (CCSS) defining grade-by-grade goals for mathematics achievement have been adopted by forty-five states, four territories, and the District of Columbia, so the chances are very good your child will be required to meet the new higher standards.[1] This book exists to help parents ensure that their children are able to meet the new higher standards.

In respect to the mathematics standards, the front page of the National Council of Teachers of Mathematics (NCTM) website[2] proclaims

> Fifteen presidents of the professional societies that make up the Conference Board of the Mathematical Sciences,[3] including NCTM President Linda Gojak, have signed a statement of strong support for the Common Core State Standards for Mathematics. The statement calls the Common Core State Standards *an auspicious advance in mathematics education.*[4]

The adoption of the new standards could be the best thing that ever happened to your child's educational experience in mathematics. Or, as the following quotation describing student outcomes shows, there could be cause for serious concern:

> **One-half of all high school graduates will take at least one remedial course in college (most often in mathematics). Fewer**

---

[1] For exact current information consult the website: http://www.corestandards.org/in-the-states

[2] See http://www.nctm.org/

[3] The CBMS includes all the major American professional societies in mathematics. A list of member societies may be found at: http://www.cbmsweb.org

[4] See more at: http://www.nctm.org/#sthash.myW3P67a.dpuf

than one-fourth of these students will earn any postsecondary credential.

Uri Treisman, *Iris M Carl Equity Address: Keeping Our Eyes on the Prize*, NCTM, Denver, April 19, 2013.[5] (Hereafter, UT 2013.)

To every parent and every child, the most important fact is that the child completes each stage of the educational process ready for the next. That half our children on completion of high school are **not ready for post-secondary** is a tragedy for those so affected. But many children are not college bound, so we should explore what this means generally.

1.1 What Do We Mean by: Ready to Go?

A three-year study produced by the National Center on Education and the Economy (NCEE) identifies high school graduates who **are able to successfully complete the first year of a community college program as being college and work ready**.[6] [7] This study, reported on by Marc Tucker who is president and CEO of the NCEE, found that students meeting this criteria are able to complete their two-year program on time, and either, successfully transfer to a four-year college, or proceed to their chosen career. Since the students in the study represent about half of high school graduates, it is a reasonable sample on which to base policy analysis.

From the perspective of a parent, knowing your high school graduate is college and work-ready would be satisfying. The alternative leads to a future of broken dreams and excessive debt associated with additional years of schooling that is seldom successful (see UT 2013). No parent would wish this for their child. Moreover, I can tell you as an educator, to be a young adult confronting this situation is devastating.

It may be thought that this problem is faced only by those less privileged who have attended public schools. But as reported in the New York Times, a recent study on ACT college admissions' test results found that

> ... **only a fourth of those tested were ready for college**. And that was among motivated students who want to go to college, from all sorts of schools, not just public school students.[8]

---

[5]The address in a variety of formats can be found at: http://www.nctm.org

[6]See http://www.ncee.org/marcs-blogs/tuckers-blogs-archives/#MAY13 *What Does it Really Mean to be College and Work Ready?* Marc Tucker, May 2013. (MT 2013)

[7]For an alternate definition based on SAT scores,
see http://www.huffingtonpost.com/2013/09/26/sat-results-2013_n_3991523.html

[8]See *The Common Core and the Common Good*, Charles M. Blow, NYT, 21 August, 2013. (CMB 2013)

At this point we have a clear statement of what it means to be **Ready to Go**. We also know that at least half of students are not ready to go. What we don't know in any real sense is Why?

1.2 Why Most High School Graduates Are Not College- and Work-Ready

Again we turn to Tucker's blog on the study by the NCEE on the meaning of college- and work-ready. The comments that are particularly relevant here are the following:

> Most of us take it more or less for granted that as a student progresses through the grades, that student does 8th grade work in 8th grade, 9th grade work in the 9th grade, and so on until, in the first year of community college, that student is doing 13th grade work. But it seems that that is not the case at all. A very large fraction of 12th graders leave the 12th grade to do 8th or 9th grade work in community college. And that is not the end of it, because about a third of our high school graduates show up at the community college unable to do work at the 8th or 9th grade level. Many of the rest, apparently, those who are admitted to credit-bearing courses at their community college, have only the shakiest command of 8th and 9th grade mathematics, reading and writing. (See MT 2013.)

The data underlying these comments show there are achievement deficits in mathematics, reading and writing evident by Grade 9. To consider things prior to Grade 9, we examine scores on the National Assessment of Educational Progress test instruments in mathematics.[9] According to data at their website, the average score on this assessment of mathematics achievement by Grade 8 students in 2011 was 284. The score that represents proficiency is 299 (See UT 2013). According to data at the NCES site, only 35% of all Grade 8 students are proficient.[10] Again, the reader may be tempted to conclude that this result is because the sample includes **all** students. But even when the sample is restricted to private school students, more than half the students fail to achieve the proficiency score.

We can take one further step back because similar assessments are done in Grade 4.[11] In Grade 4, 40% of all students were found to be proficient in 2011. Performance

---

[9]This data is available from the National Center for Educational Statistics (NCES) website: http://nationsreportcard.gov/math_2011/.

[10]See http://nationsreportcard.gov/math_2011/summary.aspx and click on Grade 8 in **Proficient** paragraph.

[11]See http://nationsreportcard.gov/math_2011/summary.aspx and click on Grade 4 in **Proficient** paragraph.

by students at private schools was better, but still more than half are not proficient.[12]

Some may believe that the reason student performance appears dismal is due to the difficult nature of the test questions and that only mathematically gifted students could be expected to perform well. To address this possibility, the reader should look at some test questions. While complete test instruments are not available, sample questions can be found for all test levels at the NAEP website.[13] The questions demand little more than recall and in my opinion these questions are so basic that every child ought to be able to answer all questions correctly.

In briefest summary, the data shows a majority of students begin accumulating knowledge deficits in mathematics by Grade 4 and this accumulation continues throughout the school career until graduation from high school. At this point, the knowledge deficits confront realities of college entrance and job markets that are unforgiving and demand competence.

1.3 What Does This Mean to Parents and Students?

Because this information has such profound consequences for parents and students, we review it again. Sixty percent of Grade 4 students are not proficient in math. By Grade 8, the number has reached 65%. And because of the nature of mathematics as a discipline, it would be expected that the number of students deemed proficient would continue to decrease as students progress through the system. This accounts for at least part of why the ACT study shows only 25% of students tested were **college-ready** (CMB 2013).

Finally, we can deduce one last thing. Students who are not proficient are for the most part being passed on in spite of their deficiencies. How do we know this? Because if failure rates in the general school system reflected this reality, there would be universal outrage on the part of affected parents and demands that the system respond.

The above data establishes the negative consequences of a lack of proficiency at math. But are there benefits associated with proficiency?

---

[12]One of the ideas current in respect to education is that if only we involve the private sector, the education system will dramatically improve. Hence the charter school movement. Before a parent jumps on this band wagon, or indeed invests in any private school, I recommend you read Chapters 4 and 5 of: **The Public School Advantage** by C. & S. Lubienski, available in paperback from Amazon.

[13]See http://nationsreportcard.gov/ltt_2012/sample_quest_math.aspx

1.3.1 What Do Studies Show About Math Skills and Success?

Given the prevalence of mathematics in our culture, we should expect that analysis of data on math achievement and subsequent success in various forms are linked. That is, we would expect more math skills to be associated with more success. Here are some of the things we know about young people and adults that have high achievement levels in math on completion of high school:[14]

1. students with high achievement in math are more likely to go to college;

2. students with high achievement in math are more likely to succeed in post-secondary;

3. students with high achievement in math who choose not to go to college have an easier time finding their first job;

4. students with high achievement in math who choose not to go to college, should they become unemployed, spend shorter times being unemployed;

5. students with high achievement in math earn higher salaries;

6. students with low salaries show low math achievement;

7. the effect of high math achievement is independent of the type of school system - public/private;

8. the effect of high math achievement significantly reduces the effects associated with low economic status of the child.

It seems incredible that one academic subject which appears to have such a small place in actual work activities could have such profound consequences on an individual's economic success. How could this be so?

How children learn and how their brains develop is the subject of profound research efforts. One aspect of this research is how children learn to solve problems effectively. A 2014 study shows learning math to the level of recall creates internal changes in a child's brain that produce enhanced problem-solving skills that persist into adulthood.[15] For this reason there are simple things every parent can do as their child proceeds through elementary school to ensure their child reaps the benefits identified above and we discuss these below.

---

[14]This information can be found in the report: *Mathematics Equals Opportunity* prepared for Richard W. Riley, U.S. Secretary of Education, 1997. See http://mathpl.us/docs/mathemat.pdf (MEO).

[15]http://www.nature.com/neuro/journal/vaop/ncurrent/full/nn.3788.html

1.4   The Motivation for Common Core State Standards in Mathematics

Beginning in 2000 and every three years since, the OECD[16] has conducted a program of testing to evaluate the education systems of individual countries on a comparative basis.[17] The testing series was given the acronym PISA which stands for: Programme for International Student Assessment. Perhaps the most objective between-country comparisons arising from the PISA data sets are mathematics related because essentially identical questions can be administered in every country. For this reason, when the US was ranked 25 th in the PISA results in mathematics (UT 2013), it was viewed by the education establishment as a matter of great concern. Indeed, this ranking led to the conclusion that to maintain its international competitiveness, the US would have to substantially raise standards in its school mathematics programs. In short, a tipping point had been reached.[18]

As a result, a process was put in place to identify mathematics performance goals that would return the US to a place of superior competitiveness. These goals became the CCSS-M and here is some of what the developers wrote about their task:

> For over a decade, research studies of mathematics education in high-performing countries have pointed to the conclusion that the mathematics curriculum in the United States must become substantially more focused and coherent in order to improve mathematics achievement in this country. To deliver on the promise of common standards, the standards must address the problem of a curriculum that is a mile wide and an inch deep. These Standards are a substantial answer to that challenge.
>
> CCSS for MATHEMATICS, p. 3 (CCSS-M).[19]

The criticism *mile wide and an inch deep* applied to North American school mathematics curricula is a familiar one. It has been repeated about each revision of the school mathematics curriculum in the US and Canada for the past 20 years. But what does it mean?

---

[16]OECD is the acronym for the Organization for Economic Co-operation and Development.

[17]For more information visit: http://www.oecd.org/pisa/home/.

[18]For a more complete description of what led to the standards go to: http://www.corestandards.org/about-the-standards/development-process/.

[19]The CCSS-M document is available at http://www.corestandards.org/Math and can be obtained by anyone.

1.5 Curricula and Your Child

The CCSS-M is intended to provide standards that will lead to a **coherent and focused** curriculum. In order to explain why this is important in a fashion useful to parents, we need to briefly review some of the key facts derived from comparative studies of school curricula in mathematics. Much of what we have to say is extracted from studies performed by researchers located at the Education Policy Center (EPC) at Michigan State University.[20]

1.5.1 Teachers or Curricula?

One finding by the EPC group based on a TIMSS[21] data set collected in 1997, was that the education systems in various countries have a similar structure. At the top is a department of education which specifies what is to be taught through setting an **intended** curriculum. This department may be nation-wide in a small country like Singapore, or state- or province-wide in a large countries like the U.S and Canada.

While there may be some intermediates, for example a county school board, the delivery system for the intended curricula is the teacher. The curricula that is actually taught in classrooms and learned by students is called the **enacted** curricula, and if you think about it, what is actually taught and learned in a particular classroom may, or may not, closely reflect what is intended.

Thus when you consider the situation in respect to your child, what your child learns has two components: the teacher (as presenter) and the enacted curriculum (as topics presented).

Here is a quick summary of what parents ought to know about these two components as reported in a paper by Schmidt, *et al.* in 2002:[22]

1. what students learn from the enacted curricula is what is in the intended curriculum;

2. some intended curricula are much more effective than others in respect to measurable student outcomes;

3. effective intended curricula have common features in respect to number of topics presented each year, order and duration of presentation, etc.;

4. successful intended curricula are narrowly focused and coherent;

---

[20]For more information visit: http://education.msu.edu/epc/.

[21]TIMSS is the acronym for the Third International Math and Science Study.

[22]*A Coherent Curriculum*, Wm. Schmidt, H. Houang and L. Cogan, **American Educator**, Summer 2002, available on-line. Hereafter, ACC.

5. North American intended curricula extant in 2002 tended to be ineffective in respect to outcomes precisely because they lacked focus and coherence, hence they were termed *mile-wide and inch-deep.*

The first thing to understand about these conclusions is they are universal and apply to all school systems operating under the auspices of a Department of Education; this would include all public schools but not necessarily private or charter schools.[23]

Thus, your child's teacher is teaching your child what is in the intended curriculum set out by your state's Department of Education, and that teacher is likely to be doing a good job of teaching that curriculum as measured by tests. That's the upside.

The downside is that some curricula are much less effective in producing positive student outcomes, so that even though the child has a great teacher, the constraints imposed by an ineffective intended curriculum lead to reduced learning outcomes.[24] In short, the curriculum set out by your state's Department of Education really does matter when it comes to your child's long-term success because that is what your child's teacher is going to teach. For this reason, it is important for you to have a fair idea of how to recognize an effective curriculum.

1.5.2   A Model Coherent Curriculum

Before proceeding further with this discussion we need to make clear what the difference is between coherent curriculum and a typical mile-wide North American curriculum in terms of student outcomes. Arriving at such an assessment starts with asking students from different countries to perform the same, or equivalent, tasks - this is easy to do in a discipline like mathematics.

Within a country students' results are rank ordered by percentile. Thus, if your child is ranked in the 100 th percentile in the US, it means no US student answered more questions correctly than your child. If your child was ranked in the 75 th percentile in Canada, it means 25 % of Canadian students answered more questions correctly, while 74 % answered less correctly. Finally, if your child was ranked in the 25 th percentile it means that 75 % of students answered more questions correctly. Percentile rankings provide no information as to how many actual questions were correctly answered by any student. They only indicate how a particular test score by one child ranks relative to the test scores generated by all children in the given country. Percentile data in one form or another is generally what is available on public websites.

---

[23]Here again we refer readers to Chapters 4 and 5 of **The Public School Advantage**.

[24]The EPC surveyed teachers on exactly this point and found the principal determinant of what that teacher did was what was in the intended curriculum. More information can be found at the EPC website in Working Paper 33.

Countries can also be rank ordered based on the relative performance of that countries' students in comparison to other countries — recall, the US was ranked 25 th on PISA (UT 2013).[25] Within each country there is a percentile/raw-score table which informs researchers exactly how many correct answers were required to achieve a given percentile ranking. Because the same test instrument is used by each country, the table can be used to determine exactly what an individual raw score in one country means as a percentile in any other country. Thus for example, the researchers at EPC found that the raw score of a US student in the 75 th percentile was ranked below the 25 th percentile in Singapore (ACC). Another way to think of this is that 75 % of students being taught under Singapore's A+ curriculum achieve at levels that only 25 % of US students achieve at using a US curricula! This is the functional difference in individual student outcomes between learning from a coherent curriculum and the typical mile-wide curricula prior to CCSS-M. This difference is the reason why parents need to understand exactly what constitutes an A+ curriculum. So let's get back to that task.

As noted above, Schmidt **et al.** found that effective mathematics school curricula have a number of common features. The table that follows is derived from the effective curricula identified by Schmidt **et al.** in their 2002 paper (ACC) and that they term A+. The A+ curricula illustrates the features that make effective curricula effective.

Our table faithfully reproduces the data reported in ACC for A+ curricula but has been truncated to omit the data from Grades 7 & 8 because our focus is elementary school. As well, our table has been rearranged so that like topics are presented as a group, e.g., numbers and operations on numbers are presented together.

Solid circles (•) indicates that the topic at the left was taught in the grade at the top by 100% of high-performing countries, i.e., countries that met Schmidt's criteria for being A+ (ACC). A blank indicates the topic was taught by fewer than two thirds of the A+ countries; most likely, the topic was not taught. A careful read of the Schmidt study leads to the conclusion that what is not taught is as important as what is. In other words, more is not better. Rather, fewer, carefully selected topics need to be taught in a concentrated fashion so that students learn to mastery and retain what they have learned over the long-term.

Focus for a moment on the first group of topics in the table related to numbers and operations. Notice that within this group each consecutive line of the table represents more difficulty and/or sophistication. For example, the first line of the table is concerned with the meaning of whole numbers and reflects the fact that the computational procedures of arithmetic depend on the place-value system of notation. The intention is for children to learn to count and through this process learn about

---

[25]This is not a percentile. It simply means that the students of 24 other countries were judged to perform better on this series of tests.

the Arabic (place-value) notation system (see §2.6.1). On the next line the topic is whole number operations. The first operation studied here is single-digit addition for which there is a finger counting procedure that is based on counting. More advanced topics like two-column addition are founded on the simpler topic combined with a deeper knowledge of place-value.

TOPIC & GRADE:	1	2	3	4	5	6
Whole Number Meaning	●	●	●	○	○	
Whole Number Operations	●	●	●	○		
Common Fractions			□	●	●	○
Decimal Fractions				○	●	○
Relationship of Common & Decimal Fractions				○	●	○
Percentages					○	○
Negative Numbers, Integers & Their Properties						□
Rounding & Significant Figures				○	○	
Estimating Computations				○	○	○
Estimating Quantity & Size				□	□	

TOPIC & GRADE:	1	2	3	4	5	6
Equations & Formulas			□	○	○	○
Properties of Whole Number Operations				□	○	
Properties of Common & Decimal Fractions					○	○
Proportionality Concepts					○	○
Proportionality Problems					○	○

TOPIC & GRADE:	1	2	3	4	5	6
Measurement Units	□	●	●	●	●	●
2-D Geometry: Basics			□	○	○	○
Polygons & Circles				○	○	○
Perimeter, Area & Volume				○	○	○
2-D Coordinate Geometry					○	○
Geometry: Transformations						○

TOPIC & GRADE:	1	2	3	4	5	6
Data Representation & Analysis			□	□	○	○

This table is adapted from the A+ curricula identified by ACC. Topics have been reorganized to reflect domains identified in the CCSS-M and only grades 1-6 are shown. Topics identified with a ● are in the intended curricula of all the A+ countries in the grade shown; ○ identify 80% of A+ countries; □ identify 67% of A+ countries. Topics not on this list are **not** in the intended curricula of A+ countries in Grades 1-6!

A+ countries expect children to have mastered operations with whole numbers by the end of Grade 4.[26] Indeed, 20 % of the A+ curricula complete the learning of operations by the end of Grade 3. To ensure children succeed at this, A+ curricula limit the topics taught in Grades 1 and 2 to whole numbers, addition and subtraction and making measurements, as with a ruler.

In A+ countries there is little or no study of common fractions before Grade 3, and a full third of the A+ curricula delay any discussion of fractions until Grade 4. In making this decision, A+ countries recognize the children must master operations with whole numbers first because operations with fractions utilize whole-number operations.[27] Again, introducing more topics in earlier grades does not produce better results.

Another feature of the A+ curricula is the avoidance of topics that derive from algebra and data analysis in the early grades. Rather, the entire focus in Grades 1-3 is on number, operations with numbers and measurements related to the geometry of the line. The reader may wonder, if students are dealing with fewer topics, why don't they learn less? The answer is that students learning from an A+ curriculum are expected to have a more extensive knowledge of what they do learn. For example, the place-value system of notation (§5.1.2) connects to our system of computations (§6.4) and A+ curricula expect that students will come to terms with these relationships.

We can summarize the import of this section for parents as follows: the best way to help your child is to provide learning support in respect to the limited number of topics identified in the A+ curriculum. Thus, for example, in Grades 1 and 2 the issue for your child is coming to terms with the number system and how to add and subtract. In Grade 3, the focus is on multiplication and extending the knowledge of the number system, particularly in respect to place value. In the long run, ensuring your child learns the short list of topics well is what will provide the best foundation leading to success in the workplace and/or post-secondary.[28]

---

[26]My memory of my own public schooling is that we had finished long division by the end of Grade 4.

[27]Ideally, one would like children working with fractions to deal with whole-number operations on a recall basis as discussed in §1.7.1. To see exactly why, see §14.3.3.

[28]Recently, a friend asked my advice about a child who was having trouble with the math taught at the start of Grade 4. I told him to make sure his child learned her addition and times tables well. At the end of the year his child got the prize as the best math student. The point is: small amounts of knowledge learned really well go a long way in arithmetic. There is also huge gratification to a child who knows they know the answer.

1.6 Assessing Your Child's Progress

The learning target for children in an A+ curriculum is that while only a few critical topics are taught, these topics are learned to the point of **mastery**. The teaching/learning process must account for the fact that children are unique in respect to their learning rates. To achieve success, what is absolutely essential to every parent, child and teacher is knowing whether that child is successfully mastering the required curriculum elements on a continuing basis so that prompt action can be taken where and when necessary. Clearly, the only way to make this determination is through some form of testing. Since testing is an area of sensitivity, we discuss it in some detail.

1.6.1 The Nature of the 3 R's

As we all know, in primary and elementary grades children learn three critical life skills, reading, writing and arithmetic. This learning is of a profoundly different nature than much of that which occurs in most other subjects. The difference is that these subjects all involve

doing.[29]

Consider what it means to learn to read. At a successful conclusion of learning to read, we expect a child to be able to read and understand written material they have never seen before. In other words, they can **do reading**.

Similarly, for a child who has learned how to add whole numbers, we expect to be able to give the child a sum to perform and they should be able to correctly find the sum, even if they have never seen that particular sum before.[30] This is what it means to be able to **do addition**. And, we could make the same observation about writing.

While all subjects involve some level of doing, none are so substantively about doing as mathematics, and while this type of learning continues at all levels, it is most intense in primary and elementary. What is critical to realize here is that **doing** involves **new behaviors** and successfully acquiring new behaviors requires practice. In this sense, learning the behaviors required for reading, writing and doing arithmetic are just like learning the skills required to succeed at sports or playing a musical instrument. Helping children get the required practice is an area where parents can provide critical support.

To succeed children must have feedback and respond to that feedback in effective ways which is what we consider next.

---

[29]Learning a new language would be a further example of a subject that requires **doing**.

[30]Fluidly execute the standard algorithm in the words of the CCSS-M.

1.6.2 Formative Assessment and Your Child

The importance of assessment/feedback as a component of the CCSS is described in a position paper on formative assessment at the NCTM website.[31] In particular, the paper recommends:

1. The provision of effective feedback to students

2. The active involvement of students in their own learning

3. The adjustment of teaching, taking into account the results of the assessment

4. The recognition of the profound influence that assessment has on the motivation and self-esteem of students, both of which are crucial influences on learning

5. The need for students to be able to assess themselves and understand how to improve

(quoted from NCTM-FA).

The purpose of formative assessment is that students would immediately know whether they have successfully acquired a body of material and, if not, that information would be available in a timely fashion enabling an immediate response to correct deficits. Without doubt, using assessments to correct deficits in the manner described is a desirable outcome for student, parent and teacher and would prevent the kind of knowledge deficits prevalent by Grade 4 under the current system.

1.6.3 How the Process Works

The implied theory of formative assessment as described in the five points involves three steps: teachers will identify difficulties, communicate those difficulties in a suitable manner to learners, and the learners will take responsibility for fixing the problem. Consider the following example from Grade 2.

Step 1

A child takes an addition test consisting of five sums of two digit numbers and gets the first two questions right and the last three wrong. Specifically, the child gets this, and similar problems right:

---

[31]See **Formative Assessment** A position of the NCTM, found at: www.nctm.org/uploadedFiles/About_NCTM/Position_Statements/Formative%20Assessment1.pdf (NCTM-FA).

$$\begin{array}{r} 23 \\ +45 \\ \hline \end{array}$$

but gets this and similar problems wrong:

$$\begin{array}{r} 27 \\ +45 \\ \hline \end{array}$$

The fact that the child gets the first and similar problems correct suggests two things to the teacher: the child knows all single digit sums less than 10, and the child knows the basics of the standard algorithm for addition (see §6.4).

The fact that the child gets problems of the second type wrong suggests a problem either with single digit sums more than 10, or with the carrying procedure. To determine which, the teacher considers the answers given. For example, here are three possible wrong answers: 73, 62 and 61. Compared with the correct answer, 72, the first answer is wrong in the *ones* place but correct in the *tens* place. Thus, the child is able to correctly execute carrying; but the error suggests the student believes $7 + 5 = 13$. The second answer, 62, suggests that the student has an incomplete knowledge of place value and carrying while the third answer suggests the student has problems with single digit sums that have a two digit answer and also has difficulty carrying. Confirming the exact source of each error may require asking the child:

Exactly how did you arrive at this answer?

In any case, at the conclusion of evaluating the child's responses, the teacher should have a fair idea of what the child knows, what the child doesn't know and where more practice is needed.

Step 2

Communicating test results to the students and their parents is complicated because these results are loaded with success/failure connotations.[32] Point 4 above recognizes that discussing test results with students/parents is delicate. We all need to recognize that the purpose of testing is to determine those areas in which the student needs more practice and to act on test results accordingly. Testing is not about finding fault or failure. It is about identifying future required action.

---

[32]Test results are, and have been, wrongly used to make judgements about student abilities, teacher abilities, school quality, and the list goes on. We will keep our focus on the child.

Step 3

The intended response to test results is that the child will undertake sufficient additional practice in the identified area to correct the difficulty. This step is critical because once a wrong fact becomes fixed in memory, e.g. $7 + 5 = 13$, it becomes almost impossible to unlearn.

For a successful response to be effected, all parties, teachers, parents and students, must believe that essentially every child has the intellectual capacity to learn arithmetic to the CCSS-M standard. Without such a belief, one of the parties will subvert the learning process and the only party that can conceivably be excused is the child who may be frustrated by the need for additional work. The point is that all must believe that success can be achieved.

The reason I say that all children can achieve at the required level is that the skills of arithmetic that the CCSS-M expect children to master are algorithmic. By this I mean there is a fixed solution procedure which may also require knowledge of a small data base, e.g., the addition table in the example above. And even the items in that data base are found by procedure, namely counting, so if a required fact is not remembered, it can be constructed on the spot by the child. For this reason, achieving mastery is one of practicing procedure and/or recalling facts to reach the CCSS-M standard.

Once we all agree that the issue is one of practice, we have to consider whether it is reasonable to expect a child in Grade 2 to act responsibly in this matter. My answer is that ensuring an individual child in elementary school undertakes the practice requires adult supervision[33] and in providing this supervision, teachers and parents should be allies.[34]

1.6.4   Meeting the Standard

The most important point I want to emphasize is that ensuring a child meets the standard set by the CCSS is **labor intensive**. It further seems unlikely that governments will provide the necessary additional resources in the form of money and qualified personnel. Indeed you can already find evidence of this on Internet news

---

[33]In my experience, there are many members of extended families who are willing to take this responsibility on.

[34]I recall an article in *Scientific American* on why the children of immigrants tended to out-perform local children in their school work. The answer was that such children did homework around the kitchen table after supper as a regular practice. In other words, there was parental supervision on a regular basis. This practice was dropped in the 2nd generation and as a result the grandchildren of immigrants performed like local North American children.

sites.[35]

Given these realities, it seems plausible that the educational system may continue to fail for many children and unless responsibility for ensuring the child's success is taken on by that child's parents. What is truly encouraging is that survey data by the EPC of parents in CCSS adopting states indicate that parents are willing to undertake this responsibility.[36] This book exists to help parents who want to help their children by acting as mentors. It does this by providing the detailed knowledge of procedures mentors will need in a framework they can understand. The book has been written with the specific intent that it should be accessible to all parents, even those who consider themselves weak at math.

1.7   What a Mentor Needs to Know

An adult who contemplates undertaking this responsibility may be reluctant on the grounds that it is absurd to expect the average person to be able to help a child through the entirety of the school math curriculum. Moreover, recent studies seem to confirm this idea.[37] These studies suggest that once a child enters middle school, parental efforts can become counter productive and actually cause their child's test scores to decrease. An identifiable cause is that

> parents may have forgotten, or never truly understood, the material their children learn in school (see The Atlantic, p. 85).

So we are clear, based on more than forty years of teaching in universities, the only portion of the curriculum you need to concern yourself with is primary and elementary and there are things you can do that are guaranteed to help your child (see §1.7.1). During primary and elementary children are at an age most amenable to help from parents and if a child completes this portion of the curriculum with no deficits, the child will do just fine on the rest of the school math curriculum. So in terms of math content, this book confines itself to topics from arithmetic identified by the CCSS-M. As already noted, all of math at this level is procedural and the book presents procedures in an easily understood manner. For this reason, with the

---

[35] A search of Huffington Post has more than 50 pages of articles on Common Core. Some focus on resource/training requirements and whether such resources/training will be available in particular states. See, for example, http://www.huffingtonpost.com/stephen-chiger/to-improve-teaching-get-s_b_3655190.htm

[36] See the EPC Working Paper 34, on-line.

[37] ... And Don't Help Your Kids With Their Homework, Dana Goldstein, **The Atlantic**, April, 2014, p. 85

aid of this book, even parents who consider themselves weak at math will be able to help their children.

In respect to helping your child, you need to know various things:

1. Is my child performing up to standard on a grade-by-grade basis?

2. You need knowledge of the underlying content to be able to intervene and help your child where necessary.

3. You need to know where to find additional practice materials to support your intervention.

Information regarding each of these points is included.

1.7.1   Rote Learning

At the end of the Preface we identified two websites reporting on how math skills learned in early childhood affect internal brain development. The first thing to know is that these brain developments **last a lifetime** and continue to effect skills.

In respect to these skills, here's what you need to know. Cognitive psychologists identify two kinds of problem-solving. The first is **procedure-based** which simply means the problem solver uses an external procedure to solve the problem, for example, counting on one's fingers to add 5 and 4. As discussed, arithmetic is completely procedure-based. The second is **memory-based** problem-solving which means the problem solver recalls the solution from memory. Obviously, memory-based problem-solving is more efficient. It is also the case that children learning math transition from procedure-based to memory-based problem-solving. This is why it is essential that correct procedures are the basis for the transition, because unlearning a memory-based skill in next to impossible.

The research we are bringing to your attention found that the transition to memory-based problem-solving involves changes in your child's brain and that these changes continue to enhance the individual's skill at problem-solving as they become adults.[38]

So here are two things that every parent/mentor can do that are guaranteed to help a child succeed. First, ensure the child knows the addition table from memory by the start of Grade 2, and second, ensure the child knows the multiplication table from memory by the end of Grade 3. Of course there are many other things parents can do, but these are guaranteed to have immediate benefit to your child.

---

[38]See http://www.nature.com/neuro/journal/vaop/ncurrent/full/nn.3788.html and http://www.medicaldaily.com/math-skills-childhood-can-permanently-affect-brain-formation-later-life-298516

1.8 The Content

The remainder of this book is about numbers, their representation and the operations on numbers; these topics comprise the **number strand** in the CCSS-M. It is about these topics because as indicated above, in the experience of the author, students who can do and understand these arithmetic topics have no difficulty getting through courses in calculus and statistics of the type required for a career in business, the sciences, engineering or economics. It's also the case that if a student can't do arithmetic they are not only blocked from the above, but also from many industrial trades having a substantial math component, e.g., the electrical trades. Finally, the book is about topics children learn in primary and elementary because the test data tells us that this is where deficits begin to accumulate.

1.8.1 Using the Content

In previous sections we have made the point that the computations of arithmetic can all be performed using straight-forward mechanical procedures. For example, I think most would accept that the most feared topic in elementary school arithmetic is **fractions**. In §14.3.3 a three-step procedure is presented for adding fractions that for common fractions involves only computations with whole numbers and in all cases always produces a correct answer, even for algebraic fractions. Once a child has learned this procedure, that child is set for life when it comes to adding fractions. At a minimum, every parent can familiarize themselves with this procedure and make sure that their child knows how to use it.

At this point you may be saying something like: If it's that simple, why is there a problem with math? The key here is that the nature of mathematical knowledge is hierarchical — new procedures use and incorporate previous procedures. Thus, unless a child has mastered the previous procedures, that child will be unable to effectively learn the dependent procedure. As test results show, a mile-wide, inch-deep curriculum makes this type of mastery learning impossible for all but a few.

The following point by point outline presents the key procedures of arithmetic in a fashion that illustrates their hierarchical relationship. The only knowledge that is not procedural is the abstract notions related to number. These items must be mastered in the order shown:

1. the concept of counting number as measure of the relative size of a collection and conservation of counting numbers — Chapter 2;

    (a) the process of counting by rote;

    (b) the place-value naming system for counting numbers — Chapter 5;

(c) ordering of counting numbers;

2. addition of counting numbers — Chapter 6;

   (a) procedure for single-digit addition based on counting — addition table;

   (b) procedure for two two-digit sums using addition table;

   (c) procedure for two multi-digit sums using addition table;

3. subtraction of counting numbers — Chapter 7;

   (a) procedure for subtraction as take-away;

   (b) procedure for subtracting two-digit numbers;

   (c) borrowing;

4. multiplication — Chapter 8;

   (a) multiplication of single-digit numbers as repetitive addition — multiplication table;

   (b) multiplication of two-digit numbers by single-digit numbers;

   (c) multiplication of multi-digit numbers by single-digit numbers;

   (d) multiplication of two-digit numbers by two-digit numbers;

   (e) multiplication of multi-digit numbers by two-digit numbers;

5. division — Chapter 11;

   (a) concepts of multiples and factors;

   (b) procedure for division;

6. concept of fractions — Chapter 13;

   (a) making measurements;

   (b) concept of unit fractions;

   (c) notation for fractions and the Notation Equation;

   (d) the Fundamental Equation governing the behavior of unit fractions;

7. computations with fractions — Chapter 14;

   (a) multiplication of fractions;

   (b) addition of fractions with same denominator;

(c) addition of fractions with different denominator;

8. exponentiation and notation for decimals — Chapter 17;

9. operations with decimals — Chapter 18.

The most important thing to take from this outline is how each procedure builds on and/or incorporates previous procedures. This is why mastery of prerequisite procedures is essential both for you as mentor and for your child who is learning. In my experience, mastery of the topics outlined above will enable your child to succeed in higher level math courses precisely because knowing how to do the underlying computations will not be an issue. Finally, thorough knowledge of these procedures is the foundation on which **mathematical understanding** rests.

1.8.2   Understanding as a Requirement of the CCSS-M

You may wonder why the topics listed in the outline above require 17 chapters and 400 pages. Indeed, you may believe that that is the reason why children cannot learn math. The procedures in the outline require only a small portion of the total pages to explain. The additional chapters and sections contain information on the underlying ideas, for example, the Distributive Law and why it must be satisfied by our number system. Explanatory material is included that enables parents to achieve the **same level of understanding expected of their children** as set down by the authors of the CCSS-M. This material has been included so that parents have the knowledge required to help and support the learning of their child.

So the reader is clear as to what is being discussed in respect to **understanding**, we quote from the CCSS-M. The item discussed, which applies the Distributive Law, would occur in a middle school course in algebra, hence outside the immediate focus of this book. We include the quotation here because it defines the intent of the new standards and also because it provides a context to explain how procedural knowledge is the foundation on which understanding rests.

> These Standards define what students should understand and be able to do in their study of mathematics. Asking a student to understand something means asking a teacher to assess whether the student has understood it. But what does mathematical understanding look like? One hallmark of mathematical understanding is the ability to justify, in a way appropriate to the student's mathematical maturity, why a particular mathematical statement is true or where a mathematical rule comes from. There is a world of difference between a student who can summon a mnemonic device to expand a product such as $(a+b)(x+y)$ and a student who can explain

20

where the mnemonic comes from. The student who can explain the rule understands the mathematics, and may have a better chance to succeed at a less familiar task such as expanding $(a+b+c)(x+y)$. Mathematical understanding and procedural skill are equally important, and both are assessable using mathematical tasks of sufficient richness. See CCSS-M p. 4.

The intent described in this paragraph demands a higher level of learning than what is in a pre-CCSS-M school curriculum. It may well be that you would choose to leave material like this to schools and focus your efforts on procedures. Without doubt, that is a valid strategy that will provide significant help to your child. However, should you want to go further, we have included material in the book to enable you to confidently take the next step.

There are three laws that drive our understanding of numbers: the Associative Law, Commutative Law and Distributive Law. The first two laws applied to addition reflect the fact that any way of combining numbers using addition must produce the same answer. Similarly, the first two laws applied to multiplication tell us that any way of combining numbers using multiplication must also produce the same answer. Why this must be so is explained in Chapters 6 and 8, respectively. Children experience these laws functionally; for example, each time they actually compute

$$5 + 3 \quad \text{and} \quad 3 + 5$$

and observe the two sums are the same, they witness the Commutative Law in action.

Consider the Distributive Law which asserts:

$$a \times (b + c) = (a \times b) + (a \times c).$$

Again, children learn the truth of this law when they compute

$$2 \times (5 + 3) \quad \text{and} \quad (2 \times 5) + (2 \times 3)$$

and get the same result. This is how procedural knowledge produces a deeper knowledge of principle.

In respect to the notion of *understanding* in the quotation above, children learning algebra are confronted with computations like:

$$(a + b) \times (x + y) = ?$$

There are two ways we could teach children to find the answer. The first is a common mnemonic labeled: FOIL method.[39] The second approach is to get children to perform

---

[39]See wikipedia.org/wiki/FOIL_method entry for a complete description.

this computation as two applications of the Distributive Law:

$$(a + b) \times (x + y) = ((a + b) \times x) + ((a + b) \times y)$$
$$= (a \times x) + (b \times x) + (a \times y) + (b \times y).$$

To learn through this approach, children must first recognize that the letters stand for numbers and as such behave no differently than numbers. Second, they must have a functional familiarity with the Distributive Law gained by working with numbers in elementary school. Moreover, a child who learns through this approach has little trouble dealing with the computation required in

$$(a + b + c) \times (x + y) = ?$$

If you want to take the next step, it begins with the Associative, Commutative and Distributive Laws and these are are explained thoroughly in a manner that an adult who knows the procedures can understand.

1.8.3 Using this Book

In summary, the purpose of this book is to provide parents with the mathematical information they need to ensure their children master arithmetic. There are three aspects to this:

1. providing the procedural knowledge on which computations are based so parents are able to help their child master the computations of arithmetic;

2. providing the information required to make judgements about whether the curriculum their child is being taught meets CCSS-M standards;

3. providing detailed information as to what they should expect their child to know on completion of grade levels so that they can assess for themselves whether their child is making adequate progress;

4. suggesting specific activities that help and encourage their child in the learning process so as to ensure their child achieves appropriate levels of mastery.

The first thing you need to know about this book is that it is written for adults. It is not for children. The author assumes the reader has enough knowledge of arithmetic to perform calculations with the aid of a calculator. The only other requirement on the part of the reader is a desire to learn what you need to know to help your child. We will explain what is in the CCSS-M curriculum in respect to arithmetic. What we discuss is what your child is expected to learn. No more and no less.

1.8.4   What Parents Need to Know Again

In our first remarks on this, we listed three items:

1. You need knowledge of the underlying content to be able to intervene and help your child where necessary.

2. You need to know whether your child is performing up to standard on a grade-by-grade basis.

3. You need to know where to find materials to support your intervention.

There are two kinds of help your child may need. The first is how to do a specific type of calculation. For example, your child may need help coming to terms with the procedure for finding

$$\begin{array}{r} 27 \\ + \;\; 84 \\ \hline \end{array}$$

Responding to this situation is fairly straightforward in that you go over the examples in the appropriate section, in this case §6.4 to make sure you know how to correctly perform the procedure, and then work through some similar examples with your child. Working through the examples will enable you to assess whether your child has the prerequisite knowledge to perform the calculation; for example, does the child know $7 + 4$ as a matter of recall?

   The second type of help your child may need is in developing an understanding of why procedures work. It is in expecting children to know the *Why?* that the CCSS-M fundamentally raises the standard. Working through examples with your child will enable you to form judgements about the degree to which your child understands particular elements of a computation. (For example, why does the 7 have to be in the same column as the 4 in the setup?) Explaining a *Why?* generally involves significant prerequisite knowledge. In the example above,

> the simplest explanation of *Why?* is that the addition procedure requires us to add *ones* to *ones*, *tens* to *tens*, and so forth. Since both the 4 and the 7 are *ones* they are placed in the same column to ensure we add *ones* to *ones*.

This type of discussion can lead to a deeper discussion of **place** in the Arabic System. So the second type of help requires a deeper understanding on your part and an expectation that you will have read the prerequisite material. All the information required to deal with questions of *Why?* is in the book in the appropriate sequence with complete explanations.

As stated in the content overview, each chapter concludes with grade-by-grade goals based on the standards. For example, by the end of Grade 2 it is expected that your child can add two four digit numbers using the standard procedure. Whether your child achieves this goal is easily verified: you simply ask your child to perform some calculations. If you know the goals in advance and monitor your child's progress throughout the year, by the end of the year there should be no question that the goals are met. Basically, it all comes down to practice.

Testing

Testing has something of an evil name, certainly among students. The fact is, there is only one way to determine whether anyone knows something and that is to ask them.

In the opinion of the authors, the only legitimate purpose to testing in mathematics is to:

> **determine whether a student has learned the material that has been taught.**

This may seem obvious, but often times test questions seem to have another purpose; for example, finding out if a student is clever. The results of tests should be used only to inform parents and teachers whether a child has assimilated the knowledge required to proceed.

Here is what we recommend in respect to testing. Each year your child should have a notebook devoted solely to tests in arithmetic. As each new test is added to the collection, you and your child should go over the test question by question. Going over all questions provides an opportunity to hand out praise. Going over questions that are wrong provides the opportunity to find out if your child knows why the answer was wrong and whether there are issues in respect to the question that need to be addressed. We stress: **going over a test is not an adversarial process**. It is about helping your child learn what is needed.

There may be rare situations in which neither you nor your child knows what is wrong and what the correct solution is. In that case, ask the teacher for the correct solution with an explanation.

Finally, this notebook will provide pointers as to what sorts of additional practice your child needs to master a topic. The next section shows where you can get help with this.

1.9 Math Websites for Mentors

There are any number of websites that have been developed to support the CCSS-M. Some sites offer worksheets that are available for download at no cost. We provide a sample list of such sites with descriptive material taken from the site. We also provide a list of subscription sites.

The website

http://www.achievethecore.org/parent-community-common-core/parent-resources/

has more information on the Common Core movement, including detailed information on expectations.

The website

http://www.pta.org/advocacy/content.cfm?ItemNumber=3552

contains numerous items on the Common Core movement including a number of math videos on specific topics, for example, *coherence*. Parental guides from PTA can be found at http://pta.org/parents/content.cfm?ItemNumber=2910

The website

http://everydaymath.uchicago.edu/parents/

has information for parents on curriculum topics at all grade levels. Of particular interest to parents is the material on alternative algorithms for performing standard computations. Short videos demonstrate the use of these methods, for example, the partial sums method, for performing computations. It also provides access to on-line learning games such as **Bunny Count** and **Connect the Dots**. We will refer to this site as EDM.

1.9.1 Free Websites Containing Worksheets

The website

http://www.math-aids.com/

contains the following content description:

> The website contains over 72 different math topics with over 847 unique worksheets. These math worksheets may be customized to fit your needs and may be printed immediately or saved for later use. These math worksheets are randomly created by our math worksheets, so you have an endless supply of quality math worksheets at your disposal. These high

quality math worksheets are delivered in a PDF format and include the answer keys. Our math worksheets are free to download, easy to use, and very flexible. These math worksheets are a great resource for Kindergarten through 12th grade. A detailed description is provided in each math worksheets section.

The home page at this site has a list of topics on the left. Clicking on a topic produces a worksheet page that begins with a description of the types of worksheet. The *Kindergarten* button produces worksheets on a variety of entry-level topics. Across the top is a list of buttons, one of which is a site map which is very useful. There is also a button leading to links to other resources, most of which are commercial. We will refer to this site as M-A.

The website

http://www.superkids.com/aweb/tools/math/

has an interactive engine that will produce worksheets on a fairly complete list of computational topics from arithmetic. It also covers pre-algebra and exponents. Access is free. The following is from the site description:

Have you ever wondered where to find math drill worksheets? Make your own here at SuperKids for free! Simply select the type of problem, the maximum and minimum numbers to be used in the problems, then click on the button! A worksheet will be created to your specifications, ready to be printed for use. We will refer to this site as SKids.

The website

http://www.softschools.com/

provides math worksheets and interactive games on an extensive list of topics. There is a lot of interactive stuff on counting, so it's good for Pre-K. The following is from their site description

SoftSchools.com provides free math worksheets, free math games, grammar quizzes and free phonics worksheets and games. Worksheets and games are organized by grades and topics. These printable math and phonics worksheets are auto generated. There are many counting games at Pre-K level. We will refer to this site as SS.

The website

http://www.kidzone.ws/

has **free** printable worksheets. Unfortunately, the items covered are limited in scope to counting, the basic operations, and word problems related to these computations. The grade levels covered are Pre-K to Grade 5. The site does not have material on fractions.

The website

> http://www.pbs.org/parents/education/math/

has lots of general education stuff. There are on-line games for children and their parents. It is probably more valuable to new parents and parents of the very young.

1.9.2 Subscription Websites Containing Worksheets

The website

> http://www.adaptedmind.com/Math-Worksheets.html?type=hs

has worksheets organized by grade level. The list of topics is extensive and parents will have to pick and choose. There is useful material on place value in topics for Grades 1 and 2. Not free: $10/mo.

The website

> http://edhelper.com/math.htm

offers extensive list of worksheets and activities covering K-12 at a cost of $19.99 per/year for a limited subscription and $39.99 for a complete subscription.

The website

> http://themathworksheetsite.com/

has worksheets available on all aspects of arithmetic. A subscription is $25.

The website

> http://ca.ixl.com/

contains practice problems that are interactive. It is comprehensive and geared to the Canadian curriculum. For this reason parents will need to be choosey about which activities they ask their child to undertake. As long as parents focus on curriculum elements from the CCSS-M, there's lots of great interactive material to provide practice for your child. Unfortunately, access is not free, but twenty free worksheets can be had from http://www.math-drills.com/ which appears to be a subsidiary of IXL. Other subsidiary sites are

> http://www.mathsisfun.com/worksheets/
> http://www.dadsworksheets.com/
> http://www.mathworksheetsland.com/

Chapter 2

Collections and Numbers

Chapter Overview. The order of presentation of material in this book is the same as the order in curriculum guides. This order, as far as we know, reproduces the historical development of mathematical knowledge. Thus the material in this chapter can be found in the Pre-K curriculum and reflects the earliest development of mathematics, namely the concept of **counting number** as an attribute of a physical collection. A critical issue is that of determining whether two collections have the same number of objects, or that one has more than the other. A procedure for determining when two collections must be assigned the same counting number will be given. Learning goals for Pre-K derived from the CCSS-M are presented and discussed.

2.1   What Numbers Are and Why They Exist

People have always been interested in answering the questions:

How much? How many?

as they pertain to collections of things in the world. The most primitive cultures, mathematically speaking, answered these questions with either: one, two, or many. Indeed, there are still some cultures in the world today whose mathematics is limited to one, two, or many. Over time, limiting the answers to one, two, and many, proved inadequate and more complete systems for answering these questions were developed. This required the development of numbers to give precise meaning to the question: How many are in a collection?

As treated in this document

numbers are things that tell us the size of something.

Historically, the first numerical questions posed by our ancestors were probably applied to collections. For example, a wife might want to know how many birds were brought home for supper, or how many clams had been collected at the beach, or in another vein, how many days until the next full moon.[1] Such questions involve identifying groups of objects and assembling them into collections. So in the first instance, our focus will be on numbers that arise as a response to the question:

How many objects are in this collection?

The answer to this question is a fact about the real world that is determined by counting the number of objects in the collection. In this sense, the answer is the result of an experiment and there is a universally held belief that any two competent counters will get the same result. Every time you count a group of objects as part of your normal activities you are performing a counting experiment that witnesses the truth of this universally held belief.

The number of objects in a collection is referred to as an **attribute** of the collection. Numbers that capture this attribute of a collection are called **counting numbers**, or **cardinal numbers**.

Obviously, there are other kinds of numbers that answer questions about objects, for example, the **length** of a soccer pitch, the **area** of a plot of land, or the **weight** of a cow, etc. All of these are **numerical attributes** of real-world objects. While we will deal with such numbers later, as stated, our initial focus will be on the counting numbers associated with collections.

However, before continuing our investigation of what numbers are, we will concern ourselves with the fundamental question:

Why should numbers exist at all?

2.2   Conservation

The most important and deepest laws in the natural world are conservation laws. Conservation laws have application in every area of science and indeed, one such law is at the heart of arithmetic. To see what it is, consider the following **thought experiment**.

---

[1]The earliest known potentially mathematical object is a baboon's leg bone with 29 notches in it dated from 35,000 years ago. (see Wikipedia: Prehistorical mathematics)

Suppose you have a pile of buttons. Count them, place them all in a jar; seal the jar. Place the jar on the shelf. Now suppose you come back at a later time and observe that your seal is intact. How many buttons will be in the jar? More specifically, do you need to count the buttons in the jar to determine your answer?

Obviously, you do not need to recount the buttons. Without counting you know that the number of buttons remains the same as when you placed them in the jar. If the seal is not broken, no buttons can be removed, and none can be added.

Consider still another experiment:

Unseal the jar of buttons and carefully pour the buttons into another empty jar making sure all the buttons are transferred. How many buttons are in the new jar when you are finished? Do you need to count?

Again, we do not need to recount the buttons so long as we are sure that all the buttons were transferred to the new jar and none were added.

Consider one last experiment:

Suppose we have some empty jars. Unseal the jar containing buttons and distribute some of the buttons among the empty jars, possibly retaining some in the original jar. How many buttons are in **all** the jars? Do you need to count?

In all of these cases there is no need to count the buttons. We started with a fixed number of buttons, and unless there is a source of new buttons or a sink for the old ones, the number of buttons is unchanging in time.

These ideas are illustrated in the following diagram:

A drawing showing a box containing six buttons on the left. These buttons have been distributed between the two boxes on the right. There is no need to count to know the number of buttons in the two boxes is the same as the number we started with, because buttons are conserved.

In summary:

Conservation Principle (CP). The number of things in an isolated collection is unchanging in time.

The effect of this fact about the world is that the cardinal (counting) number of a collection is a **stable** attribute. We refer to this observation about the world as **Conservation of Number**. The reason underlying the principle is that buttons, and physical things more generally, are not spontaneously created or destroyed by the world. This observation may seem obvious, or, trivial. It is neither.

The age at which children come to terms with this principle in its various forms was studied by a famous cognitive psychologist, Jean Piaget.[2] His results are complex and show children's understanding of this conservation principle develop and change over time as the brain matures.[3]

To see that conservation is absolutely essential to all our reasoning, ask yourself how the world would work if the principle weren't true. Among other things, this principle is what enables you and your banker to agree on the contents of your bank account.

We prefaced this discussion with the question:

Why can there exist numbers at all?

Now we have our answer.

Numbers can exist because we can form collections having constant membership, that is, that have a numerical attribute that remains constant over time. We will see how this is applied in discussions about numbers below.

This fact that isolated collections have constant membership permits us to have standard collections against which we can test any others. In this respect, such test collections would be like the cylinder of platinum and iridium in Paris which is the standard kilogram.

2.3   What are Collections?

Since our focus will be to develop numbers as a response to: How many are in a collection?, it is important that we know what collections are. To be precise:

a **collection** is any identifiable group of objects in the real world.

---

[2]Cognitivist (learning theory) is the theory that humans generate knowledge and meaning through sequential development of an individual's cognitive abilities, such as the mental processes of recognize, recall, analyze, reflect, apply, create, understand, and evaluate. (From Wikipedia.)

[3]For example, Piaget found that realizing that the total things in a collection are conserved when the collection is subdivided, required substantial learning. In another example, Piaget found that young children believed that the contents of a tall thin container was more than those same contents transferred to a flat, wide container, even though they watched the transfer!

Since it is always useful to have a concrete idea in the back of one's mind, we could think of a standard collection as some buttons in a jar. Thus, the collection consists of all the buttons in the jar. The jar, or container, is not part of the collection, only the buttons. The buttons themselves, are referred to as **members**, or **elements** of the collection.

Consider two jars of buttons, that is, two distinct collections. In the real world, we know the buttons in the first jar are different objects from the buttons in the second. This assertion merely reflects the fact that it is not possible for a physical object to be in two places at once.

In our development, we will identify the numerical properties of collections. In doing this, we will depend on the fact that every collection considered, or group of collections considered, can, in principle, exist in the real world. As such, the same object cannot belong to two collections simultaneously.

2.3.1   Sets vs Collections

Mathematics is about abstractions. As a professor in one of my classes observed more than fifty years ago as he put some vectors on a blackboard: "These are not vectors, they are squiggles on a blackboard. Vectors do not exist in the real world."

"What does this have to do with us?" you may ask.

Almost certainly, your child's curriculum uses the word **set** in respect to collections. For this reason we need to go over how mathematicians use the word.[4]

Sets are the things mathematicians use to **model** collections.

We need to provide some explanation of what we mean by *model*. So let's start with a physical model. Perhaps when you were small you built a model airplane. It was a device that captured some of the properties of the actual object, and that could be used to study that object. Model cars, model trains and the like are ubiquitous. Dolls are also models, although we don't usually think of them that way. The point is that every child has played with physical models of various kinds.

Engineers build more realistic models to study properties like loads on a dam or lift on an airplane wing. Such models had better be accurate, or when the actual structure is built, it could fail with catastrophic consequences.

Not all models are physical. Some are entirely abstract. If you have a bank account, there is an abstract model that you regularly use. It is the record of checkbook

---

[4]When sets were introduced into the school curriculum as part of something called the **New Math** back in the 1960's, all but a few found it intimidating. Readers of the *Peanuts* comic strip from the 1960's which is being recycled in papers today, may recall seeing individual strips complaining about aspects of the New Math. Such strips were indicative of the degree of discomfort that greeted the New Math in the general population.

transactions. The squiggles recorded there are not money, which is real, nor are they numbers, which are entirely abstract. But these squiggles provide a means of keeping track of how much money is left in your account. In this sense they are a model. And correctly used, your bank book provides a perfect means of predicting what is in your account at any moment in time.

As stated above, sets are the abstractions that mathematicians study to model the behavior of collections. Because they are abstract, rather than try to say what sets are, mathematicians concern themselves with rules that describe how sets behave and how sets can be constructed. For the most part, these rules reflect what we know about real-world collections, but there are some differences.

Much of the nomenclature regarding sets is identical to that for collections. Thus, things in sets are called **members**, or **elements**, these words being used in exactly the same way as for real-world collections. Thus, if we have a set A and something in the set called b, we would say b is a member of A, or b is an element of A. The standard mathematical notation for b is a member of A is

$$b \in A.$$

where the symbol $\in$ is read *is a member of*. We will see this notation in Chapter 10 and remind the reader of this discussion at that time.

But there is a fundamental difference between *collection*, as we are using the term, and *set* as used by a mathematician. It is this. If I put a button in a collection, you cannot put that same button in a different collection. On the other hand, if I put 1 in a set, there is nothing to stop you from putting 1 in a different set, even while I still have 1 in my set. Moreover, the 1 in my set is exactly the same as the 1 in your set. They are indistinguishable.

This difference between collections and sets has consequences. For us, the most important consequence is the following. Suppose you have a collection with one button, and I have a collection with one button. If we combine our collections, the new collection will have two buttons. Alternatively, suppose you have a set having a single member, 1. And I also have a set having a single member, namely, 1. If we combine our sets, that is put the single member of your set and the single member of my set into a new set, that set will contain only the single member 1.

It is clear that sets have an added level of complexity over collections. But not to worry, as we will be working with collections. Indeed, in what follows we will spend a lot of time thinking about jars of buttons, and if you wanted, you could actually physically reproduce any part of the discussion with your own jars of buttons.[5]

---

[5]In my lifetime I have met and worked with some really fine mathematicians. All thought in the most concrete terms possible, and were not afraid to say so. Moreover, all believed that the most effective strategy for doing research begins with understanding the work that went before.

Although the terms *collection* and *set* are often used interchangeably, we will continue to apply the noun *collection* only to groups of real-world objects. Thus, any collection we will speak of could, in principle, physically exist. *Set* will be reserved for groups of abstract objects which do not exist in the world.

We adopt this approach for the expressed purpose of emphasizing that the generation of the ideas being discussed in respect to arithmetic arises from considerations about the real world. Since collections are things all of us deal with everyday, the rules collections obey are so ingrained in our thinking, we don't even know they are there. But we can identify these rules and use them as a basis for arithmetic.

2.4   First Perception of Numbers: Pairing

We continue our development by following the steps a child takes.

Consider the thought experiment:

> Suppose you and your very young child are driving by a field that contains two cows and three horses. You ask your child: Are there more cows or more horses? What I want you to consider is what your child needs to know to answer this question. Specifically does your child need to know the names *two* and *three* to draw a conclusion?

To make this more concrete, consider the schematic field shown below.

> A schematic field outlined by the frame. Horses are denoted by an H and cows by a C.

A child that knows nothing about numbers could answer the question based on proximity by noticing that two cows and two horses are feeding close together, while the third horse is far away. So by **pairing**, as explicitly shown below, all the cows with some of the horses, and noticing there is an unpaired horse, the child can draw the correct conclusion without ever knowing about numbers.

34

A schematic field outlined by the frame. One horse is not paired with a cow, so there are more horses than cows.

Thought Experiment. Alternatively, suppose you and your very young child come to another field that has three cows and three horses. You pose the same question.

Again, your child can answer the question without knowing about *three* simply by observing that there is an **exact pairing** of cows and horses as illustrated below. Mathematicians refer to an exact pairing between the elements of two collections as a **one-to-one correspondence**. We will continue to use **exact pairing** because we think pairing is more easily understood as a concept.

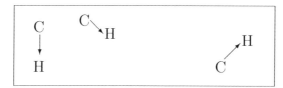

A schematic field outlined by the frame. There is an exact pairing between cows and horses, hence an equal number of cows and horses.

Here is an example of how the notion of exact pairing can be useful:

Suppose you are setting the dinner table for eight — the cousins are coming over — and your young child, who can't count, wants to help. You could lay out the place mats and tell the child to put one fork, one knife and one spoon at each place mat.

Again, pairing substitutes for knowledge of number. But doing the pairing leads to the notion of number as an attribute of collections.

2.4.1   Why Pairing Has Limited Utility

The pairing process requires proximity in space and time. Consider the following:

Suppose you and your very young child are driving by a field that contains four cows. The child says, "Look! Cows." Later, you come to another field that has five horses. The child says, "Look! Horses." And now you say, "Are there more horses in this field, or cows in the last field?"

We illustrate this situation below:

Schematic fields with four cows (above) and five horses (below).

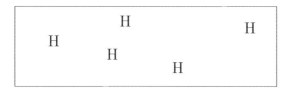

Because the fields, and hence the animals, are separated in space and time, it is no longer possible to determine whether there are more cows, or more horses using a direct pairing based on direct proximity of the animals. This requirement of the pairing process, that collections be proximate in space and time, is a severe limitation.

2.5   Counting Numbers as a Solution

To answer the question: Which has more? for collections separated in space and time, the child has to deal at a much higher level. What is required is the notion that there is something fundamental about all collections for which there is an exact pairing with a collection containing four cows. It is the abstract notion of *fourness*. Notice that to operate at this level the child must also believe that were you to return to the field with the cows, there would still be four. In other words, the child must come to terms with conservation at some level.

In a separate vein, the child must also understand that when we say a collection has the numerical attribute of *fourness*, this tells us nothing about the physical nature of the members of that collection. It only answers the question;

How many are in the collection?

Because it is an abstract idea, the attribute of fourness may be equally well applied to cows, or horses, or spoons, or whatever. Similarly, all collections containing five items have a common attribute that gives rise to the notion of *fiveness*. And so on for other numbers that answer the question: How many?

The realization that every collection has an attribute which answers the question, *How many are in the collection?*, gives rise to the abstract notion of **counting numbers (cardinal number)**.

36

Counting Numbers are the numbers we use to name the attribute of collections of things that tells us: *How many items are in the collection.*

We stress that **cardinal number** is just another name for counting number.

2.5.1 Equality Between Counting Numbers

Once we have identified counting numbers as being things we want to use, we need a fixed and robust procedure for determining when two counting (cardinal) numbers are equal. Based on the discussion above:

> **Equality Principle.** Two collections will be assigned the same counting number exactly if there is an **exact pairing (one-to-one correspondence)** of all the objects in one collection with all the objects in another, with none left over in either collection.

We illustrate the equality principle using two collections having four members each:

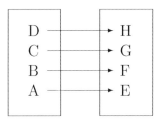

> A diagram illustrating an exact pairing between two collections. So, as a consequence of the Equality Principle, both must be assigned the same counting number as the answer to: *How many?* Note that the decision as to whether the two collections have the same counting number does not depend on what that number actually is. Nor does it depend on what the elements of the collections are. It only depends on whether there is an exact pairing between the two collections.

Thus, the procedure for determining when collections have the same cardinal number is derived from our understanding of the nature of collections in the real world. We emphasize that the determination does not depend on the nature of what is actually in the collection; nor does it require a notation for the counting number that captures the cardinality attribute because the pairing process does not mention number. As such it captures our real-world experience. I want to emphasize this point strongly. What we are doing is creating an abstract model of certain aspects of the real world. This model must exactly replicate the aspects it seeks to capture. Thus, if we have two collections for which there is an exact pairing between the objects, they must be assigned the same counting (cardinal) number.

As we have already noted, the exact pairing process has limited utility due to the proximity requirement. Thus, for counting numbers to be the solution, there must be another process for determining equality between collections. There is, but to describe that process we need to explore the deeper properties of real-world collections.

2.6 What Your Child Needs to Know

Sections like this one will provide answers to the following questions for the topics presented in each chapter.

1. How old should the child be when these topics are presented?

2. To what extent should I show the child how to find the answer?

3. Are there materials available that could help me?

Since the answers are critical to parents helping their children, we go through the questions and answers in detail.

In respect to the first question, the grade-level-goals for the CCSS-M were identified by mathematics educators using research data on what and when children can effectively assimilate various concepts. There are goals in the CCSS-M for K-12. There are no Pre-K goals, but fortunately this situation was remedied by the educators at the University of Chicago.[6]

2.6.1 Goals for Pre-K

On entering Kindergarten it is expected your child can:

1. Verbally count in sequence to 10 and beyond; develop flexibility in counting, including counting on and counting backward from a given number.

2. Count objects with one-to-one correspondence and know the last counting word tells "how many."

3. Develop an awareness of numbers and their uses; associate number names, quantities, and written numerals; recognize and use different ways to represent numbers (for example, groups of objects or dots).

---

[6]See grade-level-goals in parents' section of http://everydaymath.uchicago.edu based on the CCSS-M.

4. Compare and order groups of objects using words such as more, fewer, less, same.

5. Solve and create number stories using concrete modeling; explore part-whole relationships (for example, 5 is made of 2 and 3).

Let's go through these goals to make clear the intention and at the same time identify where you can find help.

Goal 1 says that your child should be able to count to 10 and beyond. What is being discussed is **rote counting**, which is clear at the website. This means your child could repeat in order the numbers up to 15, but would not necessarily be able to count the buttons in a jar containing 15 buttons and report that fact. It is expected that the tasks set out under this goal can be executed by rote. If you want to operationally see what's required, try the **Connect the Dots** game with numbers at the EDM website. The system of notation we use for counting numbers is discussed in Chapter 5.

Goal 2 says that your child should be able to count a group of objects in the manner that an adult would do it. This process works by assigning counting numbers in order, the last being the cardinality of the collection. Using this process represents a deeper level of knowledge. Again, these ideas are covered in Chapter 5. If you want to operationally see what's required, try the medium level **Bunny Count** game with numbers at the EDM website.

Goal 3 says that your child should be able to directly associate numerals and collections having the cardinal number named by that numeral. For example, 3 and a box with three dots (see Chapter 5). Clearly this represents a still deeper level of knowledge. If you want to operationally see what's required, try the **Bunny Count** game with numbers at the EDM website.

Goal 4 says that your child should be able to compare and order groups of objects. This begins with the processes discussed in the present chapter. If you want to operationally see what's required, try the **Bunny Count** game with characters at the EDM website.

Goal 5 says that your child should be able to write simple number stories involving operations. For example:

> Sally went to the store and bought two dolls and one model car. How many toys did she buy?

The intention of such stories is to develop a sense of the operation of addition. Addition is discussed in Chapter 6.

Returning to the issue of what and when in respect to Pre-K, encouraging a child to explore and experiment can begin very early. For example, I have observed such an interaction counting a small collection of objects between a mother and her 14 month-old child in which the child was both interested and engaged. In another context, a friend describes their 18 month-old grand child negotiating the number of episodes of a video she could watch before bedtime. The parent would offer 1, the child would respond with 5; the negotiated settlement was usually 2. The point is number concepts are being absorbed by children at a very early age, 2^+ years according to research data.[7] Obviously, the younger the child, the simpler the questions that should be posed. So for example, for a 2-year old: Are there more forks (2 or 3), than spoons (1 or 2), would be a good place to start. Or more forks (2 or 3) than fingers? The point is that this is about finding ways to encourage your child to think about the size of collections and counting.

Obviously, children are individuals, and what is appropriate for one, may not be for another, even in the same family.[8] Thus, you will have to gage the response of your child to determine how to proceed. A good rule is that learning is like play; it should be fun.

In respect to providing guidance, research shows that young children respond very differently when provided with instruction, namely, they tend to focus on the solution provided by the instructor, as opposed to exploring the situation as widely as possible.[9] So an alternative to instructing is to simply pose the problem and let the child work it out. At the point the child has a potential solution, get the child to tell you how the solution was arrived at. When you understand the child's thinking, you can suggest new problems.

There is one very important thing that should be understood. Because a young child is trying to find a solution to a problem for which **no solution is known to them**, all solutions they might propose have value. The process is one of **trial and error**. It is a fact that most trials end in error.[10] There is no fault in this. There is only the learning that arises from eliminating an incorrect solution. Trial and error is the process by which we humans managed to create culture. However, learning from those with experience is how we expand thousands of years of culture. Balancing these paradigms is the key to successful learning.

---

[7]See *Scientific Thinking in Young Children: Theoretical Advances, Empirical Research, and Policy Implications*, Science, **337**, 1623-1627 (28 September 2012).

[8]I have a colleague who has twin sons, who, according to his descriptions, are completely different in respect to their mathematical behavior and interest. While both are competent, one is clearly deeply interested in numbers, and has been so since the age of two. The other is much more interested in artistic activities.

[9]See *Scientific Thinking in Young Children: Theoretical Advances, Empirical Research, and Policy Implications*, Science, **337**, 1623-1627 (28 September 2012).

[10]See **Adapt: Why Success Always Starts With Failure**, Tim Harford, Farrar, Strauss and Giroux, 2011.

Lastly, with respect to Pre-K, there are a number of sites like EDM that offer interactive games and/or interactive worksheets. Particularly with small children you could jointly play some of the games and do some of the worksheets. Building counting skills is essential. **Counting and conservation** are the basis for all of what we do. They are the foundation of every principle that is presented in this book. All the key laws are experimentally verifiable by a counting process that depend on conservation. It is by doing the activities and worksheets that your child comes to terms with these things.

Chapter 3

Counting and Collections

Chapter Overview. In this chapter we develop five properties of collections that our system of arithmetic must capture.

In the last chapter, we identified counting numbers as a numerical attribute of collections. We implied that our arithmetic was based on the behavior of this numerical attribute. Although there are no records of how numbers were originally invented, it seems most likely it was through the study of collections. Indeed, the standard properties, used by mathematicians to develop the counting numbers, closely reflect the behavior of real-world collections. So it is important to know what these properties are.

3.1 Counting Goals for Kindergarten

Understanding the properties of collections as they relate to what your child has to learn is critical to parents. So the focus of this chapter will be on how we count. In particular we will examine collections from the perspective of the following learning goals taken from the CCSS-M for Kindergarten.

1. Understand the relationship between numbers and quantities; connect counting to cardinality.

   (a) When counting collections of objects, say the number names in the standard order, pairing each object with one and only one number name and each number name with one and only one object.

   (b) Understand that the last number name said tells the number of objects counted. The number of objects is the same regardless of their arrangement or the order in which they were counted.

(c) Understand that each successive number name refers to a quantity that is one larger.

Read the goals and then think about getting six spoons to set a table. The process requires that you understand that you need to construct a collection having a certain numerical attribute, namely, the cardinal number 6. (Although we have not yet discussed the existence of numerals for numbers, since we all know about them, we will behave as though we already have them.) The process by which we construct the required collection involves removing spoons **one-at-a-time** from a drawer and assigning the numbers 1, 2, 3, 4, 5, and 6 to the spoons. In other words, an exact pairing, or one-to-one correspondence is constructed between a collection of spoons and the set of numbers

$$\{1, 2, 3, 4, 5, 6\}$$

which exists only in our minds. We know we can stop when we get to 6 because the collection now has the right attribute. We know this works because at each step of taking another spoon, we know the collection will be one spoon larger. This process is so ingrained that we don't even think about it. But it is **the basis of all our arithmetic** which is why we study it in detail in this chapter.

3.2   Properties of Collections

We begin by noting that every property of collections discussed in this chapter can be observed in experiments.

Collections Have Elements

We defined a collection to be:

> any identifiable group of objects in the real world.

It is implicit in this description that each collection has something in it. This approach, that a collection must have something in it, seems most consistent with how the earliest humans might have thought about things. To a mathematician, a collection having something in it is referred to as being **non-empty**. Obviously, as soon as one has identified *non-empty* as an important feature of collections, the notion of **empty** collection cannot be far behind. Historically, however, the notion of empty came much later in the development of numbers and arithmetic.

In what follows in this chapter, all collections are non-empty.

Finally, we stress again that the identification of these properties is based on observations about the world. As such, the reader can recreate situations that exactly confirm each result.

3.2.1 First Property

Experiment 1. Consider the two collections of things as shown in the figure below. The collection on the right has a single member. The question we want to consider is: Which has more, the collection on the right, or any other possible collection we might consider on the left. To make the experiment concrete, we have placed four elements in the collection on the left but any number of elements would do.

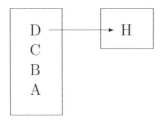

A diagram of the two collections. The one on the right has a single member. We want to know whether any other collection, which must have something in it, can have less members than the one on the right?

To answer the question, we pair elements of the collection on the left, with elements of the one on the right. The diagram shows one such pairing; since this pairing has elements in the collection on the left that are not used, we know the collection on the left has more elements than the one on the right. Moreover, we know, based on experience, that there will be elements in the collection on the left that are left-over no matter how we construct the pairing.

Since the question we are considering is whether there can be a collection that has less members than the one on the right, we might try removing elements from the collection on the left. However we perform this removal process, our experience tells us that eventually the collection on the left will be reduced to having a single element remaining, for example, as shown:

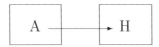

A schematic of the two collections in which three of the elements have been removed from the collection on the left.

At this point, we have an exact pairing between the two collections and, in this circumstance, the collections must be assigned the same counting number. But we want the one on the left to have less elements than the one on the right. When we remove another element from the collection on the left, the following situation results:

A schematic of the two collections in which the last element has been removed from the collection on the left, hence the collection disappears. No pairing can be constructed, since the schematic container on the left has no members!

As is obvious, removal of the last member from the collection on the left leaves an empty container. Clearly, however we were to try this with any real-world collection, for example, a jar of cookies, a box of nails, etc., if the collection has a single element to start with, when we remove that element, there is nothing left and the container is empty. The point we take from all this is:

> **There are smallest collections, characterized by the property that removal of any element leaves no collection at all.**

If we have a collection that qualifies as *smallest*, then like the collection discussed above, it must consist of a single member. This is an observable fact about the world. As we have said, all such smallest collections must be assigned the same counting number, and as we all know, that number is **one**. Anticipating future development, we use the symbol 1 as the notation, or **numeral**, for this number. This permits us to state the first essential property of counting numbers:

> **There is a smallest counting number, to which we give the name** *one,* **and the numeral** 1 **.**

In identifying this property of counting numbers, we are illustrating the direct relation between the experimental facts about collections and the properties of the abstract attributes we call counting numbers. Moreover, if there were no smallest collections, we could not possibly construct collections of a fixed size by the one-at-a-time process described above.

3.2.2 Second Property

Experiment 2. Consider two collections for which there is an exact pairing. In consequence, we know they must be assigned the same cardinal number. We picture this below for collections having six members.

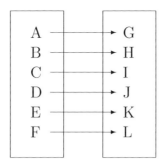

A schematic illustrating an exact pairing between two collections having six elements each. Once again, as a consequence of the Equality Principle, both must be assigned the same counting number.

Suppose we put one extra element in each collection. We want to consider the effect on the counting numbers assigned to the two collections. The collections with the extra element are illustrated below:

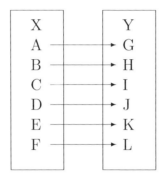

A schematic of the two collections to which a single element has been added to each.

In the second diagram the single additional element denoted by X has been put in the collection on the left. The single additional element Y has been placed in the collection on the right. Notice that since there has been no change in the original members of either collection, we are able to construct a partial pairing of the two collections based on the earlier exact pairing between the two collections, as shown. Since there is exactly one unpaired member of each collection left over, when we pair the new element on the left with the new element on the right, we will have an exact matching between the two collections, and, in consequence, the two collections must be assigned the same counting number. A bit of experimentation will convince you that whether the collections started with six members, or six thousand, the result would be the same. The two altered collections, each having one additional member, would still have to be assigned the same counting number.[1]

This observation provides us with our second property:

> **Given two collections for which there is an exact pairing, if we put a single additional element in each collection, there will continue to be an exact pairing between the members of the augmented collections.**

For the time being, we leave this as a statement about collections. However, we will revise it as a statement about counting numbers later.

3.2.3 Third Property

Experiment 3. Again we consider two collections, neither of which consists of a single element, and between which there is an exact matching. Suppose we remove one element from each collection. Will there still be an exact matching? The situation is pictured below in a sequence of two diagrams:

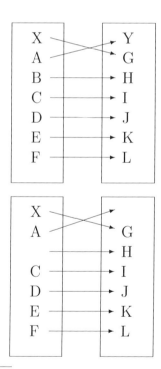

---

[1]The reader will of course know, that the counting number of the augmented collection is obtained by adding 1 to the original number.

A schematic of the two collections, each having seven elements (previous page), followed by a schematic in which one element has been removed from each (this page).

In the second diagram, the exact pairing has been disturbed by the removal of the item **B** from the collection on the left, and the item **Y** from the collection on the right. However, the exact pairing is easily restored by pairing the item **A** with item **H**, as shown below.

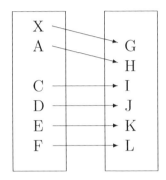

A schematic of the two altered collections with the exact pairing restored.

The reader could experiment with this by taking the original collections and, after removing one item from each, checking that an exact pairing will always exist.

The discussion above can be repeated for all collections except those consisting of a single element. If the two collections each have a single element, removal of that element results in an empty container in both cases. Experience tells us that if we have two empty jars, when we ask the question: How many are in the jar?, we get the same answer in either case, namely, none. Thus, even when we start with collections having only a single element each, when we remove one element from each collection, we get two entities (either two collections, or two empty containers) that are equivalent in respect to: How many?

What we have observed can be stated as our third property:

> **Given two collections for which there is an exact pairing, after removal of a single element from each of them, either it is possible to construct an exact pairing between the two remaining collections, or both collections cease to exist.**

3.2.4   Fourth Property

Experiment 4. Consider two collections, the first, which we call A, having some unknown number of elements, and the second, which we call B, having **exactly one more element than the first collection**. For concreteness, think of A as a jar containing three buttons and B as a jar containing four buttons. We ask the question: Is there a third collection, which we call C, having the following two properties **simultaneously**:

1. the third collection has more elements than A;

2. the third collection has less elements than B?

If you try this with actual collections, you will quickly conclude that because B has exactly one more element than A, you cannot construct a collection having more elements than A but less elements than B.

To make everything above explicit, we provide a schematic based on A having three members.

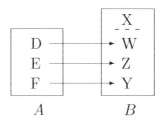

A schematic illustrating the result of trying to pair elements of A with elements of B. Since A has three elements, while B has exactly one more, there will always be one left over, as indicated by the unpaired X.

The purpose of this experiment is to find out whether there can be a collection, C, having more members than A, but less members than B. We will present an argument, based on pairing, to show that no such C can exist, because if C has more members than A, it must have at least one more! You could give your child a jar with three buttons and another jar with four buttons and ask your child whether they can make a collection of buttons with the two properties described for C and discussed below.

Suppose there were such a collection, C. What would C have to look like?

49

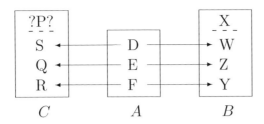

A schematic showing the relationship between the hypothetical C and previously discussed A which has three elements and B which has four. C is required to have more elements than A and less than B.

The diagram constructed assumes that C satisfies the first requirement, namely has more members than A. So, it must be the case that any attempt to pair elements of A with elements of C leaves something in C left over. This is indicated in the diagram by the **?P?**. We have placed question marks around the **P** to indicate that we do not know whether we are dealing with a single item or multiple items. Now, simply remove from C any items paired with an item in A, and remove from B any items paired with an item from A. This will produce:

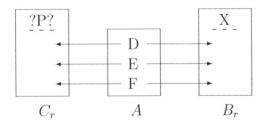

A schematic showing the collections C and B after removal of elements paired with something in A. The revised collections have been renamed, C_r and B_r, respectively.

What we can now say with certainty is that the revised collection, B_r (see diagram), consists of a single element, because we removed only elements paired with an element in A. Hence, B_r is an example of a smallest collection.

What we can also say with certainty is that C_r, the revised collection resulting from removal of the items in C paired with items in A, is that it has not been destroyed. Hence by the First property of collections, C_r must have at least as many elements as B_r.

Returning the missing elements to B_r and C_r shows C must have at least as many elements as B. Therefore, once it is known that C has more elements than A, it cannot be the case that C has less elements than B.

This discussion was based on a collection A that had three elements. But some thought and experimentation will convince you that the same reasoning applies to any pair of collections A and B for which B has exactly one more element than A.

We summarize this as:

> **Given two collections A and B. If B has exactly one more element than A, then there is no third collection C having more elements than A and less elements than B.**

3.2.5 Fifth Property

There is one last property of collections to be discussed. Suppose that you want to set a table for a dinner party. So you go to the drawer containing silverware and take out forks, spoons and so forth. What I want you to again consider is the exact process which was discussed in §3.1. If it is a large party, then rather than simply grabbing a handful, you will likely take the implements out one-at-a-time until you have the required number. And this is the key point: collections that arise in the real world can be assembled by a one-at-a-time process that starts from nothing and adds a single element to the existing collection at each step. We state this as:

> **Every collection arising in the real world can be assembled by a process that starts with nothing and adds to the collection a single element at each step.**

The reason why this is true is due to the Fourth property.

To understand the role of the Fourth Property, suppose we want to assemble a collection of buttons of a certain size. We start with an empty jar. We put in one button. Either, we have the collection we want, or, it lacks buttons. If the latter, we put in one more button. Again, either this is the size we want, or we have to add at least one more button. Because of the Fourth property, we can't skip over any possible sized collection of buttons if we only add one button at a time, so this process must lead to the collection of the size we are trying to construct. Finally, our understanding of the real world tells us that no collection of buttons is so large that we can not assemble it by adding one button at a time.

Mathematicians have a special name for collections of this type. They are called **discrete**. For our purposes, we will think of a discrete collection as a collection that can be constructed by putting in items one at a time.

In the most concrete terms, we can think of any discrete collection as being constructed by the same process as getting forks for a dinner party. We go to the fork drawer and one at a time get sufficient forks to set the table.

The **one-at-a-time** construction process is very important as it is the basis of counting and arithmetic. Your child needs to come to terms with this process as is evident from the CCSS-M learning goals.

3.2.6   Summary of Collection Properties

In the statements below, instead of referring to an **exact pairing** between the elements of collections, we say the collections have the **same number of elements**. The properties are restated as follows:

C1 There are smallest collections. A collection in this category contains a single element; removal of that element produces an empty container.

C2 Given two collections that have the same number of elements, if a single element is added to each collection, the augmented collections will also have the same number of elements.

C3 If two collections have the same number of elements and a single element is removed from each, then, either the two collections will have the same number of elements, or both collections will cease to exist.

C4 If a collection B has exactly one more element than a collection A, then there is no collection C having more elements than A, but fewer elements than B.

C5 Every collection can be realized by starting with an empty container and putting in elements one-at-a-time.

3.3   What Your Child Needs to Know

We have identified five properties of collections. They are fundamental properties that each child must come to terms with on an operational basis. The only way they can do this is by playing with collections.

3.3.1   Counting Goals for Pre-K and Kindergarten

By the end of Kindergarten is is expected that your child will be able to construct collections having twenty members and understand the process as set out below:

1. Understand the relationship between numbers and quantities; connect counting to cardinality.

    (a) When counting collections of objects, children should say the number names in the standard order, pairing each object with one and only one number name and each number name with one and only one object.

    (b) Understand that the last number name said tells the number of objects counted. The number of objects is the same regardless of their arrangement or the order in which they were counted.

    (c) Understand that each successive number name refers to a quantity that is one larger.

This means your child will know the numerals up to 20 (see Chapter 5) and understand the relation between a collection and its cardinal number. The child will understand that cardinal number is a **stable attribute** (unchanging property) of collections. The child will understand the effect of putting one more element into a collection on the cardinal number of the collection, that is, item (c) above.

In terms of learning the numerals, **Connect the Dots** at EDM is useful. In terms of cardinality and counting, **Bunny Count** again is useful. If you try these games you will know what the intention is and you can explore other activities, or indeed, make up your own using available tokens.

Chapter 4

Counting Numbers

Chapter Overview. The object of this chapter is to identify the properties of counting numbers that capture the five properties of collections. These properties will be the foundation for our arithmetic computations. Because the counting numbers are abstractions, we will be forced to adopt the language of sets. In this regard the notation $\mathcal{N}$ will be introduced to denote the ensemble of all counting numbers.

4.1   Why Are Numbers Necessary

In Chapter 2, we showed that each collection had a numerical attribute attached to it that answered the question: How many are in this collection? In Chapter 3 we studied the properties of collections related to this attribute. As the reader knows, these attributes were given the name **counting** or **cardinal numbers**.

In this section, we briefly consider the question:

Are counting numbers really needed?

In showing why each collection had a counting number, we used the notion of *pairing*. Specifically, for two collections, we could answer the questions: more?, less?, or the same?, by constructing a pairing between the members of the collections. So given we can do this, why do we need numbers?

One way to avoid the need for numbers and continue to use pairing would be if everyone carried a collection of tally sticks. For those unfamiliar with this concept, a **tally stick** is a piece of wood with markings on it, one mark for each item in a collection being tallied. For example:

54

A schematic of marks recorded on a tally stick.

Imagine that each of us carries a standard collection of tally sticks. (They don't have to be sticks, they could be made of some efficient material so we could have lots of them.) Each time we wanted to know: How many?, we would consult our tally sticks using the pairing process to find the answer. Although this process is very concrete, it is also very ineffective for dealing with all but very small collections. Clearly, a better alternative had to be found.

4.2   Counting Numbers as Abstract Entities

The most important fact about counting numbers is that they are abstract representations of an attribute of things that exist in the real world. In that sense, they are like **beauty**. To make this comparison precise, consider two observers looking at a vase containing five roses. Both observers will have ideas in their heads as to whether the flowers in the vase satisfy their own notions of what it means to be beautiful. And both will also have an idea in their head of how many flowers are in the vase. In this sense, counting numbers and beauty are alike: they are both ideas in peoples heads. But there is a fundamental difference. The two observers may not agree on whether the flowers are beautiful. But if both observers were asked to generate tally sticks showing the number of flowers in the vase, we are sure the tally sticks would both show five marks. Succinctly stated, the essential point is:

> counting numbers are abstractions that are precisely defined and about which there is universal agreement.

The author's contention from the beginning has been that the foundation of arithmetic is the behavior of collections. So, by studying collections, we can understand the essential properties of counting numbers. In the last chapter we identified five critical properties of collections as related to the concepts of more and less. The entire thrust of much of the pre-K and kindergarten math program is to teach these properties to children as they apply to counting numbers. Of course they don't explicitly tell the children this. Rather, they supply them with activities that will ensure that they "get it". My belief is that if you are going to be successful in helping your child and making judgements about where to concentrate your efforts, you need to understand how the five properties of collections are translated into the arithmetic properties of counting numbers.

To this end, we restate properties C1–C5 of collections for easy reference:

C1: There are smallest collections. A collection in this category contains a single element; removal of that element produces an empty container.

C2: Given two collections that have the same number of elements, if a single element is added to each collection, the augmented collections will also have the same number of elements.

C3: If two collections have the same number of elements and a single element is removed from each, then, either the two collections will have the same number of elements, or both collections will be destroyed.

C4: If a collection B has exactly one more element than a collection A, then there is no collection C having more elements than A, but less elements than B.

C5: Every collection can be realized by starting with an empty container and putting in elements one-at-a-time.

Using Letters as Names

In the lines above, we have used capital letters as names for collections. When mathematicians use letters this way we refer to them as **variables**. By using a variable as a name, we are able to speak about an arbitrary object, as opposed to, a specific object. For example, in what follows we will be speaking about cardinal numbers and we will often use a letter to represent such a number. Thus, we can say n is a cardinal number, and this does not convey any more information than the fact that n is a number of the type specified. By using letters in this way, we are able to speak about all numbers of a certain type. Alternatively, we could pick a particular cardinal number as a representative to work with, for example 19. We will sometimes do the latter for purposes of clarity.

As an example of a general assertion about all counting numbers, consider the Commutative Law which states that the addition of any pair of counting numbers n and m satisfies the equation

$$n + m = m + n.$$

The reader knows that if we replace n and m, respectively, by any other counting numbers whatsoever, then a true statement will result, as in

$$18 + 94 = 94 + 18.$$

We stress that in turning a general statement involving variables into a particular example using numbers, all instances of each letter must be replaced be the **same** number. (In the example above n is replaced by 18 and m by 94.)

The CCSS-M begins introducing this type of mathematical notation as early as Grade 2 and expects students to be completely comfortable with the use of variables by Grade 6.

4.3 Properties of Counting Numbers

In what follows, we will develop the abstract notion of **counting numbers** ($\mathcal{N}$) and the operation of **successor**. Our intention is that counting numbers and successor (adding 1) can serve as the basis for arithmetic. To be successful, this abstraction, $\mathcal{N}$ and successor, must correctly capture the behavior of the numerical attribute of collections that we are calling **cardinal number** and the process of counting. It turns out to be harder to specify appropriate properties for $\mathcal{N}$ and successor than it was to identify C1-C5 in the first place. Rather than trying to cope with the underlying logic, some readers may want to go over the properties N1-N6 and then proceed directly to Chapter 5 which is geared more directly to what is appropriate for children.

The ensemble that consists of all the attributes that we have been referring to as cardinal numbers for collections that exist, or could exist, in the real world, will be called the set, $\mathcal{N}$. Members of this set will be referred to as **cardinal** or **counting** numbers.

The set $\mathcal{N}$ has for members things that only exist in our minds. Indeed $\mathcal{N}$ only exists in our minds. This is why we use the descriptor **set** as opposed to **collection**. But we want to make our thinking as concrete as possible. To this end, we will imagine a standard collection of tally sticks. The idea is that the standard collection of tally sticks will contain exactly one tally stick corresponding to each counting number (cardinal number). Throughout the remainder of this chapter we will refer to this collection as the **standard collection**.

Our task is to articulate the properties that counting numbers and $\mathcal{N}$ must have if they are to model the five properties of collections identified in the last chapter and listed above as C1-C5. We will identify these properties by considering:

What do C1-C5 tell us about the tally sticks in the standard collection?

4.3.1 Property 1

N1: 1 is a member of $\mathcal{N}$.

We know from C1 that there is a counting number associated with smallest collections to which we gave the name *one* and the notation 1. Further, we know the standard collection must contain a tally stick corresponding to 1. This tally stick must look like the stick pictured below:

The tally stick associated with smallest collections as set out in C1.

Since 1 is the name of the counting number associated with the tally stick pictured above, and $\mathcal{N}$ includes all counting numbers, 1 must be a member of $\mathcal{N}$. N1 merely confirms this fact.

But a question remains:

How do we know the tally stick pictured above is the one associated with smallest collections?

We know this because it has a single mark and if we remove that single mark then we will have a tally stick that is unmarked and is no-longer a tally stick.

When we think about these requirements in the context of the ensemble $\mathcal{N}$, we are thinking about an entity which is an abstraction and is composed of counting numbers, each of which is also an abstraction. N1 tells us that there is a counting number in $\mathcal{N}$ having the name *one* and the notation 1. But, that's all N1 tells us. In particular, N1 doesn't tell us that 1 is the smallest counting number. Making sure the element we identify as 1 has all the correct properties is why things get hard. The only way we have of ensuring the counting number we are calling 1 actually is what we intend it to be, namely the counting number associated with the tally stick pictured above, is to specify its **behavior**. That is, we make it do the things we know it has to do according to C1–C5. In this respect, our primary target is C5. Specifically, we want the process of

adding 1 to a counting number to correspond to putting a single additional element into a physical collection.

4.3.2   The Operation of Successor

Next consider C2, which tells us that if we have two collections for which there is an exact pairing, and we put a single new member in each collection, there will be an exact pairing between the augmented collections. We want to translate this into a statement about counting numbers. To do this, we analyze C2 which is about physical processes taking place in the world.

Suppose we have two collections, A and B for which there is an exact pairing. This means A and B have the same size (cardinality). Call this cardinal number n. C2 now says that if we add a single member to each collection, we get new collections A' and B', and there will be an exact pairing between A' and B'. This means that A' and B' also have the same cardinality, and hence the same counting (cardinal) number. We call this new counting number which gives the size of A' (B'), the **successor** of n, and give it the notation $n+1$. We also call n the **predecessor** of $n+1$.

For clarity, suppose we had two jars containing five buttons each. Place one additional button in each jar. Both jars now have six buttons each. What's more, we know that if anyone else in the world repeated the process of placing one more button in a jar containing five buttons, after the addition, the jar would contain six buttons.

We can think of **successor** as a real-world **operation** (procedure) that takes us from n to $n+1$ that is defined by:

> Given a counting number n, find a collection, A, that has n members. Form B by adding a single element to A. The successor of n is the number of members in B.

We call this procedure an operation because C2 guarantees that however we construct the collection A so that it has n members, and whatever **single** element we add to make B, we will always arrive at the same counting number as the successor to n. For this reason, we are able to give this number the unique name $n+1$ and know that there will be universal agreement as to which tally stick it corresponds to.

We can picture this process using tally sticks:

> The **process** diagram depicts the successor procedure. Start with 6, construct a collection with 6 members as indicated by the tally stick on the left, move to the tally stick on the right by adding one additional element. We call the associated counting number the successor of 6 and give it the notation $6+1$.

The successor operation is at the heart of counting. (To see why we say this simply re-read §3.1 on the counting goals for Kindergarten.) So our number system and the notation we use for counting numbers must capture this process. It will require several mathematical assertions about the behavior of members of $\mathcal{N}$ to capture all the important facts in relation to the successor procedure as derived from the process for collections.

4.3.3 Property 2

N2: If n is a member of $\mathcal{N}$, then the successor of n, namely $n+1$, is a member of $\mathcal{N}$.

N2 tells us that the number we are identifying as the successor of a counting number is again a counting number. Thus, whenever a number n is in $\mathcal{N}$, there has to be something in $\mathcal{N}$ that is called the successor of n. But as with N1, which says there is something in $\mathcal{N}$ that plays the role of 1, we still don't know that the counting number we have labeled $n+1$ is the number identified by the successor procedure. Since the successor of n in the real world is obtained through addition of a single element to a collection of n elements, it seems clear that for $n+1$ to have the right properties, 1 has to have the right properties. Property 3 will guarantee that.

4.3.4 Property 3

Let's go back to C1. It says there has to be a smallest collection. However, we know that smallest collections cannot be obtained by starting with a collection and adding a single element. This is because all collections have something in them and smallest collections have only a single element. To make this concrete, consider the figure illustrating the successor process using tally sticks (see §4.3.2). What we are saying is that the tally stick associated with a smallest collection cannot occur on the right side of the arrow in this figure. Since 1 is intended to be the smallest counting number, let's use the fact that its tally stick cannot be on the right side of the arrow in the figure as an essential property of 1:

N3: 1 is not the successor of any counting number.

In simplest terms, N3 says 1 has no predecessor and in this sense it is like all collections that have only one member. N1, N2 and N3, do not by themselves make 1 the smallest counting number, but N3 does give 1 a key property that 1 has to have. If we can ensure that 1 has to be the least counting number, then this will in turn force $n+1$ to have the property required by the process diagram above.

4.3.5 Property 4

N4: If n and m are any two counting numbers such that $n = m$, then $n+1 = m+1$.

N4 completes the capture of C2 for counting numbers, and makes the process $n \to n+1$ (illustrated in the process diagram) always result in the same counting number as required by C2.

4.3.6 Property 5

N5: If n and m are any two counting numbers such that $n + 1 = m + 1$, then $n = m$.

N5 tells us that different counting numbers have different successors. This captures C3.

4.3.7 Property 6

The last property we need corresponds to C5, which says we can construct every possible collection by starting with nothing and adding elements one-at-a-time. As described in §3.1, this process is the basis for counting. Replacing *collection* by *counting number* in C5 gives us:

N6: Every counting number, n, can be obtained by starting at 1 and applying the successor operation repetitively.

Let's see how N1-N6 apply to constructing the standard collection. This collection has to have exactly one tally stick corresponding to each counting number. We start with an empty jar which naively we label **standard collection**.

Following N1–N3, the first thing we do is put a tally stick corresponding to 1 into the jar. This tally stick looks like:

By properties N1–N3, we have to put this tally stick into the jar. Note this is the only tally stick which cannot occur on the right side of the process diagram.

That was easy, but now the real work begins because we have to follow N2. N2 tells us that whenever we find n in $\mathcal{N}$, we must also be able to find $n + 1$ in $\mathcal{N}$. So we apply this to the standard collection in the following way. Since a tally stick corresponding to 1 is in the standard collection, by N2, there must also be a tally stick corresponding to $1 + 1$, the **successor** of 1, in the standard collection. Then, since the tally stick corresponding to $1 + 1$ is in the jar, we have to place a tally stick corresponding to $(1 + 1) + 1$, the successor of $1 + 1$, into the jar. And then we have to put a tally stick corresponding to $((1 + 1) + 1) + 1$, the successor of $(1 + 1) + 1$, into the jar, and so forth.

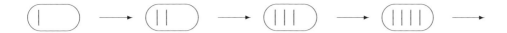

As soon as we put a tally stick in our standard collection, we have to add the tally stick corresponding to its successor. This is a process that never stops!

Does the process ever stop? No, never! We have to continue adding successors forever. Obviously, it is impossible to create a standard collection with a complete collection of tally sticks, one for each counting number, in the real world.[1] To do so would require generating an **infinite** collection. But such collections don't exist in our world. That's why $\mathcal{N}$ is a set and not a collection. Finally, the effect of N6 is that $\mathcal{N}$ consists only of things that can be obtained from 1 by the application of successor. There are no other counting numbers, just as there are no collections that cannot be constructed by the one-at-a-time process.

At this point we have what we need to ensure that the abstract notion of counting numbers captures the required properties of collections. It is possible to generate all of arithmetic from these properties. But that process is beyond the scope of this book. Instead, our approach will be to set down some assertions which we will take to be true. (Mathematicians would refer to these assertions as **axioms**.) We will justify the assertions using conservation. We will then use these axioms as the basis for our arithmetic.

4.4 What Your Child Needs to Know

The properties discussed in this chapter are aimed at identifying and capturing the key properties of the principal numerical attribute of collections, namely their size. In sum total what the properties come down to is that counting numbers let us count!

Your child must learn how to count, not by studying what is in this chapter, but by counting how many things are in collections found in the real world. The counting experience should be encouraged at every turn. This shouldn't be hard, since Primary school teachers I've spoken to all report that children love to count.

In the next chapter we will discuss the Arabic system of numeration. Coming to terms with this system is a multi-year process for the child. That said, one essential

---

[1]A Kindergarten teacher told the following story about one of her students. The child reported with great pleasure counting to a very large number. A bit later the child returned and reported counting to an even larger number. This was repeated several more times, each time with less happiness. Finally, the child returned a last time and with tears in his eyes said: "There is no largest number, is there?" A great discovery!

pre-K goal is that children will be able to count collections of twenty or so items with facility.

Let's be precise about the expectation. This means for a jar of twenty, or so, marbles, the child can count the marbles by choosing a first marble, assign it 1, choose a second marble, assign it 2, choose a third, assign it 3, and continue this process of choosing and assigning a number, until the last marble is assigned a number. At the conclusion of the process the child knows that the last number assigned is the number of marbles that are in the jar. The activities generated by **Bunny Count** at EDM develop these skills.

4.4.1   Counting Goals for Pre-K and Kindergarten

Goals for Pre-K were set out in §2.6.1. Because of the importance of the counting process, we reiterate the Kindergarten goals listed in §3.1 and §3.3.1.

By the end of Kindergarten is is expected that your child will be able to construct collections having twenty members and understand the process as set out below:

1. Understand the relationship between numbers and quantities; connect counting to cardinality.

    (a) When counting collections of objects, say the number names in the standard order, pairing each object with one and only one number name and each number name with one and only one object.

    (b) Understand that the last number name said tells the number of objects counted. The number of objects is the same regardless of their arrangement or the order in which they were counted.

    (c) Understand that each successive number name refers to a quantity that is one larger.

Chapter 5

Making Counting Numbers Useful

Chapter Overview. The main purpose of this chapter is to introduce the Arabic System of notation for counting numbers. A thorough knowledge of this system is essential because it supports all our numerical computations. Because this system is so powerful, it requires multiple years in the curriculum for children to master. You will need to understand this system in detail if you are to help and support your child's learning. The study of this system will be facilitated with a concrete realization that we refer to as the **Button Dealer System** which will be used in later chapters to provide concrete illustrations of addition and subtraction. Finally, we will describe grade related expectations tied to Core Standards that will enable you to assess your child's progress.

There are two distinct ways in which we use counting numbers as applied to collections. The first is to record the number of objects in a collection found in the world. This, it seems most likely, is the objective that led to the recognition and generation of counting numbers. The second is, given a counting number, to construct a collection having exactly that number of members. These uses are illustrated below:

A diagram showing a box containing a number of buttons on the left. To get beyond tally sticks, we must have a **name** and a **numeral** for the cardinal number of each such collection. In this case, the name is *six* and the numeral is 6. But the number itself is an abstraction that exists only in our minds.

A diagram showing a box containing a numeral on the left. The requirement is to construct a collection containing the number of buttons identified by this numeral. The process by which this collection is constructed is **counting**. The constructed collection is displayed on the right.

As the reader can see, both of these uses require a way to represent numbers, in other words, a **system of numeration**.

In the scenarios developed in Chapter 2, we found it was easily possible to compare the size of collections with no direct knowledge of numbers, as long as the collections were not separated in time and or space.

One solution to the separation problem is the use of a **tally stick** as a means of recording numbers. For example, consider a scout for a tribe of early hominids that wants to convey information to the chief on the number of interlopers crossing the boundary into their territory. By observation the scout could make a tally stick that contains one mark for each interloper and send this to the chief.

A schematic of an early hominid tally stick that simply matches marks to members of the collection of interest.

Or, consider the child with cows in one field and horses in another trying to judge whether there are more cows or horses. Again a tally stick with one mark for each cow in the first field solves the problem. The child can use the tally stick as a standard against which to compare the number of animals in any other field.

While a tally stick will solve both problems, a little thought will quickly lead to the conclusion that tally sticks are not a good solution.

Consider the second process, namely, constructing a collection with a specific number of elements. A very common example of this arises when the cousins come to dinner and you want to set the table. First you need to know how many cousins there are. Now you could have a tally stick for cousins, but clearly this would lead to closets chock full of tally sticks for cousins and all the other sundry things you need to keep track of. But, given a tally stick for cousins, it would be possible to set the table simply by creating one place setting for each mark on the stick. Indeed, putting out place mats and telling a child to put one set of silverware on each place mat amounts to using a tally stick.

What is clear from the above is that the mere fact that each collection can be associated with a unique counting number has little utility unless counting numbers can be accurately transmitted through time and space, and from one person to the next. The solution that enables men to transmit a culture of ideas through space and time is **language**. And so it is with numbers. What is required is a language for describing numbers. At a minimum, this language must contain a name for every counting number, a tall order given there are an unlimited number of counting numbers, as we concluded in the last chapter.

To solve the problem of language, we have to agree on a name and notation — way of writing the number — for each number we assign to collections. In addition,

the name of each number must be unique.

In other words, that different numbers must have different names and, different names must identify different numbers.

Coming up with a system for naming the counting numbers was a challenge because, as we have seen, there are an unlimited number of numbers, and the name and notation for each individual number must be **universally recognized**.

To see why having to name each member in an unlimited collection causes difficulties, consider using tally markings based on groups of five (shown below) which we have used to denote the number eighteen:

A schematic for a more sophisticated tally stick based on groups of five.

This system of notations does produce a unique representation for each counting number. In addition, any notation constructed in this system will, almost certainly, be universally recognized and correctly interpreted. But the notations are unwieldy to say the least. For example, the notation for one hundred, which is not a particularly large number, requires twenty groups of five.

But a system of notations based on tally markings has a greater failing than merely being unwieldy.

To see why, consider again the problem of the scout. Suppose the scout observes the number of interlopers shown on the tally stick represented above. How can he communicate this to the chief? If the only system available to the scout for communicating numbers is the tally stick, he must show the tally stick directly to the chief to transfer the information. This is a severe limitation on the communication process. Thus, to realize the full utility of numbers, we must have not only a notation for each number, but also a verbal name.

One way around the length problem associated with tally marks is to introduce a unique symbol for each counting number. Presumably, a verbal expression would correspond to each symbol. The problem with this approach is that such a system would require an unlimited number of different symbols. Moreover, as ever larger numbers are needed, new symbols would have to be created. How would these new symbols become universally accepted in a commercially active world that demands universal acceptance of mathematical symbols at all times in order to function? This is simply not possible if new symbols have to be made up each time a larger number is needed.

The solution to the number notation/naming problem was a major step forward in human intellectual development. Aside from tally sticks, several candidates were tried and discarded before the current system came into use. We will study this system of numeration in detail because it is the heart of our computational system. Mastery of this system is the second great intellectual step your child will take during the study of arithmetic. (The first is, of course, learning to count.)

5.1   The Arabic System of Numeration

In the system we use for counting with which we are all familiar, each number has been given a **name** and a **numeral**, as follows: *one* or 1, *two* or 2, *three* or 3, *four* or 4, *five* or 5, and so forth.

To be clear, names are the words we use in language. Numerals are the symbols (notations) we use in computations. Neither is the number they represent anymore than your name is you. Thus, the name *one* is the name of the number for which we employ the symbol (numeral) 1. The number denoted by this name and numeral is an abstraction and does not exist other than in our minds. That being said, in common usage we speak of *the number* 1 instead of the *number denoted by the numeral* 1.

Below we display diagrams showing numerals for the first nine counting numbers together with a collection containing that number of items, that is, having the same cardinality. Complete knowledge of the relationship between each numeral and the associated collection is a key learning goal to be achieved by the end of Pre-K.

A series of schematics showing the nine single digit numerals associated with collections, together with the collections they enumerate. Notice that as we proceed through the collections from smaller to larger, each collection contains one more item than its predecessor, thus illustrating the one-at-a-time process for generating collections which is the basis for the counting process.

5.1.1 The Roles of Zero

There is one special number having a single digit numeral which has not been listed because it is not associated with a real-world collection. Recall that collections, as we are using the term, **have to have members**. It took human beings a long time to realize that the result of removing the last member from a collection resulted in something that needed a numerical description. What we are saying is,

nothing also needs a number.

The idea that nothing also needs to be counted is really abstract, much more abstract than counting numbers themselves. The evidence for that is that it took human beings an extra 2000 years to discover that nothing needed a number even though they were using 0 in its other role as a **place-holder** in the number system.[1]

The use of zero as a place-holder will be discussed at length when we consider the Arabic System of numeration in detail. So we leave the second use for the time being and return to considering zero as a number.

If you ask yourself, How do I think about nothing?, you begin to see the problem with arriving at zero as a number that represents a quantity. What seems most direct is the notion of **empty set**, that is, a set with nothing in it. While a collection in the real world ceases to exist when we take out the last element, in our minds we can imagine the **empty set** as being what is left. And if we are asked how many elements are left, we would say: None. Assigning this a number leads directly to the

---

[1]According to Wikipedia, zero as a number was identified in India by the 9th century AD. The difficulty of recognizing zero as a number is evident from the fact that the requirement for a zero-like place holder was known to Babylonians 2,000 years earlier!

idea that a set with nothing in it has **zero** members. The numeral for zero is: 0 , and a graphic descriptor analogous to the ones above is:

A schematic of an empty set on the right and the numeral of its counting number on the left.

The numeration system we now use is the **Arabic System**. It uses ten digits, which in order starting from zero are:

$$0, 1, 2, 3, 4, 5, 6, 7, 8, 9$$

Corresponding to each symbol is a verbal name, which for completeness in corresponding order is:

zero, one, two, three, four, five, six, seven, eight, nine.

The symbols 0 – 9 are the only symbols that occur in the Arabic notation for any counting number. Since there are an unlimited number of numbers, these symbols may have to be used more than once in the expression for a particular number. How this is done is one of the really clever features of Arabic notation. Other systems, for example Roman numerals, also use the same symbol multiple times, but not nearly as effectively, and not in a way that connects to the system of computation. The connection to computations is, perhaps, the key reason why Arabic notation became universally accepted. Understanding the Arabic System of notation and developing fluidity with its computational schemes is an essential goal of the CCSS-M.

5.1.2 A Concrete Realization of the Arabic System

To understand arithmetic and to be able to help your child, you need a thorough understanding of the Arabic numeration system. For this reason, we develop an example, based on a mythical button supplier that provides a physically based realization of the Arabic System of numeration. We will use this realization as the starting point for all our explanations of how the computational procedures work. These explanations will ultimately trace all computations back to counting! So it is important that you thoroughly understand this example. It is also the case that many of the manipulatives used in pre-K–Grade 1 to explain the Arabic System to children incorporate many of the same ideas.

Button Dealer's System. The button supplier has a very large supply of buttons. Customers show up, tell the supplier how many buttons they need, and he fills their order. To do this efficiently, the button supplier keeps his buttons in jars labeled with one of the numerals 1, 10, 100, 1000 and 10000. Each jar contains the number of buttons identified by its numeral. So a jar with 100 on the front contains one hundred buttons. The jars are stored on shelves according to the following system he has devised:

> on the first shelf, jars labeled with: 1;
> on the second shelf, jars labeled with: 10;
> on the third shelf, jars labeled with: 100;
> on the fourth shelf, jars labeled with: 1000;
> on the fifth shelf, jars labeled with: 10000.

Suppose a customer shows up who requires 6038 buttons. To satisfy this order the dealer looks at the notation, 6038, and proceeds as follows. Since Arabic is read right-to-left, the dealer starts with the right-most digit. This digit is an 8, so he goes to the first shelf and gets *eight* jars marked with a 1. The second digit, one to the left of the 8, is a 3, so he goes to the second shelf and gets *three* jars, each one of which is marked with a 10 and puts them with the jar marked 8. The third digit is a 0, so he gets no buttons off the third shelf. The fourth digit is a 6, so he goes to the fourth shelf and gets *six* jars, each marked with 1000 and puts them with the other jars. Since there are no more digits in 6038, he combines the buttons from the seventeen jars into one big jar and gives the buttons to the customer.

Given that the various jars contain the number of buttons specified, it is clear that this simple process will produce a jar containing exactly 6038 buttons. We want to consider why this process will work.

5.1.3   Designing the Button Dealer's System

To understand why the Button Dealer's System system works, we need to analyze the process of designing this system for supplying buttons. We will assume that the designer knows about the numerals 1 to 9 and the cardinal numbers they represent as indicated in the next diagram.

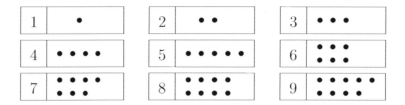

The diagram presents the exact relation between each of the symbols used by the Arabic System and the numerical attribute these symbols represent.

For purposes of discussion, we will assume the designer has to meet the following criteria:[2]

1. the system should use the symbols 1 — 9 and as few others as possible;

2. the system should require a minimum of counting to fill an order;

3. the system should be easy to use and to learn how to use;

4. the system should be perfectly accurate;

5. the system should be able to fill any order for an amount of buttons up to some maximum size.

With these criteria in mind, setting up a system to serve customers where the maximum order size is *nine* is straightforward. The designer simply puts jars containing a single button each on a shelf. To serve a customer, the server has to know and understand the contents of the previous diagram perfectly in order to get the correct number of jars off the shelf. The server never has to count to more than *nine*, since that is the maximum order size. In addition, the server has to know that a collection of the required size can be constructed by taking one jar off the shelf for each dot in the collection associated with the numeral specified by an order, and combining all the buttons from the various jars into one jar which is given to the customer.

---

[2]It would be nice to think our ancestors were clever enough to actually plan things out. History appears to suggest the Arabic System was the result of a long trial and error process that took thousands of years to complete.

Ordering More Than 9

Major design questions arise when the system has to accommodate orders of more than *nine* buttons. To be accurate, the system must not skip numbers, which means that the first number that has to be considered carefully after nine is the successor to nine. Obviously an order for *ten* buttons could be accommodated by simply filling an order for *nine* buttons and getting *one* more jar containing a single button off the shelf. Notice that counting to *ten* means our server must come to terms with a diagram that looks like:

where the question mark indicates an appropriate, but unknown, notation for the successor of *nine*. In other words, as soon as the Button Dealer wants to fill an order for *ten* buttons, he has to have a notation for the number *ten*.

Because our system is comprised of jars of buttons, it is reasonably easy to imagine simply creating a second shelf on which we place jars, **each of which contains ten buttons**. This would satisfy the minimal counting criteria because to fill an order for *ten* buttons we only have to count to *one*. Because *ten* is the successor of *nine*, we know we didn't skip a number. This is the easy part.

The hard part is figuring out a suitable notation for this number, that is, what to write on the jars on the new shelf.

We could, for example, come up with a new, single-digit notation, as in:

It would be possible to create a system based on this idea that would look something like Roman Numerals. While such a system could provide notations for numbers, it would not support computational procedures in the way the Arabic System does.

The really critical insight that our designer came up with was the realization that when a second shelf with jars containing *ten* buttons was created, the label on these jars must convey **two** pieces of information:

1. get no jars off the first shelf;

2. get one jar off the second shelf.

The question is:

> How can we communicate these two pieces of information using the numeration system?

Here is where the system designer got incredibly clever.

First, the designer realized that **communicating two distinct pieces of information would require two symbols**, not one. Further, since one of the instructions was *get no jars from the first shelf*, a new symbol was required that would instruct the user to:

$$get\ no\ jars$$

off a particular shelf. The result of this realization was the creation of the "do nothing" symbol, namely:

0	

A schematic of the new symbol, 0, that counts the items in an empty jar.

At this point the designer had all the symbols that are needed to convey the two pieces of information identified above. The question left was:

How should these symbols be displayed?

It seems a relatively small step at this point to simply say that the order in which the symbols are read will be the order in which the shelves are accessed. Thus, the first symbol read tells what to get off the first shelf, and the second symbol tells what to get off the second shelf. This being the case, why should

$$10$$

be read as the instruction:

get no jars off the first shelf;
get one jar off the second shelf,

and not the other way around? The answer to this question is that European languages are written and read left-to-right, while Arabian languages are written and read right-to-left. Thus, to the Arabic speaking designer of our button system, the first symbol in 10 is the 0 and the second is 1. Understanding this fact is an essential feature of the Arabic System of notation.

Thus, the notation for the successor of 9 in the Button Dealer system is:

$$10,$$

and it is understood by the user to mean:

> *get no jars off the first shelf,*
> *get one jar off the second shelf.*

Finally, the designer will realize she needs a name for 10, and makes up the word *ten* to correspond to 10. This discussion is summarized in the following digram:

In the diagram, the 9 jars containing one button each are combined with 1 jar containing a single button. The result is equivalent to a single jar containing ten buttons shown on the right-hand side of the equality and identified with the numeral 10.

Finally, we note that the notation 10 is the numeral for the smallest counting number that cannot be expressed with a single-digit numeral.

Given the designer invented 10 and its corresponding instructions, she will know that 11 has to mean:

> *get one jar off the first shelf,*
> *get one jar off the second shelf,*

12 means:

> *get two jars off the first shelf,*
> *get one jar off the second shelf,*

and finally that 19 means:

> *get nine jars off the first shelf,*
> *get one jar off the second shelf.*

So customers requesting 1 – 19 buttons can now be served. We illustrate serving a customer wanting 17 buttons:

An order for *seventeen* buttons is made up by combining *seven* jars containing one button each from the first shelf and *one* jar containing *ten* buttons from the second shelf.

By making similar diagrams, the reader can verify that orders corresponding to each counting number less than *twenty* can now be filled with perfect accuracy. This truth is entirely due to the fact that each new numeral is a notation for the successor of the previous number for which a numeral had been constructed.

The next question would be what to do about a customer needing the successor of 19 buttons. To deal with this, all the designer has to do is to notice if she combines two jars containing 10 buttons each, that is, two jars from the second shelf, she will have the required number of buttons, as shown below.

A diagram illustrating that *twenty* is realized as two groups of *ten*. To verify that *twenty* is in fact the successor of *nineteen*, construct 19 via the process illustrated above for 17. Then observe that the successor has the same diagram as shown for 20.

This is a critical insight, because it forces the notation for the successor of 19 to be 20, which then translates to the instruction:

> *get no jars off the first shelf,*
> *get two jars off the second shelf.*

This interpretation of 20 exactly extends the meaning of the notation previously created. Moreover, it is clear that we can now represent any number of buttons from 10 — 99, inclusive.

To see why, first observe that numbers of buttons equivalent to

> *ten*, *twenty*, *thirty*, *forty*, *fifty*, *sixty*, *seventy*, *eighty*, *ninty*,

can be obtained by using the notations

$$10,\ 20,\ 30,\ 40,\ 50,\ 60,\ 70,\ 80,\ 90$$

which mean

> *get no jars off the first shelf,*
> *get* 1 — 9 *jar(s) off the second shelf,*

respectively. These are all the numbers of buttons that can be obtained using a single digit descriptor of numbers of jars from the second shelf and no jars from the first shelf.

Missing numbers of buttons are obtained by getting the correct number of jars off the first shelf, as in the case of 17, 27, 37, etc., all of which require *seven* jars from the first shelf.

No numbers are missed because, a two-digit number whose numeral ends in 0 is the successor of a number whose right-most digit is 9, as 90 is the successor of 89. This is a consequence of the fourth property of collections (C4) identified in Chapter 3. It is also why the successor idea as the **next** counting number is critical.

The naming scheme in English adopted for two-digit numbers greater than twenty is particularly simple, once we have names for twenty, thirty, etc. It simply amounts to reading left-to-right. For example, for 25, we simply say **twenty-five**; in other words, the name is built from the digits in the expression reading left-to-right.

Once the designer has figured out how to deal with numbers from 1 - 99, the design methodology is established. It simply repeats itself. For example, since all possible two-digit combinations of the symbols 0 — 9 are used in making the notations for 0 — 99, the successor of 99 will require using a third shelf and a third digit to communicate this extra piece of information. What goes on the third shelf is jars containing a number of buttons equal to the **successor** of 99 which is the next counting number. The notation for this number is:

$$100$$

which reading right-to-left translates into:

> *get no jars off the first shelf,*
> *get no jars off the second shelf,*
> *get one jar off the third shelf.*

Because of its importance, we reiterate the following. Since 10 is the successor of 9, a collection of 10 buttons can be obtained by combining *ten* jars containing a single button each. But this also means that 100, the successor of 99, can be obtained by combining the contents of *ten* jars, each of which contains 10 buttons. While this is something you surely know, you may not have thought about it in quite this way.

If we now pick any three digit number, it is clear what the instruction will be using the Button Dealer's scheme. For example, 836 instructs:

> *get six jars off the first shelf,*
> *get three jars off the second shelf,*
> *get eight jars off the third shelf,*

where the digits are read right-to-left. Combining the buttons from the all the jars into one produces a jar containing exactly 836 buttons. We remind our readers that this is a direct consequence of conservation (**CP**).

At this point, the Button Dealer can now fill any order for $1 - 999$ buttons.

Again, there has to be a new name for jars on the third shelf. As you know, it is **one hundred**. This name, for the cardinal number that is the successor of 99, has a different flavor to it. To be specific, it identifies *hundred* as the essential name of the successor of 99 and the *one* tells us that we want exactly *one* unit of this size. So for example, the name associated with 500 is **five hundred** and specifies *five* units of *one hundred*. Compare this with **fifty**, which corresponds to 50. The name *fifty* is a single unit and it is much less transparent that this name is telling us to get *five* units containing *ten* each, as opposed to the numeral 50 which literally specifies getting *five* units containing ten each.

Using this naming scheme for three-digit numbers, in speaking of the request for 836 buttons, the Button Dealer would say that

eight hundred thirty six

buttons were supplied. Again notice the right-to-left interpretation of 836 when getting the actual buttons, as opposed to the left-to-right interpretation when speaking English.

Recall, when we reached 9 we needed a new name and numeral for the successor of 9; similarly, when we reached 99, we needed a new name and symbol for the successor of 99. For the same reason, namely, when we get to 999, we will have used all possible three-digit combinations of our symbols, and so we will need a new shelf, a new name and a new symbol for the successor of 999. Each jar on this new shelf will contain the combined contents of *ten* jars of 100 buttons each. As we know, the numeral on the jars is 1000 and its name is **one thousand**. The naming scheme at this point follows that used for hundreds. For example, 7000 is named **seven thousand**, and corresponds to the instruction to the button dealer to:

get no jars from the first shelf;
get no jars from the second shelf;
get no jars from the third shelf;
get seven jars from the fourth shelf.

Introduction of **thousands** on the fourth shelf permits the Button Dealer to accommodate all orders up to 9999 buttons, including our original example:

6038.

Once again, the designer will need a new numeral and name for the successor to 9999. The notation is 10000, and the name is **ten thousand**, which is a combination of the previous names *ten* and *thousand* and reflects their placement in the numeral $10,000$, where we have inserted a comma to emphasize the point.

It is now possible to provide names and notations for all counting numbers up to 99999. It is obvious that to a button server, a request for

$$75756$$

buttons, gets translated to:

get six jars from the first shelf;
get five jars from the second shelf;
get seven jars from the third shelf;
get five jars from the fourth shelf;
get seven jars from the fifth shelf.

A little thought will convince you that the Button Dealer could extend this system to accommodate arbitrarily large orders simply by adding more shelves as needed to the system. The required notation is built in, although there would have to be some new names created.

Let us recall the design criteria specified at the beginning of this section to see if we satisfy the requirements:

1. the system should use the following symbols, 1 — 9 and as few others as possible;

2. the system should require a minimum of counting to fill an order;

3. the system should be easy to use and to learn how to use;

4. the system should be perfectly accurate;

5. the system should be able to fill any order for an amount of buttons up to some maximum size.

The completed system uses one additional symbol beyond the symbols 1—9. Since there has to be a symbol associated with *getting no buttons*, and all the other symbols are associated with getting some number of buttons, any system will have to have this additional symbol, 0, whence this addition cannot be viewed as a failure of the first requirement. A user has to count at most *nine* jars on any shelf, which is minimal.

In order to use the system, one needs to know only two things. The first is the relationship between each single digit numeral and its standard collection, in other words, how to count to nine. The second is how the position of a digit in a numeral specifies a shelf to go to.

5.1.4   The Problem of Accuracy

There are two aspects to the problem of accuracy:

1. numeral to collection;

2. collection to numeral.

The first aspect is that given any cardinal number and its numeral, the size of a collection of objects produced described by that numeral must always be the same. The test of this is whether two collections generated from the same numeral always admit an exact pairing between them. That such a pairing should always exist is a consequence of **CP**. Another way of stating this aspect is that the process that takes numerals to collections of the specified size is **reproducible**.

The second aspect of accuracy is more complicated. Consider that we have a pre-existing collection of buttons, A, that has less than $10,000$ buttons in it. For our system to satisfy the accuracy requirement, it must be the case that there is a numeral which when given to the Button Dealer will produce a collection, B, that has the same size as A. The test of this again is whether there is an exact pairing between the members of B and the members of A.

Determining whether a the Button Dealer System satisfies these two conditions is a matter of experiment. That we all believe that it does and that the Arabic System does as well is a matter of experience. Ultimately the truth of these assertions comes down to conservation of cardinal number.

In terms of instructing a child, a system like the Button Dealer can easily be modeled using counters. But you might not want to use $10,000$ as the maximum!

5.2   Base and Place in the Arabic System

We want to summarize the key components of the Arabic System.

1. A short list of symbols and names for an initial set of counting numbers;

2. a symbol and name for the number associated with the empty set;

3. a recognition that the symbol combination 1 followed by one, or more, zeros has to be the notation for the successor of the largest number that can be written using fewer symbols.

In the actual Arabic System of numeration, the symbol for the largest counting number having a single digit notation is 9. Thus, the initial list contains nine individual symbols denoting numbers associated with the first nine collections, namely, the collections having $1-9$ members. We also have a special symbol for zero. Thus, in the Arabic System, the short list of symbols has ten members.

The number *ten* is referred to as the **base** of the Arabic System. Alternatively, we speak of the Arabic system as a **base ten** system. You can think of the base as the number of symbols in the short list. Ten is also the successor of nine and the smallest number for which there is no single digit notation.

Consider now a multiple digit number in Arabic notation, say:

$$42027.$$

If we read this as an instruction to the Button Dealer, we know it means:

get seven jars from the first shelf;
get two jars from the second shelf;
get no jars from the third shelf;
get two jars from the fourth shelf;
get four jars from the fifth shelf.

As we know, the position of a digit tells us which shelf to use, starting at the right and proceeding to the left.

The only difference between Button Dealer interpretation and the Arabic System interpretation is that the position or place of a digit now directly conveys a quantity associated with that position as illustrated below:

4 — ten thousands

2 — thousands

0 — hundreds

2 — tens

7 — ones

Thus reading from right-to-left, the first digit tells us how many ones to use, the second digit tells us how many groups of ten to use, the third tells how many groups of one hundred to use, the fourth how many groups of one thousand to use, and the fifth how many groups of ten thousand to use.

As before, the symbol 0 is essential as a **place-holder** when no groups are used. Thus, 0 has two functions. It is the numeral for the counting number of the empty set, and it prevents other digits from being assigned the wrong value based on their place (position) in multi-digit numerals. The second function is really subsumed by the first once we understand that we must say that we want no hundreds, in the numeric expression for *forty two thousand twenty seven*, 42027.

Lastly, each new group, tens, hundreds, thousands, and so forth, specifies the successor of the largest number that can be written in fewer symbols. Thus,

10 is the successor of 9;
100 is the successor of 99;
1000 is the successor of 999;

and so forth. In each case, the smallest number representable by each new grouping, that is, a 1 followed by some number of zeros, is the successor, of the largest number expressible with one fewer symbols. This fact ensures no numbers are missed by the notational scheme and is an essential feature of the system.

There is one additional feature of the base ten system that needs to be emphasized. It is that each place value numeral, *ten*, *one hundred*, *one thousand*, and so forth, is made up of 10 units of the next lowest place value. We express this as statements about collections:

- a collection having 10 members is comprised of *ten* collections having 1 member each;

- a collection having 100 members is comprised of *ten* collections having 10 members each;

- a collection having 1000 members is comprised of *ten* collections having 100 members each;

- a collection having 10000 members is comprised of *ten* collections having 1000 members each;

and so forth. We will recall these facts in Chapter 8 on multiplication.

The discussion to date has not included arithmetic computations. That will happen in successive chapters, and at that point we will see the amazing utility of the Arabic system of numeration.

5.2.1 Equality Between Numerals

In §2.5.1 we gave a specific experimental procedure for determining whether two counting numbers were equal (see Equality Principle). At that time, we did not have numerals for counting numbers available, so the procedure did not depend in any way on the notation for counting numbers. Now we have the Arabic System of notation and it is easy to say when two Arabic numerals denote the same counting number.

To be clear, given an Arabic numeral for any counting number, we know from the discussion in this section how to construct a collection having the cardinal number denoted by the given numeral. Thus, given any two such numerals, we can construct the two collections specified in §2.5.1 and check whether there is an exact pairing. Clearly this would be a tedious process to use on an every-day basis. A simple method for determining equality is the first clever feature of the Arabic System we shall identify.

Two numerals denoting counting numbers will denote the same counting number exactly if:

- the digits in the two numerals, starting at the left and taken in pairs moving to the right, are identical.

Alternatively stated, different numerals denote different numbers.

5.3 The Utility of Names

Wherever possible mathematicians tried to make the names of things convey meaning beyond merely providing an identifier. For example, the name *square root of two* identifies a certain number. But more than this it tells us exactly what property that number has, namely that

$$\sqrt{2} \times \sqrt{2} = 2.$$

So when you come across a new name, try to ask yourself: What additional meaning does this name convey? If there is additional meaning, it can be very helpful and we will see this in much of what follows.

5.4 What Your Child Needs to Know

The focus of this chapter has been on understanding the Arabic System of Numeration. Coming to terms with the Arabic System is the **bedrock** on which everything else rests. Achieving this should be a matter of great pleasure for your child.

There are two principal ways we use the names and notations for counting numbers. They are:

1. given a collection, we find the name of the counting number that tells us the cardinality of the collection;

2. given a numeral for a counting number, we construct a collection having that counting number as its cardinality.

Children need to attain complete comfort and facility with both these uses. To do this, your child needs to know and be able to use the information in the following diagrams:

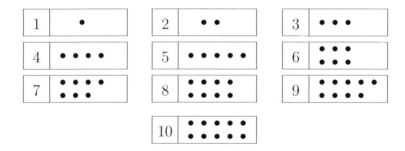

A critical feature of these diagrams is the fact that the numbers are in sequence, that is, 2 comes after 1 , 3 comes after 2 , and so forth. This means your child understands that each counting number is followed by a **next** counting number, what we know as the **successor**. Assimilating these ideas is a critical goal of Pre-K and we direct your attention to activities like the **counting up to** games at the SS website, the number recognition worksheets (click on *Kindergarten* button) at the M-A website and Bunny Count at the EDM website for building these skills.

Here is a list of specific things your child ought to be able to do by age 4–5:

1. given a numeral from 1 — 10 , form a collection with that number of members;

2. given a collection having ≤ 10 members, identify the correct numeral associated with that collection;

3. given multiple representations of collections having the same size, recognize that they have the same total number of members;

4. understand for numbers less than 10 that the next number in the sequence is associated with a collection containing a single additional element; for example, the collection associated with 6 has a single more dot than the collection associated with 5 ; (This is how your child comes to terms with the **successor** process.)

5. understand that to count a collection of objects we select the objects one-at-a-time and assign a number starting with 1, then 2, then 3, and so forth, until the objects are exhausted, with the largest number so assigned being the number of objects in the collection;

6. count collections containing as many as twenty objects;

7. starting at a given counting number, count a designated further amount.

The third point is illustrated below:

$$5 \;\; \boxed{\begin{smallmatrix} \bullet \;\; \bullet \\ \bullet \\ \bullet \;\; \bullet \end{smallmatrix}} \quad = \quad 5 \;\; \boxed{\bullet \; \bullet \; \bullet \; \bullet \; \bullet}$$

Two representations of 5. The pattern on the left is what appears on dice used in board games. Playing such games is a fun way for children to learn to instantly recognize these patterns and practice counting skills. Many of these games are designed for children who have not yet started school.

The fourth point, finding the next number after 8, is illustrated below:

$$8 \;\; \boxed{\begin{smallmatrix} \bullet \; \bullet \; \bullet \; \bullet \\ \bullet \; \bullet \; \bullet \; \bullet \end{smallmatrix}} \quad + \quad 1 \;\; \boxed{\bullet} \quad = \quad 9 \;\; \boxed{\begin{smallmatrix} \bullet \; \bullet \; \bullet \; \bullet \; \bullet \\ \bullet \; \bullet \; \bullet \; \bullet \end{smallmatrix}}$$

The successor of 8, namely 9, is obtained by adding one element to an existing collection having 8 elements. Again, board games using dice and requiring counting are a great way for your child to come to terms with these ideas.

The fifth point is illustrated by counting a collection of containing *seven* dots as follows:

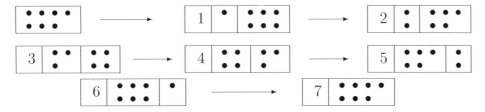

The step-by-step process of counting a collection containing *seven* objects is illustrated. As each object is counted, the object is moved to the left and the numeral is incremented by *one*. Arrows indicate the order that is followed. When all objects have been counted, the last numeral indicates the number of objects in the collection.

To achieve the sixth goal, you need to encourage your child, as early as possible, and before they start school, to learn to count. The purpose of this is not merely so the child learns their numbers. Equally important is that the child develops an intuitive sense of:

How much is five, or six, or eleven, etc?

Attaining this sense of the quantity attached to a number requires experience.

The seventh goal can be achieved by posing questions like: Find the number that is five more than twelve. This requires the child to start at 12 and count 13, 14, finishing at 17. Tasks like this anticipate addition, in this case finding the sum of 12 and 5.

5.4.1   Goals for Kindergarten

The following list is based on the Core Standards:

1. Count to 100 by ones and skip-count to 100 by tens (see M-A *Kindergarten* for worksheets).

2. Write the numbers from 0 to 20. Write the counting number of a given collection of objects up to 20. Represent a count of no objects by 0.

3. Fully understand the relationship between counting numbers and size of collections (cardinality).

4. When counting, say the numbers in proper order.

5. Understand that successive counting numbers correspond to quantities that are one larger.

6. Understand that the size of a collection is independent of its arrangement – conservation of size attribute.

7. Compare (larger or equal) single digit numbers based on their written numerals.

8. Compose and decompose numbers from 11 to 19 into 10 *ones* and some further number of *ones*. For example, 12 is 10 *ones* and 2 *ones*, where this can be expressed with dots, or physical objects.

Counting to higher numbers is initially learned by **rote**. The websites M-A, SS and EDM all contain activities designed to achieve this and the other learning goals. Skip-counting by tens as in 10, 20, 30,... etc. is a good way to begin to come to terms with 10 as the base of our number system.

5.4.2 Goals for Grade 1

By the end of Grade 1 your child will be able to do the following in respect to place-value in the Arabic System.

1. Extend rote counting skills to 120 starting from any number less than 120.

2. Read and write numerals less than 120 and represent numbers of objects with a written numeral.

3. Know that a two digit number represents groups of *ten* and groups of *one* and that places in the Arabic System are determined starting at the right and working to the left. Understand the following special cases:

   (a) 10 can be thought of as a bundle of 10 *ones* that we call *ten*;

   (b) Be able to completely describe the place-values assigned to digits in the numbers 11 – 19, e.g., 18 is 1 *ten* and 8 *ones*;

   (c) know the meaning of 10, 20,..., 90 as 1, 2, ..., 9 groups of *ten*, respectively, and be able to describe the role of 0 in each numeral.

4. Compare two-digit numerals using <, =, and > (See §9.4 Ordering Integers).

5.4.3 Goals for Grade 2

By the end of Grade 2 your child will be able to do the following in respect to place-value in the Arabic System.

1. Understand that the three digits of a three-digit number represent amounts of *hundreds*, *tens*, and *ones*; e.g., 706 equals 7 *hundreds*, 0 *tens*, and 6 *ones*. Understand the following as special cases:

   (a) 100 can be thought of as a bundle of 10 *tens* called a *hundred*;

   (b) the numbers 100, 200, 300, 400, 500, 600, 700, 800, 900 refer to 1, 2, 3, 4, 5, 6, 7, 8, or 9 *hundreds* and 0 *tens* and 0 *ones*.

2. Count within 1000; skip-count by 5 s, 10 s, and 100 s.

3. Read and write numbers to 10,000 using base-ten numerals and number names; identify the digits in each place of a four-digit number, and know the value of the digit.

4. Be able to correctly compare two three-digit numbers based on the value of the various digits in their numerals and to correctly record this information using the $<$, $=$ and $>$ symbols. (See §9.4.)

The reader who compares the CCSS-M grade-level goals with ours will discover that our list of goals is significantly shorter. There are two reasons for this. First, our goals are listed in the appropriate Chapter. Thus, goals involving addition are listed at the end of the chapter on addition.

The second and more important reason is that our focus is on developing skills that will ensure your child can succeed in Algebra I and ultimately, to end up college- and work-ready. This is the target. While there are many additional topics covered in the CCSS-M curricula, in the view of the author they are peripheral. Our focus is on those essential topics that cause students to fail in post-secondary. Ensuring children properly learn these topics ensures they will be able to learn the other things they need to know along the way. For this reason, the topics that have been identified as being essential need to be known at the **mastery level**. The list provided lets you know where to focus your efforts.

Ultimately, your child must come to terms with the Arabic System of notation. Achieving this takes time. Recognition of this fact is implicit in the by grade goals listed above. There are many activities you can find on the websites we have listed that will aid in achieving these goals. After reading this chapter, you should have a good idea of the basics of the Arabic System. That will enable you to pick and choose activities suited to your child. But you should understand that coming to terms with this system of notation will take time. As can be seen from the goal structure, a four-year process is contemplated.

Chapter 6

Addition of Counting Numbers

Chapter Overview. In this chapter we will define the operation of addition on any pair of counting numbers in terms of a real-world process on collections. This definition will be reformulated in terms of the **successor** operation defined in Chapter 4. The addition definition will be extended to include zero. The Commutative and Associative Laws will be stated and their essential truth as consequences of a real-world process on collections discussed. The standard procedure for addition computations supported by the Arabic System of numeration will be presented. Sample computations will be presented in detail. Grade-by-grade curriculum goals for addition will be presented which will enable parents to evaluate their child's progress.

Let us take a moment to review where we are in the scheme of things.

In Chapter 2, we studied real-world collections and observed we could compare collections in respect to **more** or **less**. Most importantly, we discovered that the number of items in an isolated collection does not change, so that **counting number is a conserved attribute of collections (CP)**. Indeed, it is the fact that isolated collections in the real world have constant membership that gives rise to the very concept of counting number as an attribute of collections.

We also discovered that there were smallest collections characterized by the property that removal of any member destroyed them. Such collections contained a single element, and were used to define the cardinal number one (see Chapter 3).

Lastly, we saw that collections could be constructed by successively adding single elements to the collection being constructed. We referred to this as the **one-at-a-time** process, which we know is the basis of counting.

In Chapter 4, we translated the properties of real-world collections into specific statements about counting numbers. These identified 1 as the notation for the least counting number, that is, the counting number associated with smallest collections. By analogy with adding a single element to a collection, we defined the operation of forming the **successor** of a number by the addition of 1 to that number. An important fact about this operation was that there was no counting number strictly between a given counting number and its successor. Thus, given a counting number, there is a **next largest** counting number.

In Chapter 5, we developed the Arabic System for naming counting numbers. This system has **zero** as the number denoting the cardinality of the **empty set**. The system also uses place as a means to differentiate the value assigned to a single digit in multi-digit numerals. The discussion surrounding the Arabic System was based on an analogy of a Button Dealer in the real world. In what follows, it will be convenient to refer to that analogy as a means for demonstrating how and why various procedures work.

Our purpose in the remainder of this book will be to develop the properties of arithmetic that you will need to help your child succeed. So let's begin by considering what arithmetic is.

> **Arithmetic consists of numbers, operations on those numbers and the procedures for performing the operations.**

We have already identified a set of numbers, namely, the counting numbers denoted by $\mathcal{N}$ on which to define arithmetic operations (see §4.3). We start with this set because, so far as we can tell, it was the set on which human mathematics was founded.

We need to be clear about what we mean by an **operation**. Any procedure that takes numbers as input and produces a number as an output, and **always produces the same output when given the same input** will be considered to be an **operation**. In other words, the result of an operation is universally reproducible.

We already have seen an example:

> Given a counting number as input, construct a collection having that number of elements, add a single element to the constructed collection, and record the counting number of the new collection as output.

The reader will recognize that if the input counting number is n, then the output counting number is the successor of n, namely, $n + 1$ (see §4.3.2). The property C2 of collections guarantees that given the same number as input, the result will always be the same. So this procedure defines an **operation**. Since the input is a single counting number, we refer to this as a **unary** operation.[1]

---

[1] The reader will likely be familiar with the square root operation on a calculator which is another example of a unary operation that takes a single number as input.

Our entire system of arithmetic can be developed starting with the set $\mathcal{N}$ and the operation of successor. We will take a mildly different approach, the first step of which is to generalize the operation of successor to the addition of any pair of counting numbers.

6.1 What is Addition

Addition is a **binary operation** on the counting numbers. It is **binary** because it takes as input two counting numbers. To be an operation, we need a procedure.

> **Addition Procedure.** Given two counting numbers, n and m, construct a collection having n elements and a collection having m elements; combine the two collections into a single collection; count the number of members that are in the combined collection; the resulting counting number is $n + m$.

The quantity $n + m$ is called the **sum** of n and m. The inputs n and m are called **summands**, or **addends**. A concrete example of the addition procedure is pictured below.

A diagram showing *seventeen* is the sum of *nine* and *eight.*

This process works because cardinal numbers are **conserved (CP)**.

To apply conservation, suppose we start with a collection, say a jar containing some number of buttons and we also have an empty jar. Place some number of buttons from the first jar into the second jar. Count the buttons remaining in the first jar and call that number, n. Count the buttons in the second jar and call that number m. Put the buttons remaining from the first jar into the second jar, count the total, and call that number $n + m$. What we know from **CP** is that so long as we are careful, the total number of buttons will be constant, namely, the same number as we started with initially in the first jar. Thus the sum, $n + m$ is fixed. This is the process captured in the diagram below.

90

The original jar, on the left, has 17 buttons. A number are taken and placed in the first jar on the right, 9 as shown. The remainder are placed in the second jar on the right, 8 as shown. **CP** demands that however the buttons are divided between the two jars on the right, the total number of buttons on the right-hand-side (RHS) of the equality must be the same as the original number on the left-hand-side (LHS).

Viewed in this way, finding the sum of two counting numbers is an experimental process. It is because the cardinal numbers of collections are **conserved** that we know we will always come up with the same result, and that result is obtained by **counting**.

The diagrams and ideas presented above should be the bed-rock foundation on which every child's thinking about addition is based.

6.1.1 Use of CP

As we work our way through the theory of arithmetic you will see the conservation of cardinal number **CP** given as the reason why something must be true. Indeed, we have already used this principle above as the reason why the quantity $n + m$ is fixed and unique. Appealing to **CP** as a primary reason why something must be true does not make the given rationale a **mathematical proof**. Rather it is a statement that for any system of arithmetic we might come up with to have value, it must agree with this fact about the real world.

Arithmetic must agree with your experience of the real world.

Arithmetic is an abstract model of the behavior of counting numbers as they apply to real-world collections. Models make predictions. Thus, we can think of $n + m$ as the predicted value of the counting number that will be observed when a collection having n members is combined with a collection having m members. Since we can check this by combining two real-world collections of appropriate size, we have an experimental procedure to verify our abstract model.

6.1.2 Why Use Collections Instead of Sets in the Addition Procedure

There is one subtle aspect of this that we need to emphasize. It is the point concerning the difference between sets and collections discussed in §2.3.1. Consider two collections, for example, two jars of buttons in the real world. Suppose we pour the buttons into a third jar. There is no possibility that two of the buttons coalesce to become one. In other words we cannot have a single button that occupies both jars at once in the real world. This is a physical fact about the world.

Because mathematics is abstract, this is not so in the mathematical world. For example, we can have two sets, A and B, each having 3 members as shown:

$$A = \{1, 2, 3\} \text{ and } B = \{2, 3, 4\}.$$

When we combine the two sets by forming their **union**, we get[2]

$$A \cup B = \{1, 2, 3, 4\},$$

which has only four members, not the six we would expect from two physical collections of buttons, each of which has three members. This is one reason why dealing with sets is confusing. But the folks who invented our arithmetic did not have these abstract ideas around. They understood the focus was on real-world collections and what such collections had to say about arithmetic. This is why we have spent so much time emphasizing that our ideas about arithmetic should be guided by the real world.

6.2   Equality Properties

The equality relation is an essential part of arithmetic and has an important place in the CCSS-M. For this reason, we need to discuss its important features.

In §2.5, we provided a procedure using pairing for determining when two collections had to be assigned the same counting number. This procedure is the entire basis for equality of counting numbers. Thus, any properties we might state must be consistent with this procedure.

Further, if we have a general statement about counting numbers that involves equality, if we replace any variables by particular counting numbers, we can perform an experiment with collections and counting that will verify any instance of the equation. For example, using $n = 12$ and $m = 3$, the equation

$$n + (m + 1) = (n + m) + 1$$

becomes

$$12 + (3 + 1) = (12 + 3) + 1.$$

---

[2]The symbol $\cup$ denotes taking the **union** of two sets. The result is a new set containing all members in either, or both, sets.

To witness the LHS combine a jar with 12 buttons with a jar containing $3+1=4$ buttons and do the count. To witness the RHS combine jars containing 12 and 3 buttons, respectively. Then add a single button and count. Both counts must produce the same cardinal number, in this case 16. Does anyone doubt that the counts will agree? Of course not. But the fact that we know that both processes will give the same count is the basis for our choice of what properties we must take to be true in our formulation of arithmetic.

We state the following four Equality Properties using the word *quantity* to denote any mathematical object, in particular numbers. We let A, B, C and D be any mathematical quantities:

E1: $A = A$;

E2: if $A = B$, then $B = A$;

E3: if $A = B$ and $B = C$, then $A = C$;

E4: if $A = B$ and $C = D$, then $A + C = B + D$.

The following are examples using counting numbers:

E1: $5 = 5$;

E2: if $3 = 2 + 1$, then $2 + 1 = 3$;

E3: if $4 + 3 = 7$ and $7 = 5 + 2$, then $4 + 3 = 5 + 2$;

E4: if $2 = 1 + 1$ and $3 = 2 + 1$, then $2 + 3 = (1 + 1) + (2 + 1)$.

Notice the use of parentheses to indicate groups of symbols that are to be treated a single quantities. Thus, $(1 + 1)$ is treated as a unit or **term**.

For clarity, we consider how the principles are used. In our minds, we can think of

the equals symbol means: *is the same as.*

As well, we can think of a statement like $5 = 5$ as expressing a mathematical fact. Similarly, $5 = 4 + 1$ expresses a mathematical fact about Arabic numerals. What E2 tells us is that given the mathematical fact $5 = 4 + 1$, we immediately know that $4 + 1 = 5$ is also a mathematical fact about Arabic numerals. At the deepest level, what is being asserted by the equality $5 = 4 + 1$ is that

5 and $4 + 1$

93

are names for the same counting number in the Arabic System of numeration. That said, you can operate successfully by thinking of equality at the highest level of meaning, namely, *is the same as*.

Again, E1-E4 can be demonstrated by constructing collections. For example, for E2, consider two counting numbers m and n. We know they are equal exactly if when we construct a collection A having m elements, and a collection B having n elements, then there will be an exact pairing of the members of A with the members of B. If such a pairing exists between A and B, a similar pairing will exist between B and A. (Just reverse the arrows.) Thus, if $m = n$, then $n = m$ for any pair of counting numbers.

Similar constructions are possible to support the properties E1, E3 and E4.

We restate the Equality Properties in words:

E1: every quantity is equal to itself;

E2: if one quantity equals a second, the second also equals the first;

E3: two quantities equal to the same thing are equal to each other;

E4: equals added to equals are equal.

E4 will have the consequence that:

E5: equals multiplied by equals are equal.

6.3 Properties of the Addition Operation

There are three properties of addition that will be used repetitively to develop the rules for arithmetic. They are stated as equations A1-A3 and are likely to be very familiar to the reader.

> **A1-A3, tell you everything you need to know about the theory of addition of counting numbers**.

So A1-A3 need to be completely understood.

6.3.1   Addition of 0

Addition was defined in §6.1 as a binary operation on the set counting numbers by appealing to properties of collections and counting. As you know zero is not a counting number, since counting numbers give the size of real-world collections and such collections must have members. Nevertheless, we know zero is a perfectly good number and we must be able to give an answer when we add zero to a counting number. The equation in A1 tells us what the sum of n and 0 will be for any counting number n.

A1: Let n be any counting number, then

$$n + 0 = 0 + n = n.$$

Because 0 satisfies this equation, it is called the **additive identity**. The adjective *additive* refers to the operation, which is addition. The noun *identity* refers to the fact that when this number is combined with any other number, say 5, using the operation of addition, the value returned is 5.

In Chapter 9, we will take the property defined by A1 as the defining property of zero. Namely, zero is the number which when added to any other number gives the original number as the answer. Thus, $0 + n = n + 0 = n$. Proceeding in this manner would be the mathematician's approach. At this point our guide is the real world in the form of collections, so let's look there for why this equation must hold in our system of arithmetic.

Experience with collections together with **CP** should readily convince you that only doing nothing to a collection has the property that it leaves the attributes unchanged. Any process that adds or removes members changes the attribute of size. For this reason, we should expect that zero is the **only** cardinal number which will act as an additive identity. In respect to this property (behavior), zero is **unique**.[3]

6.3.2   The Commutative Law

A2: Let m and n be any two members of $\mathcal{N}$, i.e., counting numbers. Then

$$n + m = m + n.$$

As the reader knows, this equation is called the **Commutative Law** of addition.

To see why it must be true, again we turn to the real world. Consider a jar with n buttons in it, and a second jar with m buttons in it. Ask yourself:

---

[3]Mathematicians consider zero to be a cardinal number. But it is not a counting number. There are many cardinal numbers that are not counting numbers, but that do not occur in arithmetic.

Will it make a difference to the total number of buttons whether the buttons from the first jar are poured into the second jar, or the buttons from the second jar are poured into the first?

Experience tells us that so long as we are careful not to lose buttons, the total number of buttons will be the same in either case. So the equation above must be true, again as a consequence of conservation **CP**.

What every child should know about the Commutative Law is that the order in which summands are added has no effect on the sum. This applies no matter how many summands there are.

6.3.3 The Associative Law

A3: Let m, n and p be members of $\mathcal{N}$, i.e., counting numbers. Then

$$n + (m + p) = (n + m) + p.$$

This property is called the **Associative Law** of addition.

To see why it must be true, we again turn to the real world. Consider that we have three jars of buttons. The first contains n buttons, the second m buttons, and the third p buttons. The parentheses in the equation are used to tell us in which order the operations are carried out. So the left-hand side instructs us to first combine the jar containing m buttons with the jar containing p buttons to obtain a jar containing $m + p$ buttons. Only when this is done do we complete the task by combining the result with the jar containing n buttons to obtain a jar containing the total, $n + (m + p)$ buttons. The right-hand side forces us to do things in the other order, namely first combine the jars containing n and m buttons, respectively, to obtain a jar containing $n + m$ buttons. Only then add in the p buttons contained in the third jar. **CP** tells us that the number of buttons in the combined collections cannot be affected by the order in which the collection was assembled. So the equation above is true in the real-world, and so we must make sure it is true about our arithmetic.

Notice that if any of m, n and p are 0, then the equation reduces to an identity. For example, if $m = 0$, then the equation becomes $n + p = n + p$ after we apply A1 above.

What a child needs to understand about the Associative Law is that it tells us that the process that is used to combine summands in any computation involving addition cannot affect the sum.

As you can see, the theory underlying addition is small, in the sense that only three equations are required. As we shall see, these three equations have powerful effects, which is why they are so important.

We stress that while the rationale for why these three equations must be true are not proofs in the mathematical sense, they are the essential reasons why the given statements must be facts about arithmetic. They are all derived from our understanding of counting and conservation and that is how they should come to be understood by children. It is almost certainly the case that studying the behavior of counting numbers as applied to collections is how humans discovered there was such a thing as mathematics and that numbers could provide useful information about the world.

6.3.4   Addition From Successor

As we have seen, the one-at-a-time process was critical to the Arabic System of numeration. We want to explore this notion in the context of addition. Given the number n, there is a **next largest counting number**, $n + 1$ which we called the **successor** of n. This fact enables us to define how we add any other counting number to a given counting number n. This process is so fundamental, we explain it in detail. As always, our reasoning is based on collections.

The following diagram illustrates the process of counting a collection containing *seven* dots:

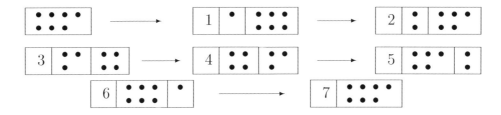

The step-by-step process of counting a collection containing *seven* objects is illustrated. As each object is counted, the object is moved to the left and the numeral is incremented by *one*. Arrows indicate the order that is followed. When all objects have been counted, the last numeral indicates the number of objects in the collection.

We can take this diagram and interpret it in a different way. Suppose we want to find the sum of 4 and 3. According to the concrete procedure we would construct two collections, one with 4 members and one with 3. We would then combine the two collections and count the result to obtain 7. This process is illustrated below:

The step-by-step process of adding $4 + 3$ by starting with a collection containing 4 members and a collection containing 3 members and transferring members one-at-a-time from the collection with 3 members to the collection that initially contains 4 members. As each object is moved to the left, the numeral is incremented by *one*. Arrows indicate the order that is followed. When all objects have been moved, the last numeral indicates the sum, in this case 7.

This diagram illustrates the following sequence of computations:

$$
\begin{aligned}
4 + 3 &= 4 + (2 + 1) = 4 + 1 + 2 = 5 + 2 \\
5 + 2 &= 5 + (1 + 1) = (5 + 1) + 1 = 6 + 1 \\
6 + 1 &= 7
\end{aligned}
$$

where we have made use of the Commutative and Associative Laws.

6.4   Making Addition Useful

While the properties of addition discussed above are interesting, particularly to mathematicians, they are not very helpful to folks who simply need numbers to keep track of things. For example, think of a rancher who has 175 cattle on the north forty, another 203 cattle on the high pasture, and who needs to know the total, so he can order winter feed. Even in a situation where the rancher has available an effective naming scheme for numbers, like the Arabic system, the equations A1-A3 above, tell him nothing about how to actually find the required sum. And waiting until he can combine the two herds so he can find the sum by doing a count may take too long. What the rancher needs is a procedure that connects the operation of addition to the naming scheme, namely, the Arabic number system. That there is a connection is one of the amazing facts about the Arabic system of numbering.

To make this connection, we will make use of our Button Dealer analogy. So let us give the rancher's problem to the Button Dealer, namely, find the sum of 175 and 203.

How the Button Dealer Does a Sum. To fill an order for 175 buttons, the Button Dealer goes to the first shelf and gets *five* jars marked with

1, goes to the second shelf and gets *seven* jars marked with 10, and goes to the third shelf and gets *one* jar marked with 100. To fill an order for 203 buttons, the Button Dealer goes to the first shelf and gets *three* jars marked with a 1, gets no jars marked with 10 from the second shelf, and goes to the third shelf and gets *two* jars marked with 100. Since the Button dealer knows that addition corresponds to combining collections and counting, she proceeds as follows. She counts the jars from the first shelf, and determines there are *eight*. She observes that she still has *seven* jars marked with 10. Lastly, she counts the jars marked with 100, and finds there are *three*. So in the Button Dealer's naming scheme, this corresponds to 378.

Look how simple this process is! To determine the total, the Button Dealer only has to count up, and record, the number of jars coming from each shelf, performing the addition process one shelf at a time. One part of this process is absolutely critical, namely that we only combine jars from the same shelf. In other words, we only count up jars labeled with 1, followed by counting up jars labeled with 10, followed by counting up jars labeled with 100. This is the key that makes the process work.[4]

Let's see if we can adapt this idea to the Arabic system for which the Button Dealer system is an analogy.

Example 1

Suppose you are confronted with finding the sum of: 23 and 54. How are you to find the answer? First we write down 23 and then under it we write down 54, as shown below.

$$
\begin{array}{r}
23 \\
+54 \\
\hline
\end{array}
$$

When we do this, we have to remind ourselves that in the Arabic system, the **position of a digit in the numeral determines its value**. So when we write down the 54 under the 23, we must make sure that the right-most digit in 54 lines up under right-most digit in 23. Thus, we have lined up the *ones* digits in one column and the *tens* digits in another column. We refer to this process as **setting up** the problem.

Once the problem is setup, we are ready to perform the addition. All we have to do is find the sum of the numbers in each column, starting with the *ones* column. These sums involve single digits.

---

[4]It is always the case that we can only add like things. Later, we will refer to this as the **apples-to-apples** principle.

The required sum of the digits in the *ones* column is: $3 + 4$. To find this sum, we create two collections, the first having three members, the second four members, and combine their contents. The size of the resulting collection is the sum. Finding this sum is illustrated in the diagram below.[5]

An equation implementing the addition procedure for $3 + 4$. The M-A website has worksheets based on adding dots.

The result is 7 and this numeral is entered in the *ones* column below the line, as shown below:

$$
\begin{array}{r}
23 \\
+54 \\
\hline
7
\end{array}
$$

The same procedure is used to find the sum of the digits in the *tens* column, namely, $2 + 5$. Again, the procedure is illustrated below:

A graphical illustration of finding the sum $2 + 5$.

The result is again 7, and the result is recorded below the line in the *tens* place as shown:

$$
\begin{array}{r}
23 \\
+54 \\
\hline
77
\end{array}
$$

The procedure described above is one of the great beauties of the Arabic system for naming numbers in that it supports procedures for performing the operations of arithmetic that are so simple they can be mastered by elementary school children. Consider that in order to add any two counting numbers, a person only has to know the sums of the digits in pairs, and mechanically follow the rules regarding place. While the thought of having to know all these sums may seem intimidating, there are very few facts to be learned in comparison to any other subject. Not only is this procedure so simple it can be mastered by children, it is so simple it can be built into a machine that can be held in the palm of your hand.

---

[5]Finding a sum such as $3+4$ is where being able to answer the question: What counting number is four more than three? is particularly useful. Simply start at 3 and continue counting four more steps, ending up at 7.

6.4.1 The Addition Table

Even though the procedure described above is straight forward to implement, constructing diagrams each time we want to add some numbers would be tiresome and time consuming! Indeed, even the best counting procedures are really inefficient. But here is where another one of the clever features of the Arabic system comes into play. There is only a short list of single digits, and so there is only a short list of pairs having sums we need to know. This list of pairs and their sums is usually presented in the form of a table, as shown below:

+	0	1	2	3	4	5	6	7	8	9
0	0	1	2	3	4	5	6	7	8	9
1	1	2	3	4	5	6	7	8	9	10
2	2	3	4	5	6	7	8	9	10	11
3	3	4	5	6	7	8	9	10	11	12
4	4	5	6	7	8	9	10	11	12	13
5	5	6	7	8	9	10	11	12	13	14
6	6	7	8	9	10	11	12	13	14	15
7	7	8	9	10	11	12	13	14	15	16
8	8	9	10	11	12	13	14	15	16	17
9	9	10	11	12	13	14	15	16	17	18

The **Addition Table**. To find the sum $7 + 6$ go to the row having 7 at the far left. Follow this row across to the column headed by 6. The table entry in this cell is 13. Notice as we move across the row, we get the sums $7 + 0 = 7$, $7 + 1 = 8$, $7 + 2 = 9$, and so forth, in order as cell entries, thereby illustrating the process of counting six more starting at 7.

Requiring children to commit the facts in the table to memory may appear intimidating. Let's consider this. Since there are ten symbols in our short list, there are 100 possible pairs $m + n$, where both m and n correspond to a single digit. So in principle, to completely master all possible sums requires learning one hundred items. Recall, that

$$n + 0 = 0 + n = n,$$

so this reduces 19 of the entries to a triviality. Also, we know that $n + m = m + n$. How does this show up in the table? Pick any cell not on the main diagonal, that is, a cell having a different row and column number. Check the value in the cell. Now observe that if you interchange the row header with the column header, the value in the new cell having this row and column header will be the same. This is

101

how $n + m = m + n$ shows up in the table. In terms of facts to be learned, the Commutative Law **reduces the total number to less than fifty**.

Think of it! Knowledge of fifty addition facts, together with an understanding of the role of place in the addition process, enables anyone to master the addition of an infinite number of pairs of counting numbers. In the marketplace of learning, this seems like a pretty good bang for the buck! The various websites provide plenty of different ways to encourage your child to commit the facts in this table to memory.

Reading the Table

Table (cell) entries are the numbers to the right and below the double lines. Each cell entry is labeled with a row header followed by a column header. Row headers are to the left of the double lines. Column headers are above the double lines. The first row is the row having 0 at the far left. The tenth row is the row having 9 at the far left. Similarly, first column has a 0 at the top, and the tenth column has a 9 at the top.[6]

To find the sum $n + m$, simply find the cell at the intersection of the row having n as its row header and the column having m as its column header. So the cell entry in row headed by 5 and column headed by 3 is the sum of 5 and 3, namely, 8.

Each entry in the body of table results from a computation of the form:

A diagram verifying the entry in the table above for $7 + 6$. The sum can be directly found by counting the total **dots** on both sides.

Learning the Table

Your child needs to know the sums in this table as a matter of recall. Achieving this means transitioning from procedure-based problem solving to memory based problem-solving. To help your child, you need to understand that the learning starts with a procedure. This procedure needs to **always produce a correct answer**. Without a procedure children will simply **guess**. Guessing never works and children need to learn this right away. If your child doesn't know the answer, you should encourage

---

[6]Entries in arrays of this type are usually referenced by row and column number. In this context, because of the row and column headed by 0, the entry in the intersection of the fourth row and fifth column would not be the sum of $4 + 5$. For this reason we choose to label rows and columns with the headers and not in the usual manner.

him/her to say so because it is at that point you can go over the procedure that leads to the answer. Having a procedure that works is key because unlearning incorrect information is far harder than learning it in the first place. In addition, the procedure should reinforce the conceptual understanding of the operation. Let's recall:

> In the real world addition corresponds to combining two collections of the required sizes and counting the total to find the sum.

Because we understand this as the process, we know that it doesn't matter which collection is poured into the other in the combining process. The result will be the same. In other words the process is commutative.

We can turn these ideas into an effective procedure for adding two single-digit numbers from the Table. The procedure is based on the process of **counting on**[7] which your child should become familiar with in Kindergarten. It works as follows:

Step 1: choose the larger of the two numbers;

Step 2: beginning the count with this number, have your child **count on** raising one finger as each successive number is counted;

Step 3: stop when the number of fingers raised matches the smaller number;

Step 4: the last number counted is the required sum.

As an example, consider finding $4 + 9$. To apply the procedure, the child starts counting at 9 and continues:

$$10, \quad 11, \quad 12, \quad 13.$$

As each new number is counted, the child raises one additional finger. At 13 your child will raise his fourth finger and the count stops with 13 as the answer. This procedure always produces the right answer and is the most efficient because, as given, the count always starts with the larger number.

Once you know your child can correctly use the procedure, you can work on transition to memory-based problem solving. Like everything else in life, it comes down to practice. Here, flash cards can be helpful. You can buy them, or you can make then using index cards. Five or ten minutes of practice each day should make your child an expert in no time.

---

[7]Exercises involving counting on can be found at the websites.

6.4.2　More Addition Examples

Example 2

For a second example we find the sum of 143 and 25. The setup is:

$$143$$
$$+25$$

Notice that one number has three digits while the other has only two. There is no digit at the left of 25 because we choose not to write down unnecessary zeros. Thus, 025 and 25 name the same number. We could write the required sum as:

$$143$$
$$+025$$

but choose not to because mathematicians are lazy. However, we can see that if we are going to be lazy, we have to be careful and make sure the columns are properly aligned so that there is no confusion as to which column a digit is in. It may be helpful for your child to set up the problem as above when starting out.

Again, we simply add the columns starting at the right with the *ones* column. Using the table, we find $3 + 5 = 8$, and this is recorded below the line in the *ones* place:

$$143$$
$$+25$$
$$\overline{8}$$

The next step is to add the digits in the *tens* column. The required sum is $4+2 = 6$, which is found in the table and recorded in the *tens* place below the line:

$$143$$
$$+25$$
$$\overline{68}$$

The final step is to sum the digits in the *hundreds* column, one of which, as discussed above, is 0. In effect, all that has to be done is to write the single *hundreds* digit below the line as:

$$143$$
$$+25$$
$$\overline{168}$$

which completes the process.

It is evident that the efficiency of this process is vastly increased if the sums of the single digits are available from memory. Another easy way to accomplish this goal with your child is to use flash cards which make learning a co-operative process between you and your child.

6.4.3 Carrying

Let us review what we know. In the examples above, the procedure for adding two numbers, n and m, was to write one down above the other so that the *ones* digit in the numeral for n was directly above the *ones* digit in the numeral for m, the *tens* digit in the numeral for n was above the *tens* digit in the numeral for m, the *hundreds* digit in the numeral for n was above the *hundreds* digit in the numeral for m, and so forth. The sum of the digits in each column was then found using values obtained from the table (or memory) to produce $n + m$. We know this process worked because addition corresponds to counting combined collections as shown by the Button Dealer analogy.

In all the examples considered so far, the sum in any column could be written as a single digit. This leaves us to wonder if the process still works when the sum is a two digit number, as in $9 + 7 = 16$? The answer is yes, but we have to revise our procedure. The revision is referred to as **carrying**.

Once again we use the Button Dealer analogy to find the sum of 28 and 54. As before the Button Dealer proceeds as follows:

> **How the Button Dealer Carries.** From the first shelf take *eight* jars marked with 1 and from the second shelf take *two* jars marked 10, to obtain 28 buttons. Then from the first shelf take *four* jars marked with 1 and from the second shelf *five* jars marked 10 to obtain 54 buttons. To find the total number of buttons, the Button Dealer starts by counting the jars marked with 1 and finds *twelve*. What the Button Dealer knows is that *ten* jars marked with 1 contain the same number of buttons as *one* jar marked with 10, so the Button Dealer simply replaces *ten* of the jars marked with 1, with *one* jar marked with 10. At this point, the button dealer has *two* jars marked with 1 and a total of *eight* jars marked with 10. Hence, there are 82 buttons in the combined collection.

Two jars of 10 come from 28, and *five* come from 54, for a total of $2 + 5 = 7$ jars marked with 10. The extra jar, comes from the fact that $8 + 4 = 12$ which in Button Dealer terms is *two* jars marked 1, and *one* jar marked 10. The process, by which we turn *ten* jars marked with 1 into *one* jar marked 10, is called **carrying**.

The above example in the context of the Button Dealer shows us how, even when we have to carry, the addition procedure comes down to counting the contents of two collections. We now consider several examples of the procedure that illustrate what children are taught and how the process is supported by the Arabic System of numeration. We start with the sum we just found.

Carrying: Example 1

As before, the setup consists of writing one numeral above the other, making sure the columns are properly aligned, *ones* above *ones* and *tens* above *tens*:

$$\begin{array}{r} 28 \\ +54 \\ \hline \end{array}$$

The sum of the numbers in the *ones* column is 12, and this is recorded in the following intermediate way:

$$\begin{array}{r} \boxed{1} \\ 28 \\ +54 \\ \hline 2 \end{array}$$

In the above, the *ones* digit from the sum of the first column, namely 2, is recorded in the *ones* place in the answer. The *tens* digit from this sum, which can only be a 1 when summing any pair of single digit numbers, is recorded in a new row in the *tens* column as shown. (This 1 has been placed in a box for clarity and corresponds to the 10 *ones* being combined in a single extra jar marked with 10 in the Button Dealer example.) We then sum all the numbers in the *tens* column, namely, $(1 + 2) + 5 = 8$ and record the result in the *tens* column below the line:

$$\begin{array}{r} 1 \\ 28 \\ +54 \\ \hline 82 \end{array}$$

This example illustrates why the columns are summed right-to-left. It is because when we carry, we will have to add a 1 in the adjacent column to the left. This fact is best understood in the context of the Button Dealer example.

Carrying: Example 2

We find the sum of 999 and 1, which we know is the successor of 999. Finding this sum will further illustrate how the Arabic System supports computations. The setup is shown below where there is more space between columns for exposition purposes:

$$\begin{array}{ccc} 9 & 9 & 9 \\ + & & 1 \\ \hline \end{array}$$

Since the sum of the digits in the *ones* column is 10, we write 0 in the *ones* place in the answer and carry a 1 into a new row at the top of the *tens* column as shown below. Note this 1 has been placed in a box and it is important to understand that it arises as the sum of 10 *ones* in the computation $9 + 1$.

$$
\begin{array}{cccc}
 & & \boxed{1} & \\
 & 9 & 9 & 9 \\
+ & & & 1 \\
\hline
 & & & 0
\end{array}
$$

The sum in the *tens* column, which now has a carried 1 at the top, is again 10. So we write 0 in the *tens* place in the answer, and carry a boxed 1 to the top of the *hundreds* column, as shown below.

$$
\begin{array}{cccc}
 & \boxed{1} & 1 & \\
 & 9 & 9 & 9 \\
+ & & & 1 \\
\hline
 & & 0 & 0
\end{array}
$$

The sum in the *hundreds* column is now 10, so we write 0 in the *hundreds* place in the answer. This time there is no *thousands* column, so we create one by carrying a boxed 1 to the top of the empty *thousands* column in the same row as the other carried 1 s, as shown.

$$
\begin{array}{cccc}
\boxed{1} & 1 & 1 & \\
 & 9 & 9 & 9 \\
+ & & & 1 \\
\hline
 & 0 & 0 & 0
\end{array}
$$

Summing the newly created column produces:

$$
\begin{array}{cccc}
1 & 1 & 1 & \\
 & 9 & 9 & 9 \\
+ & & & 1 \\
\hline
1 & 0 & 0 & 0
\end{array}
$$

which we recognize as the correct notation for the successor of 999.

The procedures described provide an ability to perform the operation of addition on all counting numbers, and indeed, on all decimal numbers after that extension has been made. This is truly remarkable for several reasons. First, it is not at all clear why there should exist a good system of notation for counting numbers that corresponds so well to the one-at-a-time process for constructing collections. Second, it is even more remarkable that this system of notation should mesh so completely with the arithmetic operation of addition. Not only does it mesh, it is so inherently simple it can, and is, successfully taught to children around the world.

6.5 What Your Child Needs to Know

In respect to this question, there is good news and bad news. The good news is there isn't all that much your child needs to know. The bad news is, what your child needs to know, your child needs to know perfectly at the level of **instant recall**.

Let's deal with the good news first. There are at most fifty separate items in the addition table, provided one is thoroughly aware of the Commutative Law:

$$m + n = n + m.$$

In comparison to the number of words a First Grader would be expected to be able to read, this is a very small number. The role of place in the numbering scheme and how it is used as part of the addition algorithm is more complex.

As noted, the bad news is these things need to be known perfectly. What do we mean by this? In respect to things that are in the realm of mathematical facts, for example, the sum of $5+4$, the correct value should be available as instant recall. With respect to things that can be thought of as computational processes, for example, the procedure used to find the sum of $973 + 858$, it is expected that the child would be able to correctly execute all aspects of the procedure with no hesitation, or in the jargon used in curriculum guides, fluently.

6.5.1 Why We Demand Instant Recall

Most of the parents who read this book will be familiar with computers. Very likely, they will have purchased a computer and have read the specifications. As such, they will know that processor speed is important, and working memory is important. Working memory is where the processor stores stuff for immediate access and if you don't have enough, the computer slows down, and if the memory is too over-loaded, the computer quits!

The brain is a biological computer. It also has working memory which it uses when it wants to solve problems. Suppose we are asked to solve the following problem:

> Johnny is on a hockey team that has 20 players. If each player has 3 sticks, how many sticks total do team members have?

To solve this problem, you have to get the information into your brain, process that information to figure out what, if any, computations are required, and then perform those computations. The part of your brain available to accept and process this information is limited and varies from individual to individual. What is clear is that the more brain an individual can apply to a given problem, the more likely that individual will be able to find a solution. Thus, any part of the problem solving process that can be dealt with by retrieving information quickly from long-term permanent memory releases working memory to address the given problem more effectively.

We liken this to muscle memory, which is what skilled athletes use on the field of play. For example, great skaters are not thinking about how they skate. In **Nike** terms, they *just do it!* Similarly, musicians do not think about what note to play when they sight-read music; if they had to, they couldn't possibly do it. Their responses are automatic, and that's what we want to achieve here.

So to take the above example, working memory will have to hold 20, hold 3, figure out that multiplication is required, and should then recall $3 \times 20 = 60$, as opposed to having to calculate the product by some other means. The point is that as the complexity in the information stored in working memory grows, if we still have to **figure out** what $3 \times 20 = ?$, the system eventually dies because working memory becomes over-loaded.

6.5.2   How to Achieve the Learning Goal

First we must recognize that acquiring these skills takes time and practice. This is why a famous mathematician once remarked: *Mathematics is not a spectator sport.* That said, for these skills to be well-learned by children, the process should be fun. And nothing is quite as much fun as **getting the right answer**. This is why we have stressed board games that require counting. There are other games that build memory, for example, Concentration.[8] In the beginning, playing the games is probably more engrossing for the child as opposed to generating the equivalent worksheet with 30 problems on it. The key websites we have listed, EDM, M-A, SS and SKids all offer worksheets and activities at all levels.

We have stressed counting activities for children as a learning device. Board games using a pair of dice effectively teach counting two collections up to 6, which, as we know, is addition. So the addition table up to 6 can be learned using a pair of dice. As well, use of dice makes children very familiar with the verification equations like:

$$3 \;\; \boxed{\bullet\ \bullet\ \bullet} \;\; + \;\; 5 \;\; \boxed{\bullet\ \bullet\ \bullet\ \bullet\ \bullet} \;\; = \;\; 8 \;\; \boxed{\begin{smallmatrix}\bullet\ \bullet\ \bullet\ \bullet\\ \bullet\ \bullet\ \bullet\ \bullet\end{smallmatrix}}$$

An equation verifying the addition table entry for $3 + 5$. See Learning Addition Game under Pre-K at the SS website.

---

[8]The cards from an ordinary deck are place in four rows of 13, face down. Players alternate by turning up pairs of cards. If the ranks match, for example, two 10 s, the player keeps the pair and takes another turn. This game can be simplified by limiting the number of pairs.

To enable the learning of all the items in the addition table may require the use of flash cards. These are easily made and their use directly involves parents in their child's learning. Moreover, involving yourself in this way gives you knowledge about the current state of your child's knowledge that is independent of school assessments.

6.5.3 Addition Goals for Kindergarten

1. Be able to represent the addition of numbers within 10 by combining appropriately sized collections of objects such as fingers, drawn dots, other physical objects as in diagram above. See Learning Addition Game under Pre-K at the SS website.

2. Be able to solve word problems within 10 by using objects.

3. Decompose numbers within 10 in multiple ways, for example, $8 = 4 + 4 = 5 + 3 = 6 + 2 = 7 + 1$.

4. Understand the use of the $=$ sign in this simple context.

5. Given any single digit, know what is required to make 10 as in

$$4+? = 10.$$

6. Fluently (by recall) add within 5.

6.5.4 Addition Goals for Grade 1

1. Use addition and subtraction within 20 to solve word problems involving situations of adding to, taking from, putting together, taking apart, and comparing, with unknowns in all positions, e.g., by using objects, drawings, and equations with a symbol for the unknown number to represent the problem.

2. Solve word problems that call for addition of three whole numbers whose sum is less than or equal to 20, e.g., by using objects, drawings, and equations with a symbol for the unknown number to represent the problem.

3. Know the three basic properties of addition and be able to use them in actual situations. For example,

$$5 + 7 + 3 = 5 + (7 + 3) = 5 + 10 = 15.$$

4. Relate counting to addition and subtraction (e.g., by skip-counting by 2 to add 2).

5. Understand subtraction as an unknown-addend problem. For example, subtract $10 - 8$ by finding the number that makes 10 when added to 8.

6. Work with equations within 20 specifically determining the truth or falsity of equations like:

$$6 + 5 = 11? \quad 5 + 4 = 7? \quad 8 + 2 + 9 = 19?$$

7. Determine the unknown in an equation like: $8 + \boxed{?} = 17$.

8. Demonstrate fluent knowledge (recall) of the addition table.

9. Add two digit numbers having a sum within 100 using the standard procedure.

10. Given a two digit number, be able to say what number is 10 more, or 10 less than the given number without counting.

6.5.5   Addition Goals for Grade 2

1. By the end of Grade 2, know addition table by instant recall.

2. Add groups of equals as foundation for multiplication, e.g., $4 + 4 + 4 + 4 = 16$.

3. Use addition to find total cells in rectangular arrays having up to 5 rows and 5 columns. For example, in a 4 by 3 array, each column has 4 cells, so the total number of cells is found by summing $4 + 4 + 4$.

4. Add up to four two-digit numbers using standard procedure.

5. Add within 1000 using concrete model (Button Dealer) and standard procedure.

6. Understand that in adding or subtracting three-digit numbers, one adds or subtracts hundreds and hundreds, tens and tens, ones and ones; and sometimes it is necessary to compose or decompose tens or hundreds.

7. Mentally add 10 or 100 to any three digit number.

8. Be able to explain why place value procedure works — understanding Button Dealer model addresses this.

111

6.5.6 Addition Goals for Grade 3

1. Fluently add within 1000 using standard procedure.

6.5.7 Strategies

If you look at curriculum documents, you will find prominent mention of **strategy**. For example, you may find that a child is requested to display two strategies for performing some computation. At this point you may be wondering what I'm talking about, so let me give an example in the current context.

Suppose your child is asked to add $19 + 41$. One **strategy** for doing this computation is to make the numbers simpler without changing the sum. Thus,

> 19 is almost 20 and that we could make it 20 by taking 1 from 41 which would mean we get the same sum in $20 + 40$. Then all we have to do is add $2 + 4$ and put it in the *tens* place.

The example given was chosen for maximum simplicity. Clearly, things get more complicated as the numbers change. For example, making 20 in the computation $18 + 41$, or $17 + 41$, and so forth.

There are many things in the curriculum aimed at this kind of thinking. For example, making it a focus to recognize the nearest 10, thus 16 needs 4 to get to the nearest 10, which means

$$16 + 7 = (16 + 4) + 3 = 23.$$

If we want to succeed in teaching math to children, we need to do everything possible to encourage them to think about numbers and how they work. So it is hard to oppose any activity that encourages children to think broadly about numbers. However, I believe a line is crossed when asking children to learn strategies becomes a substitute for learning the standard method.

What we know about the **standard method** is:

it always works.

That's why humans invented it. What we also know is: it is the simplest method that always works and on average it involves the fewest computational steps. Moreover there is no strategy that can possibly compete with merely recalling:

$$7 + 6 = 13.$$

Chapter 7

Subtraction of Counting Numbers

Chapter Overview. Subtraction of counting numbers will be defined in the standard manner using the concrete process of **take away**. This will be followed by presenting subtraction as the result of finding the solution to a simple equation. The standard procedure for performing subtraction using the Arabic system of numeration will be presented. Detailed examples, including borrowing, will be given. Grade-by-grade achievement goals will be presented.

In the last chapter, we discussed the binary operation of addition applied to counting numbers. We found that the addition of the counting numbers n and m corresponded to finding the total number of elements when two real-world collections having n and m members, respectively, are combined. Thus, the abstract mathematical operation of addition modelled the real-world process of finding the total members when two collections are combined.

In a like manner, the operation of **subtraction**, or **take away**, is also founded in the real world. But before continuing, we need a brief digression to introduce the **less than or equals** relation.

7.0.1 The Relation $\leq$

In Chapter 2, we used the process of pairing elements in two collections, A and B having n and m members, respectively, to determine which collection had more, or whether the two collections had the same number of members. The reader will recall that if there was an exact pairing between the members of A and the members of B, then the counting numbers assigned to the two collections had to be the same, and this was expressed mathematically by the equation:

$$n = m.$$

If the pairing process left elements in B unpaired after using all the elements in A, then we knew the collection B had more elements than the collection A. In this case, the counting number n is **less than** m and as mathematicians we express this by writing:

$$n < m.$$

What the reader should hold in their mind here is that when we write $n < m$ for counting numbers, it means that any collection constructed having n members will have less members than any collection constructed having m members. This can be verified by constructing the required collections and applying the pairing test.

Mathematicians combine the idea of **equality** with the idea of **less than** in one symbol: $\leq$. Thus, for counting numbers n and m, we write:

$$n \leq m, \quad \text{if and only if} \quad (n < m \text{ or, } n = m).$$

The statement $n \leq m$ is read:

$$n \text{ is less than } m \text{ or } n \text{ is equal to } m.$$

To summarize, given the counting numbers n and m, if we write $n \leq m$, it means that for any collections A and B that have n and m members, respectively, it will be the case that A has either less elements than B, or the same number of members as B as determined by the pairing process. More compactly, we would say B has **at least as many** members as A.

7.1 Subtraction as a Real-World Operation

Consider two counting numbers n and m with $n \leq m$. As discussed above, if we construct two collections A and B having n and m members respectively, then B will have at least as many members as A. Since B has at least as many elements as A, we can complete a process to **remove**, or **take away**, one element from B for each element in A.

The fact that $n \leq m$ guarantees that in the process of removing one element from B for each element in A, we exhaust the elements of A before, or at the same time as, we exhaust the elements of B.

There are two possible outcomes to the take away process depending on whether $n < m$, or $n = m$. We consider them in turn.

If $n < m$, then B has more elements in it than A. Thus, any constructed pairing between the members of A and the members of B leaves members of B unpaired. Consider these left-over members as a new collection C, and call the counting number for C, p. **CP** guarantees that however we take away the n members from B to obtain C, C will have the same number of members remaining! In other words, we always get the same counting number, p.

The other alternative was that $n = m$. In this case, when we take away one element in B for each element in A, both collections are exhausted simultaneously. We know this because there is an exact pairing between the members of A and the members of B. So in this case, there is no collection C. But we do have a number that is associated with the empty set, namely 0. So in the case $n = m$ we set $p = 0$. Again, **CP** guarantees that however we remove the n members from B, we will destroy the collection B. In such a case we end up with nothing and set $p = 0$ as the measure of what's left.

Recall, that to have a procedure qualify as an operation, given fixed inputs, the process must always yield the same result. Since **CP** guarantees this fact about **take away**, **subtraction** qualifies as an operation.

Using the procedure above, we say the number p is the result of performing the **subtraction**, m **minus** n, and write:

$$m - n = p.$$

The operation of subtraction (take away) is denoted by a **centered dash** as shown.

We reiterate for emphasis, taking away (subtraction) corresponds to removing items from collections. Because you can't remove something from a non-existent collection, it is clear why we require that the collection, from which we are removing items, must have at least as many items as we are trying to remove. Subtraction is illustrated in the following diagram:

A diagram showing *seven* is the result of subtracting *ten* from *seventeen*. Subtraction worksheets based on this type of diagram can be found at the M-A website under the subtraction button.

7.2   A Mathematical Definition of Subtraction

As shown, subtraction is founded in the real world. As such, it gives the appearance that subtraction is an entirely separate operation from addition. In fact, subtraction is a form of addition. To get to that place, we need a more mathematical approach.

Let n and m be two counting numbers such that $n \leq m$. Because of the requirement for subtraction that $n \leq m$, we know we can find a number that makes the equation

$$\boxed{?} + n = m$$

true. This number will either be a counting number or 0,[1] and the reason it must exist follows from the discussion in the previous section. For example, $6 \leq 9$, so the equation

$$\boxed{?} + 6 = 9$$

has a solution which we know is 3 and results in:

$$\boxed{3} + 6 = 9.$$

Thus, given $n \leq m$, if p is a number with the property that:

$$p + n = m,$$

we will write

$$p = m - n$$

and say, p is the result of **subtracting** n from m. Alternatively, we say p is the **difference** between m and n.

Approaching subtraction in this manner has the effect of giving primacy to the operation of addition. To obtain a deeper understanding of these ideas let's focus for a minute on a particular counting number, say 11. Then let's ask the question: In how many ways can we decompose 11 as a sum? Notice that each such decomposition provides two counting numbers for the following equation:

$$\boxed{?} + \boxed{?} = 11.$$

For example, we know one pair of numbers that gives us a true equation is 7 and 4, as in

$$7 + 4 = 11 = 4 + 7.$$

These equations give us two subtraction equations, namely,

$$7 = 11 - 4 \quad \text{and} \quad 4 = 11 - 7.$$

Thinking about decomposing counting numbers as sums is one of the ways subtraction is introduced to children.

---

[1] The set comprised of the counting numbers and 0 is referred to as the non-negative whole numbers.

A benefit of this approach is that the requirement to verify that subtraction is an operation disappears because the work in that regard has already been performed for addition. However, in taking this approach we lose the concrete aspects of subtraction as *take away*. But this loss will be more than made up when we discuss the integers in Chapter 9.

So that you are aware of the nomenclature, in $m - n$, m is referred to as the **minuend** and n is referred to as the **subtrahend**. One way to remember this is minuends get smaller and subtrahends subtract.

Thinking of subtraction in terms of addition as in:

$$m - n = p \iff p + n = m$$

gives us a procedure for checking any subtraction result, p. To check we are told to perform the addition, $p+n$, to see whether the sum is m. We may all remember that when we learned arithmetic in school, the method for checking whether a subtraction had been done correctly, was to perform exactly this addition.

If we think of the trivial decomposition of the counting number m into 0 and itself, we have

$$0 + m = m = m + 0$$

from which we conclude that

$$m - m = 0.$$

If we reflect on why $m - m = 0$ should be true, we know it simply corresponds to the fact that when we remove all the elements from a real-world collection, we are left with nothing. This is obvious. However, we will return to this observation later and see that it leads to a major step forward in the development of numbers (see Chapter 9).

When we think of subtraction in terms of addition, as discussed above, we are implicitly finding the solution to an equation which looks like:

$$x + n = m$$

where x is the unknown quantity and m and n are given. The use of variables like x, m and n in equations does not occur until later grades and younger children are most likely to be confronted with equations like:

$$\boxed{?} + 7 = 12$$

in early grades.

7.3 Making Subtraction Useful

The discussion above tells us what subtraction is, and how to check whether a given number is the correct answer to a subtraction problem. However, it tells us almost nothing about how to perform computations using the Arabic number system. Since, understanding the role of place is essential, we develop the subtraction procedure in detail, partly as a means to review our ideas.

Recall that solving the subtraction problem $m - n$ requires us to find an unknown number p having the property:

$$p + n = m.$$

So that these ideas are really concrete, we develop the standard procedure using the Button Dealer analogy.

Suppose the Button Dealer has just filled an order from a customer for 897 buttons. The customer then says that 534 buttons are for his wife and the remainder for his mother-in-law. Since the customer is not good at arithmetic, he asks the Button Dealer to divide the buttons. How should the Button Dealer proceed?

The original order consists of *seven* jars marked with a 1, *nine* jars marked with 10, and *eight* jars marked with 100. To solve the customer's problem, the Button Dealer needs to split up the buttons into two collections, one of which will contain *four* jars marked with a 1, *three* jars marked with 10, and *five* jars marked 100.

To accomplish the split, the Button Dealer proceeds as follows. To obtain buttons for the wife, he takes *four* jars from the *seven* marked with 1; he takes *three* from the *nine* jars marked with 10; and from the *eight* jars marked with 100, he takes *five*. This gets 534 buttons for the wife. The remainder, or **difference** are for the mother-in-law, and there are *three* jars marked with 1, *six* jars marked with 10, and *three* jars marked 100, or 363 buttons left for the mother-in-law.

How easy was that?! What's more, we are guaranteed that the result is correct because of **CP**. In mathematical terms we would express this as:

$$897 - 534 = 363.$$

The representation of this problem in Arabic notation is shown below.

$$
\begin{array}{r}
897 \\
-534 \\
\hline
363
\end{array}
$$

We can check that the result 363 is correct by computing $534 + 363$ using the standard addition procedure. Further examples will be discussed below to illustrate the subtraction process in detail.

What the button analogy shows us with clarity is, that like addition, subtraction is again a real-world process. That this must be so follows from the fact that the answer to a subtraction problem is also the answer to an addition problem.

To be specific, recall that the answer, p, to the subtraction problem, $n - m$, is also the solution to the addition equation, where for given n and m, we must find p such that

$$p + m = n.$$

Since subtraction is really a form of addition, having the addition table available for reference would be useful. (Of course, if the table has already been committed to memory, this is unnecessary. But, even in this case, its presence is useful for discussion.)

+	0	1	2	3	4	5	6	7	8	9
0	0	1	2	3	4	5	6	7	8	9
1	1	2	3	4	5	6	7	8	9	10
2	2	3	4	5	6	7	8	9	10	11
3	3	4	5	6	7	8	9	10	11	12
4	4	5	6	7	8	9	10	11	12	13
5	5	6	7	8	9	10	11	12	13	14
6	6	7	8	9	10	11	12	13	14	15
7	7	8	9	10	11	12	13	14	15	16
8	8	9	10	11	12	13	14	15	16	17
9	9	10	11	12	13	14	15	16	17	18

To explain how the table is used in subtraction, suppose we are given m and n subject to:

$$m \leq 18 \quad \text{and} \quad n \leq 9$$

and we want to find $m - n$. Observe that the conditions merely mean that m can be found as an entry in the body of the table, and n can be found in the top row of the table as a column heading. Further, we require that m can be found in a cell in the table having n as its column header. For example, $m = 17$ and $n = 9$ satisfy these conditions, whereas $m = 16$ and $n = 4$ do not.

Now to use the table to find $m - n$, simply go to the column headed by n, proceed down the column to the cell containing m, and thence to the far left to find the number heading that row. Call this number p. Because the table is the **addition**

table, we know

$$p + n = m,$$

which means $m - n = p$. So the addition table provides answers to all subtraction problems that meet the conditions listed above.

Let's do an example using Arabic notation.

Example 1.

Find $854 - 623$. As with addition the first step (setup) is to write the minuend 854 above the subtrahend 623 so that *units* are above *units*, *tens* are above *tens* and *hundreds* are above *hundreds*. The result is displayed below with a subtraction sign and a line to indicate where the answer goes.

$$
\begin{array}{r}
854 \\
-623 \\
\hline
\end{array}
$$

The only requirement here is that $n \leq m$, which we see is true. As in the case of addition, we perform the computation column-by-column starting at the right. The first subtraction we have to perform is $4 - 3$. Since 4 and 3 satisfy the requirements for using the table, using the procedure outlined, we find $1 + 3 = 4$, so $4 - 3 = 1$. This result is recorded below the line in the *units* place as shown:

$$
\begin{array}{r}
854 \\
-623 \\
\hline
1
\end{array}
$$

Similarly, we find $5 - 2 = 3$ and $8 - 6 = 2$, which are recorded below the line in the *tens* and *hundreds* places, respectively. The resulting solution is:

$$
\begin{array}{r}
854 \\
-623 \\
\hline
231
\end{array}
$$

This example is like the example solved by the Button Dealer, namely, the subtraction required in each column satisfies the condition, $n \leq m$.

7.3.1   Borrowing

Example 2.

Consider the subtraction problem setup below:

$$62$$
$$-45$$

First observe $45 \leq 62$, so we can indeed do the subtraction. However, the first subtraction to be performed, namely $2 - 5$, does not meet the criterion, $n \leq m$. Recall, addition problems required us to carry into the next column to the left when the sum became a two digit number. In subtraction, when $m < n$ **for a given column**, (that is to say, the subtrahend n is bigger than the minuend m in a given column) we borrow from the column to the left as shown below.

$$
\begin{array}{rr}
5 & \\
\not{6} & 12 \\
-4 & 5 \\
\hline
\end{array}
$$

When we borrow from the *tens* column, we reduce the *tens* digit in the number we are subtracting from by 1. This is indicated by crossing out the 6 and writing the replacement digit at the top of the *tens* column. We put the borrowed 1 to the left of the *ones* digit in the number being subtracted from, as shown. This has the effect of putting an extra 10 *ones* in the *ones* column.

The purpose of borrowing is to ensure that the required conditions on m and n are satisfied in each column, that is, $n \leq m$ and m occurs as a value in the column headed by n in the addition table. As the reader can check, this is now the case for both columns since 12 occurs in a cell in a column headed by 5. At this point, we can find the difference using the table. The first step is to perform the subtraction in the *units* column by finding $12 - 5$. The result, 7, is recorded as shown:

$$
\begin{array}{rr}
5 & \\
\not{6} & 12 \\
-4 & 5 \\
\hline
& 7 \\
\end{array}
$$

The second step is to perform the revised subtraction in the *tens* column, $5 - 4$, with the result recorded as shown:

$$
\begin{array}{rr}
5 & \\
\not{6} & 12 \\
-4 & 5 \\
\hline
1 & 7 \\
\end{array}
$$

The subtraction process is completed by first finding $12 - 5$, then $5 - 4$. Note, the 5 replaces the crossed out 6 in the calculation.

Example 3.

We consider one more example of borrowing, namely, $804 - 578$. We begin by noting that $578 \leq 804$, so we know we can perform this computation in the form of removing 578 members from a collection having 804 members. We set up the problem below:

$$
\begin{array}{rrr}
8 & 0 & 4 \\
-5 & 7 & 8 \\
\hline
\end{array}
$$

Notice, neither the *ones* column, nor the *tens* column satisfies the condition that $n \leq m$, which enables one to find of $m - n$ using the addition table. Moreover, the *tens* place in 804 is a 0 which further complicates the problem.

The problem has been properly expressed in Arabic notation. As before, start with the *ones* column at the far right. We observe that since $4 < 8$, we will have to borrow. However, when we try to borrow from the *tens* place in 804, we are confronted with a 0. While this seems problematic, we know we must be able to perform the computation, because $578 \leq 804$. The solution is to borrow from the *hundreds* column which we illustrate step-by-step below.

$$
\begin{array}{rrr}
7 & & \\
\not{8} & 10 & 4 \\
-5 & 7 & 8 \\
\hline
\end{array}
$$

In the first step, 100 is borrowed from the *hundreds* column and placed in the *tens* column as 10 *tens*.

The borrowed one hundred has now made the *tens* place non-zero, so we can borrow 1 *ten* from the *tens* place. This borrowing is the second step and produces the revised problem pictured below where we now have an extra 10 *ones* in the *ones* column.

$$
\begin{array}{rrr}
7 & 9 & \\
\not{8} & \not{10} & 14 \\
-5 & 7 & 8 \\
\hline
\end{array}
$$

In the second step, a unit of 10 is borrowed from the *tens* column and placed in the *ones* column. Note that all the columns now satisfy the requirements on m and n to perform subtraction using the table.

We can now perform the computation, starting with the column at the right. This gives:

$$
\begin{array}{r}
7 9 \\
\cancel{8} \cancel{10} 14 \\
-5 7 8 \\
\hline
2 2 6
\end{array}
$$

The subtractions are performed column-wise starting at the right: $14-8$, $9-7$, and $7-5$.

All subtraction follows this pattern, including subtraction of decimal numbers which are essential for keeping track of your bank balance.

7.4   What Your Child Need to Know

Overall, your child needs to develop an understanding of the what subtraction is and an ability to perform the computations using the standard algorithm. In accomplishing this, the goals for addition are also relevant. As illustrated in the examples, the process of subtraction has two parts:

1. setting up the subtraction in Arabic format, and figuring out what borrowing is required;

2. performing the subtraction $m - n$ for m and n, where m is in a cell of the addition table in a column headed by n.

The first step in the process is a matter of practice and an understanding of the Arabic System of notation.

The second step, performing the column-by-column subtractions, depends on thorough knowledge of the addition table.[2] Note that in learning the table for purposes of addition, the child must solve problems like:

$$4 + 9 = \boxed{?}.$$

In accessing the table for subtraction, the child is solving problems like:

$$\boxed{?} + 9 = 13.$$

This second type requires a deeper level of learning, hence a more thorough knowledge of the table. Again, flash cards may be useful and working with your child provides you with an independent means of assessing your child's progress.

Fluency with the subtraction algorithm is expected by Grade 3.

---

[2]See also the expectations for addition.

7.4.1 Subtraction Goals for Kindergarten

The primary goal in Kindergarten in respect to subtraction is for the child to develop an understanding that subtraction amounts to **taking apart** or **taking away**. As such the, the focus is on the concrete. Thus, by the end of Kindergarten your child should be able to:

1. Represent addition and subtraction within 10 with objects, fingers, mental images and drawings.

2. Solve addition and subtraction word problems within 10. For example, 8 children went to the store. If 6 were girls, how many were boys?

3. Decompose numbers ≤ 10 in multiple ways as sums.

4. For any number < 10 know what to add to make 10.

5. Fluently add and subtract within 5.

7.4.2 Subtraction Goals for Grade 1

By the end of Grade 1^3 it is expected that your child will be able to:

1. represent and solve problems involving subtraction;

2. understand the relation between addition and subtraction;

3. solve subtraction problems involving numbers ≤ 20; work with and understand addition and subtraction equations;

4. recognize that each counting number less than 20 has multiple representations, for example,
$$2 + 5 = 7 = 4 + 3 = 9 - 2;$$

5. use place value and properties of operations to add and subtract;

6. subtract multiples of 10 that are < 100 from other multiples of 10 < 100 using concrete models, for example, sticks of length 10.

It is expected that children will be able to work with equations like:
$$\boxed{?} - 4 = 7, \quad 17 - \boxed{?} = 10 \quad \text{and} \quad 13 - 9 = \boxed{?}.$$

The point here is that each equation contains a symbolic unknown, $\boxed{?}$, and that unknown can appear anywhere in the equation.

---

[3]See also Grade 1 Addition Goals.

7.4.3 Subtraction Goals for Grade 2

By the end of Grade 2 it is expected that students are able to

1. use addition and subtraction within 100 to solve one- and two-step word problems involving situations of adding to, taking from, putting together, taking apart, and comparing, with unknowns in all positions, e.g., by using drawings and equations with a symbol for the unknown number to represent the problem;[4]

2. construct equations for solving word problems;

3. determine whether a group of ≤ 20 objects has an odd or even number of members by pairing or counting by 2 s;[5]

4. fluently mentally add and subtract within 20;[6]

5. fluently add and subtract numbers ≤ 100 using the standard place value algorithm;

6. understand and be able to subtract numbers ≤ 1000 using the standard place value algorithm;

7. mentally add or subtract 10 or 100 from numbers in the range 100 to 1000;

8. solve word problems involving money using dollars, quarters, dimes, nickels and pennies and the symbols $;

9. tell time from digital and analog clocks.

---

[4]A two-step word problem would be: Jane has 20 marbles. She gave four to Bill and eight to Mary. How many did Jane have left?

[5]The reader will recall that **even** numbers are divisible by 2. A list of even counting numbers is: 2, 4, 6, 8, 10, 12, ... Numbers that are not even are called **odd**.

[6]Within 20 means the result is ≤ 20.

Chapter 8

Multiplication of Counting Numbers

Chapter Overview. In this chapter the binary operation of multiplication is defined in terms of repetitive addition. Multiplication is given a concrete realization using area and volume. The various properties of multiplication are presented with particular emphasis on the Commutative, Distributive and Associative Laws. The notion of multiplicative identity is defined. The Arabic System of notation is discussed again making full use of multiplication. The standard algorithm for performing multiplication supported by the Arabic System is presented and 7 examples illustrating its use are discussed. Finally, grade-by-grade performance goals are presented.

The operation of multiplication was known to the ancient Egyptians and Babylonians.[1] Since keeping track of land areas, which requires multiplication, was certainly of importance in ancient Egypt, it is clear why knowledge of multiplication would be useful.

8.1   What Is Multiplication?

Like addition, multiplication is a binary operation on counting numbers. Given two counting numbers, n and m, referred to as **factors**, we form a new counting number $n \times m$ called the **product** of n and m.

In fact, multiplication applied to counting numbers is a form of addition, namely, **repetitive** addition. What is meant by this is that 4×5 signifies adding 4 to itself

---

[1]See Wikipedia entry for Multiplication.

126

5 times, as in:
$$4 \times 5 = 4 + 4 + 4 + 4 + 4.$$

To give another example:
$$7 \times 3 = 7 + 7 + 7$$

whereas:
$$3 \times 7 = 3 + 3 + 3 + 3 + 3 + 3 + 3.$$

We may think of the **first counting number as specifying what is to be added**, and the **second counting number as specifying how many of the first number are to be added** together.

Because multiplication is defined in terms of addition, the properties of addition are transferred to multiplication. For example, that multiplication is a binary operation follows from the fact that addition is a binary operation. In fact, as we proceed through arithmetic, we will see that there is really only one operation from which all else is derived and that is addition. Addition can be founded on successor, and that in turn is concretely based on counting collections. Thus, in the final analysis, almost all of arithmetic comes down to counting.[2]

These ideas underlying multiplication are most easily made concrete if we consider diagrams like the following for 4×5:

In the diagram, each rectangular column contains four dots corresponding to the counting number 4. There are five columns, one for each of the five 4 s that are to be added together. The indicated operation is addition and it is clear from the diagram why the process would be referred to as **repetitive addition**. Since the process is addition, in the end it all comes down to counting. The CCSS-M expect that children in Grade 2 can perform the addition required to find the total dots in 5 by 5 arrays such as this and can relate these ideas to multiplication.

8.1.1   Multiplication and Area

The CCSS-M expends significant time on data. In the primary grades the focus is on making measurements of length. It is critical that when a child learns about

---

[2]We say *almost all* because when we discuss the multiplication of unit fractions we will have to find another interpretation of multiplication.

length, the child understands that every such measurement involves a specific **unit of measure** like **inches** or **centimeters** and a standard measuring device would be used. Coming to terms with measurements of length and the units in which they are measured is a prerequisite to understanding the concept of **area**. It is a hands-on task that is accomplished by experiment. One outcome from such experiments is that children recognize that different units of length produce different numerical measures for the same object in the world; for example, the length of a fixed table expressed as a number of centimeters is more than its length expressed as a number of inches.

A unit of area is generally thought of as a **square** having a side of length 1 unit in some unit of measure. The defining property for squares is that the four sides have the same length and the four angles between adjacent sides have the same measure. For example, the squares pictured below can each be thought of as specifying one unit of area, although these units are obviously different.

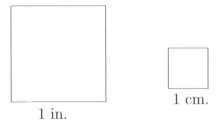

Each box satisfies the criteria of being a square and each has an area of 1 **square unit**. For the box on the left, the unit of length is inches; for the box on the right, the unit of length is centimeters.

We mentioned that multiplication was required for computing areas. To see why, consider the following grid (array):

Think of each square in the grid as 1 square unit of area. There are four units of area in each column and five columns. From the previous discussion, the total number of grid squares, hence total area, is found by computing 4×5.

Consider the combined figures:

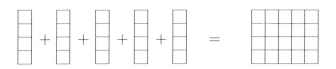

Starting with the five columns of four on the left-hand-side (LHS) of the equality, imagine removing the plus signs and moving the columns together to form the array on the right-hand-side (RHS). **CP** once again, see §2.2, guarantees the total squares on either side are the same. So area has to be **conserved**.

Rather than thinking of the last diagram as finding the sum of five 4 s, we can concentrate on the array on the RHS and think of finding the total number of cells in a rectangle of height 4 cells and length 5 cells. Considerations such as these lead to the well known formula for the **area of a rectangle** of length $\mathcal{L}$ and width $\mathcal{W}$:

$$area = \mathcal{L} \times \mathcal{W}.$$

Area is a notion that arises in the real world. Based on the above, every product of two numbers can be represented as an area. Because of **CP**, we know that area must be conserved and once again, we see that finding areas essentially comes down to counting. Understanding that the product of two numbers may be thought of as **area** is an essential part of learning about multiplication. While the formula for finding area given above does not display it explicitly, area is always expressed in **square units**, for example, square inches. The CCSS-M expect that children will know this as an essential fact about data.

It is hard to imagine a cultural requirement for multiplication that does not involve area. We know that there were procedures for performing multiplication computations known to the ancient Egyptians, Babylonians and Chinese (see Wikipedia entry on multiplication), but not why multiplication was developed.

8.2   Properties of Multiplication

Since multiplication is repetitive addition, it must be the case that all properties of addition carry over to multiplication. Two of these are:
the **Commutative Law**

$$n \times m = m \times n$$

and the **Associative Law**

$$(n \times m) \times p = m \times (n \times p).$$

where n, m and p are any three counting numbers, or 0. These laws are critical to our understanding of arithmetic. Indeed, these and a very few others are the basis for all our computations. For this reason it is essential that readers achieve comfort with their content and their use! They need to become *old friends* and we intend to make them so.

We know that arithmetic is a model for counting processes taking place in the world. Since multiplication is merely repetitive addition, and hence a counting process, we could take this to be justification for the truth of these laws. But this would not achieve our goal of deep understanding. For this reason, we examine these laws in detail.

8.2.1   The Commutative Law

Here, we show why the Commutative Law must be true for multiplication using considerations about area. The reader should remind themselves that any time we speak of area, the numerical measure will have units attached.

We consider a specific equation

$$5 \times 4 = 4 \times 5.$$

The next diagram shows the LHS of this calculation as an instance of finding the area of a figure with $\mathcal{L} = 5$ and $\mathcal{W} = 4$.

Counting the cells in the grid on the RHS tells us that $5 \times 4 = 20$. So the area of the rectangle is 20 square units.

Compare the rectangle on the RHS of the following diagram with the rectangle above. The rectangle diagramed below has has $\mathcal{L} = 4$ and $\mathcal{W} = 5$ and so has an area given by 4×5, which is the RHS of our Commutative Law equation.

Counting the cells in the grid on the RHS tells us that $4 \times 5 = 20$ as well. So the area is again 20 square units.

As the reader can see, both rectangles have the same number of cells, namely 20. While counting the cells gives us this result, we don't need to count because the RHS's are actually the same diagram! To see this, note the grid in the RHS of the current figure is the identical grid to that in the previous diagram, but rotated 90° clockwise. **CP** ensures that rotating a figure cannot change the number of grid squares! It is expected that children will understand that rotating a figure, as in the last two diagrams, cannot change the number of cells comprising the figure. Hence area is **fixed** under rotations.

In the end, every child should know that the **Commutative Law** for multiplication ensures that for every pair of counting numbers, n and m,

> **the product of n times m has the same value as the product of m times n.**

8.2.2   The Distributive Law

Before proceeding to the Associative Law for multiplication, we turn to an apparently new law, the **Distributive Law**. It states that for any counting numbers n, m and p:

$$p \times (n + m) \;=\; p \times n \;+\; p \times m.$$

A Precedence Rule

On the LHS of this equation we have used parentheses to indicate that the operation of addition is to be performed **before** the operation of multiplication. The parentheses are on the LHS to remove the ambiguity in the expression

$$p \times n + m.$$

To be clear about the nature of this ambiguity, consider the expression $5 \times 4 + 2$. Do we mean

$$20 + 2 \quad \text{or} \quad 5 \times 6?$$

The parentheses in $p \times (n + m)$ inform us that the sum $n + m$ must be computed before any multiplication occurs. Thus, p is multiplied by the sum $n + m$.

But there are no parentheses on the RHS in

$$p \times m \;+\; p \times n.$$

Parentheses are not required on the RHS because mathematicians employ what are called **rules of precedence**. Rules of precedence specify in which order operations are to be performed when there are different operations occurring in an expression, as in the Distributive Law. According to one of these rules, in the absence of parentheses, multiplication has higher precedence than addition, which means multiplication must be performed before addition. So the effect of this rule is that

$$p \times m \ + \ p \times n \ = \ (p \times m) + (p \times n).$$

That is why no parentheses are required on the RHS.

The Distributive Law Continued

To understand why the Distributive Law must be true, consider the product

$$4 \times 8 = 4 + 4 + 4 + 4 + 4 + 4 + 4 + 4.$$

We focus on the RHS and recall what the Associative Law tells us about addition. Specifically, the Associative Law says that however we introduce parentheses into the RHS of this equation we must get the same answer. We understand this as an instance of **CP**. Thus for example,

$$\begin{aligned} 4 \times 8 \ &= \ 4 + 4 + 4 + 4 + 4 + 4 + 4 + 4 \\ &= \ (4 + 4 + 4) + (4 + 4 + 4 + 4 + 4) \\ &= \ 4 \times 3 + 4 \times 5, \end{aligned}$$

because multiplication is repetitive addition. Since $8 = 3 + 5$, we we can rewrite the LHS as shown to obtain

$$4 \times (3 + 5) = 4 \times 3 + 4 \times 5,$$

which is just the Distributive Law with $n = 3$, $m = 5$ and $p = 4$. The parentheses on the LHS in $(3 + 5)$ force us to compute $3 + 5 = 8$ before we multiply, so the LHS is still 4×8.

Another instance of the Distributive Law, this time with $n = 6$, $m = 2$ and $p = 4$ is

$$\begin{aligned} 4 \times 8 \ &= \ (4 + 4 + 4 + 4 + 4 + 4) + (4 + 4) \\ &= \ 4 \times 6 + 4 \times 2. \end{aligned}$$

Since $8 = 6 + 2$, we have

$$4 \times (6 + 2) = 4 \times 6 + 4 \times 2,$$

where the parentheses still force the LHS to be 4×8.

Recall again the general statement of the Distributive Law:

$$p \times (n + m) = p \times n + p \times m.$$

For any counting numbers, we can formulate both sides as repetitive addition and recognize that the RHS is obtained from the LHS by the insertion of parentheses exactly as in the examples above. Thus, although the Distributive Law looks new, it is really only an application of the Associative Law for addition.

We can also apply the Commutative Law for multiplication to the Distributive Law. When we do this, we obtain:

$$(n + m) \times p = n \times p + m \times p.$$

Thus, on the LHS we replaced $p \times (n + m)$ by $(n + m) \times p$ On the RHS we replaced $p \times n$ by $n \times p$ and $p \times m$ by $m \times p$. Children should understand that this form of the Distributive Law is a **consequence** of the Commutative Law for multiplication.

Children should also understand why this form of the Distributive Law has to be true in the real world. To do this we consider

$$(3 + 4) \times 8 = 3 \times 8 + 4 \times 8$$

as an area computation. The last equation is represented pictorially by:

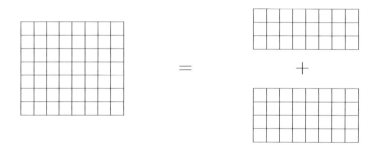

The grid on the LHS of the diagram consists of *eight* columns, each of which has *seven* cells. As such, it represents the product

$$7 \times 8 = (3 + 4) \times 8.$$

The RHS consists of two grids which are to be summed. In the top grid, there are *eight* columns of *three* cells each, corresponding to the product 3×8. In the bottom grid there are *eight* columns of *four* cells each, corresponding to the product 4×8. The total number of cells when a top column is added to a bottom column (directly below) is still *seven*. Thus, we still have *eight* columns, each of which has *seven* cells.

The grid on the LHS may be thought of as a collection of $7 \times 8 = 56$ tiles. If we split this collection into two collections, one containing 24 tiles, and the other containing 32 tiles, as is shown on the RHS, then conservation of real world objects tells us that the sum must preserve the original total. This is why the Distributive Law must be true. It is simply a version of the Conservation Principle (**CP**). So any system of arithmetic we create, as we have argued, is merely an abstract model for the process of counting real-world collections and must satisfy the Distributive Law.

Using the Distributive Law

Because multiplication is commutative, the Distributive Law is **two-sided**:

$$
\begin{aligned}
p \times (n + m) &= p \times n + p \times m \quad \text{and} \\
(n + m) \times p &= n \times p + m \times p.
\end{aligned}
$$

Ordinarily, the Distributive Law is applied in the following way:

$$4 \times (3 + 7) = 4 \times 3 + 4 \times 7 = 12 + 28.$$

In this application, multiplication by the 4 on the outside of the parentheses on the LHS is *distributed* across the addition on the inside to produce a sum of products. As a descriptive term, *distributive* seems appropriate for this process.

There is a second equally important use of the Distributive Law illustrated by the following equation:

$$4 \times a + 3 \times a = (4 + 3) \times a.$$

In this use, we start with a sum of products that have a **common factor**, in this case a as shown on the LHS. We then use the Distributive Law, **backwards** if you will, to **pull the common factor** a outside of the sum thereby producing the RHS. This second use is at least as important as the first use and needs to be recognized and understood by children early on because it is particularly important tool for manipulating symbolic quantities.

8.2.3 The Associative Law of Multiplication

Recall the general statement of the Associative Law:

$$(n \times m) \times p = m \times (n \times p).$$

We want to consider why we should take this equation as a law. We could simply appeal to the fact that multiplication is repetitive addition and addition is associative. But we want to dig deeper and look for a physical explanation that we could present to children.

Products of Three Numbers as Volume

We have already pointed out that the product of two numbers representing lengths can be taken to represent **area** (see §8.1.1). This leads directly to the question: How should we think of the product of three numbers each of which is a length?

In Figure 9.1 we picture a cube measuring 1 unit on a side. We refer to such a cube as a **unit cube** and say it has 1 **cubic unit of volume**. The property that makes this figure a cube is that each of its faces is a square. Since each face of the

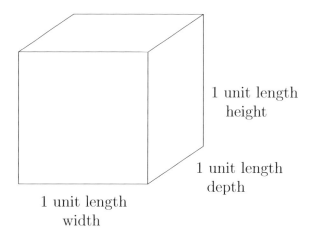

Figure 8.1: A single cube of 1 unit length on each side. Its volume is 1 cubic unit.

block is a square and the length along every edge is 1 unit of length, we know that the area of each face is $1 \times 1 = 1$. But the block exhibits more than mere areal extent of its faces because the faces extend in three mutually perpendicular directions. Thus, the block has **depth** too. Since for the unit cube, all the edges have the same length, namely, 1, the length, the width and the height of the cube are all 1, and if we multiply these three numbers together, the volume of the unit cube is $1 \times 1 \times 1 = 1$ cubic units, which is why we refer to this cube as a **unit** cube. In summary, when objects have physical length in each of three mutually perpendicular directions, we say that such objects have **volume** and like area, this is a physical attribute of objects in the world.

We can think of larger volumes as being comprised of such unit cubes in the same way we can think of areas as being comprised of unit squares. Thinking about measuring volumes by counting unit cubes comprising a whole illustrates in concrete terms why volumes involve the product of three numbers. To make this discussion in respect to the Associative Law as concrete as possible, we consider two stacks of blocks as pictured in Figure 9.2.

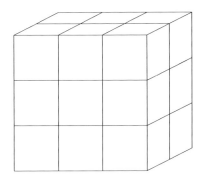 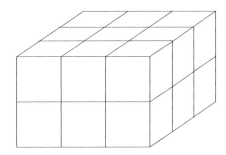

Figure 8.2: Two stacks of blocks made from 18 blocks each of which has unit volume. In the text the stack on the LHS is referred to as Stack 1, the stack on the RHS as Stack 2.

Stack 1 on the LHS is made by stacking 3 blocks to form a column. Three of these columns placed side-by-side form the front face. As the reader can see, there are a total of $3 \times 3 = 9$ blocks forming the front face. As indicated in the figure, there is a second identical row of 3 columns behind the front face which must also contain 9 blocks, so the total blocks used is $(3 \times 3) \times 2 = 18$. Now if we simply ask how many columns of blocks are in Stack 1, we see number is $3 \times 2 = 6$, which is the area of the top face. Since each column contains 3 blocks, the total blocks used counting this way is $3 \times (3 \times 2)$. Since both computations yield 18 blocks, we record this as

$$(3 \times 3) \times 2 = 3 \times (3 \times 2)$$

which is an exact instance of the Associative Law of multiplication.

Imagine that the blocks on the LHS are stuck together, and someone rolls Stack 1 so that it is now lying on its side with its front face on the top. That is the arrangement of Stack 2 on the RHS of the figure. As the reader can see, each column now contains 2 blocks. Since there are 3 columns making up the front face, we know there are a total of 2×3 blocks in the front face. But now the stack on the RHS is 3 columns deep, so there is a total of $(2 \times 3) \times 3 = 18$ blocks in the entire stack. Alternatively, the area of the top face has an area of 3×3, and each block in the top face represents a column of 2, so the total number of blocks is $2 \times (3 \times 3)$. Since the total number of blocks is fixed, these computations again are an exact instance of the Associative Law:

$$(2 \times 3) \times 3 = 2 \times (3 \times 3).$$

The reader knows that however we find the number of blocks making up these stacks, the number of blocks is fixed at 18. **CP** is the guarantor of this fact and once

136

again, we conclude that a fundamental law of our arithmetic is derived from a truth about counting in the real world.

In respect to the Associative Law, every child should understand that in any computation involving **only** multiplication, all ways of carrying out the multiplication must give the same result.

The discussion above makes clear why we consider the product of three lengths to represent volume. In respect to the stacks of blocks, we would say the total volume of the entire stack is:

$$V = (3 \times 3) \times 2 \text{ cubic units} = 18 \text{ cubic units},$$

where 1 cubic unit is the volume of one block. Every child should come to identify **volume** as one outcome of taking the product of three numbers, just as **area** is one outcome when we take the product of two numbers.

8.2.4   The Multiplicative Identity

The reader will recall the following fact about addition:

$$0 + n = n + 0 = n$$

for any counting number n. Because of this equation, we refer to zero as the **additive identity**.

The reader may wonder whether there is an analogous equation for multiplication in which we have a fixed number whose product with any other number leaves that number unchanged. In fact, there is, namely:

$$1 \times n = n \times 1 = n$$

where again, n is any counting number or 0. To be consistent with our nomenclature we refer to 1 as the **multiplicative identity**.

To see why this equation is true, we simply remind ourselves what $1 \times n$ means. The first number, in this case 1, tells us that 1s are what must be added together. The second number, in this case n, tells us how many to add, in this case, n. We could think of this in concrete terms by having each 1 correspond to a single button. We would have a total of n buttons.

Alternatively, $n \times 1$ specifies n is to be added exactly once, again giving n.

The following is an numerical example of the multiplicative identity equation:

$$1 \times 638 = 638 \times 1 = 638.$$

8.2.5   Multiplication by 0

Recall the equation that makes 0 the additive identity: $0 + n = n + 0 = n$. As we will see, this equation applies to all numbers. For our discussion here, we want to apply this fact when $n = 0$, in which case we would have

$$0 = 0 + 0 = 0 \times 2.$$

Changing the sum to a product merely uses the fact that multiplication is repetitive addition. Since the sum of any number of zeros is still 0, we can write

$$0 \times n = 0 = n \times 0$$

for any counting number n or 0. Simply stated,

when we multiply any number by 0, we must get 0.

This is a fact that every child should know.

8.2.6   Products of Counting Numbers are Not 0

Consider that we have two counting numbers n and m. As such, these numbers are cardinal numbers of real-world collections and therefore cannot be 0. Now any sum of counting numbers is again a counting number. We know this from the definition of addition as a real-world process. Since multiplication is repetitive addition, the product of these two counting numbers n and m cannot be 0. We state this as

$$n \times m \neq 0.$$

This is another seemingly simple fact. However, it has an important consequence, namely

if a product is 0, one of the factors comprising the product must be 0.

Again, this is a fact that every child should come to know as they learn arithmetic.

8.3   Making Multiplication Useful

In order to realize the utility of multiplication, we have to develop procedures for finding products of numbers written in some system of numerals. As before, we turn to the Arabic System of notation.

Let us recall the general situation for addition. The procedure really consisted of two parts, first, an ability to find the sum of two single digits, and second, a procedure for using this information to find sums of multi-digit numbers expressed in the Arabic system of notation. Developing a procedure for multiplication will involve similar steps.

8.3.1 Reinterpreting the Arabic System of Notation

Our original interpretation of the Arabic System of notation was based on the Button Dealer analogy. To be specific, given a multi-digit numeral, say:

$$8952,$$

the Button Dealer would get the following jars:

two jars marked 1 from the first shelf;
five jars marked 10 from the second shelf;
nine jars marked 100 from the third shelf;
eight jars marked 1000 from the fourth shelf.

Recall that each jar marked 1 contains *one* button, each jar marked 10 contains *ten* buttons, each jar marked 100 contains one hundred buttons, and each jar marked 1000 contains one thousand buttons. The rest of the Button Dealer process involves **combining all the buttons into one jar**, in other words addition. So let's rewrite things in terms of addition. If we do this, we find the Button Dealer gets:

$$1 + 1 = 1 \times 2$$

buttons from the first shelf;

$$10 + 10 + 10 + 10 + 10 = 10 \times 5$$

buttons from the second shelf;

$$100 + 100 + 100 + 100 + 100 + 100 + 100 + 100 + 100 = 100 \times 9$$

buttons from the third shelf;

$$1000 + 1000 + 1000 + 1000 + 1000 + 1000 + 1000 + 1000 = 1000 \times 8$$

buttons from the fourth shelf.

139

Each of the sums has been reinterpreted as a product. We express this more succinctly as the Button Dealer gets:

$$1 \times 2 \ \text{buttons from the first shelf;}$$
$$10 \times 5 \ \text{buttons from the second shelf;}$$
$$100 \times 9 \ \text{buttons from the third shelf;}$$
$$1000 \times 8 \ \text{buttons from the fourth shelf.}$$

Now we apply the Commutative Law of multiplication to rewrite things as follows:

$$2 \times 1 \ \text{buttons from the first shelf;}$$
$$5 \times 10 \ \text{buttons from the second shelf;}$$
$$9 \times 100 \ \text{buttons from the third shelf;}$$
$$8 \times 1000 \ \text{buttons from the fourth shelf.}$$

Now to complete the process, we know that in effect, the Button Dealer combines the buttons into one jar, in other words, adds the counting numbers. Thus, we have

$$8952 = 8 \times 1000 \ + \ 9 \times 100 \ + \ 5 \times 10 \ + \ 2 \times 1.$$

We remind the reader that we do not need to use parentheses on the RHS because multiplication takes **precedence** over addition and must be performed first.

This example tells us how to reinterpret Arabic notation using multiplication and the notion of **place**. In Arabic notation, we have been referring to the digit at the far right as the *ones* digit and the place as the *ones* place. This is because the right-most digit is multiplied by 1 as in our expanded form for 8952. The digit immediately to the left of the *ones* digit is referred to as the *tens* digit, and the place as the *tens* place, because in the above expression it gets multiplied by 10. The next digit to the left is called the *hundreds* digit and the place called the *hundreds* place because it gets multiplied by 100. The last digit in our four digit number, at the far left, is called the *thousands* digit and its place called the *thousands* place because it is multiplied by 1000. In a five digit number, the left-most digit will be the *ten thousands* digit and the place the *ten thousands* place, and so forth.

Applying these remarks to an arbitrary four-digit number in Arabic notation, we would have

$$n_{1000} n_{100} n_{10} n_1$$

where each of the symbols n_{1000}, n_{100}, n_{10}, n_1 is a single digit from the list 0, ..., 9 and the subscript on n specifies the **place** determined by starting at the right moving to the left. Thus, n_1 is in the *ones* place, n_{10} is in the *tens* place, n_{100} is in the *hundreds* place and n_{1000} is in the *thousands* place, in 8952:

$$n_{1000} = 8, \ \ n_{100} = 9, \ \ n_{10} = 5, \ \ \text{and} \ \ n_1 = 2.$$

The interpretation of such an arbitrary numeral is that it stands for the result of the following computation:

$$n_{1000} \times 1000 \quad + \quad n_{100} \times 100 \quad + \quad n_{10} \times 10 \quad + \quad n_1 \times 1$$

or, in equation form:

$$n_{1000}n_{100}n_{10}n_1 = n_{1000} \times 1000 \quad + \quad n_{100} \times 100 \quad + \quad n_{10} \times 10 \quad + \quad n_1 \times 1.$$

In §5.2 we pointed out an important relation that holds between adjacent place values, for example, between 100 and 10. The relation is that the

> the value of any place is obtained by summing *ten* units having the value of the adjacent place to the right.

So, for example, the *hundreds* place and the *tens* place satisfy,

$$100 = 10 + 10 + 10 + 10 + 10 + 10 + 10 + 10 + 10 + 10.$$

Obviously, this relation begs for the use of multiplication. Stating the various place relations using the power of multiplication, we have:

$$\begin{aligned}
10 &= 1 \times 10, \\
100 &= 10 \times 10, \\
1000 &= 100 \times 10, \\
10000 &= 1000 \times 10,
\end{aligned}$$

and so forth. So a digit in any place has 10 times the value of the same digit positioned one place to the right. The CCSS expect all children to come to a complete understanding of these relationships.

As the reader can see from the discussion above, this interpretation using multiplication still implements the Button Dealer analogy, but in a more compact and sophisticated way. The original interpretation was directly tied to **counting**. It becomes more abstract as we reinterpret counting as addition. The current reinterpretation, in which addition is replaced by multiplication introduces an even higher level of abstraction. But, nothing essential has changed and the notation still comes back to **counting**. That's where a child has to start, with understanding place as being groups of a fixed size coming off the shelf as in the Button Dealer analogy.

The multiplication as an operation was known to the ancient Egyptians. So in principle, it should have been possible to invent the Arabic System of notation as soon as the operation was known. But the Arabic System did not come into use until the Middle Ages. Why so long? The only possible reason is that these ideas were difficult to generate, particularly the concept of zero as a number. However, once known, the ideas are so simple, they can be universally taught to children.

8.3.2 The Multiplication Table

To make multiplication useful, we need to be able to find the products of single digits. In a like manner to addition, these products are placed in a **multiplication table**.

The multiplication table is easily constructed using the definition of multiplication in terms of repetitive addition and the addition table. It can also be constructed directly by counting areas using grids of the appropriate size.

The table is read in exactly the same way as the addition table. For example, suppose we want to know the result of 7×5. Start on the row having 7 at the far left, proceed along this row to the column headed by 5. The entry in that cell is 35, which is the required value.

There are two rows and two columns in the table that are trivial, namely those containing products where one factor is either 0 or 1. There are only a total of 36 remaining products that have to be learned. This number is reduced from 64 due to the Commutative Law.

×	0	1	2	3	4	5	6	7	8	9
0	0	0	0	0	0	0	0	0	0	0
1	0	1	2	3	4	5	6	7	8	9
2	0	2	4	6	8	10	12	14	16	18
3	0	3	6	9	12	15	18	21	24	27
4	0	4	8	12	16	20	24	28	32	36
5	0	5	10	15	20	25	30	35	40	45
6	0	6	12	18	24	30	36	42	48	54
7	0	7	14	21	28	35	42	49	56	63
8	0	8	16	24	32	40	48	56	64	72
9	0	9	18	27	36	45	54	63	72	81

The **Multiplication Table**. To find the value of 4×3, for example, find the cell in the row having a 4 at the far left and in the column having 3 at the top.

Learning the Multiplication Table

In §1.6.1 we said that every parent should make sure that their child knows the multiplication table by rote by the end of Grade 3. To achieve this, children need to start with a procedure for finding the answer that always works. Again, without a procedure, children will guess and each wrong guess creates a wrong memory, however fleetingly, that has to be unlearned. So we begin with a procedure.

The first thing your child needs to know is that:

multiplication of counting numbers is repetitive addition.

This fact is going to be the bedrock on which the child goes forward and this is why your child needs to have mastered addition within 100 by the end of Grade 2. Even if your child is very good at addition, selecting a random card from a deck of multiplication cards is not the most effective way to approach the table.

The effective way to learn the table is column-by-column starting at the left column of zeros. There are two reasons for this. First the two columns at the very left are trivial to learn. The first is nothing but zeros ($n \times 0 = 0$) and in the second, all the products are of the form:

$$n \times 1 = n,$$

since the column header is 1.

Let's go to the third column. The column header is 2, so that every product has the form:

$$n \times 2 = n + n.$$

Again, this column will be easy for children who have mastered the addition table.

Now let's skip to the column headed by 7 with the assumption that your child has learned the products in the column headed by 6. Consider how to find 8×7. Using the fact that your child knows multiplication is repetitive addition, your child can write:

$$8 \times 7 = 8 + 8 + 8 + 8 + 8 + 8 + 8.$$

So your child could simply add up the seven $8\,$s. However, since your child has already learned the products in the column headed by 6, your child can proceed as follows:

$$
\begin{aligned}
8 \times 7 &= 8 + 8 + 8 + 8 + 8 + 8 + 8 = (8 + 8 + 8 + 8 + 8 + 8) + 8 \\
&= (8 \times 6) + 8 \\
&= 48 + 8 = 56.
\end{aligned}
$$

The computation above can be viewed as an application of the Distributive Law and the fact that $7 = 6 + 1$:

$$8 \times 7 = 8 \times (6 + 1) = 8 \times 6 + 8 \times 1 = 56.$$

If your child isn't completely sure about 6×8, as long as she knows that multiplication is repetitive addition, she can go back a further column until she gets to a place where she is sure of the answer and then finish the addition from there as in:

$$8 \times 7 = (8 + 8 + 8 + 8) + 8 + 8 + 8 = (8 \times 4) + (8 \times 3) = 32 + 24 = 56.$$

Here again is where previous mastery of addition within 100 makes finding the last sum straightforward.

Once you are sure that your child can correctly find each product in the table, it is a matter of **practice makes perfect** to accomplish the transition from procedure to recall. Practice problems are available at websites. Flash cards will be helpful to make recall instantaneous.

8.3.3 The Distributive Law, Multiplication and the Arabic System

First we note that the CCSS target for children is that they will be able to multiply four-digit numbers by two-digit numbers with facility. So that our explanations are not too unwieldy, we will begin our explanations using three-digit numbers.

As discussed above, a three-digit counting number in the Arabic System satisfies

$$n_{100}n_{10}n_1 = n_{100} \times 100 \ + \ n_{10} \times 10 \ + \ n_1 \times 1.$$

To take a specific counting number as an example:

$$357 = 3 \times 100 \ + \ 5 \times 10 \ + \ 7 \times 1.$$

Now suppose we want to multiply 357 by another number, say 8. Applying the Distributive and Associative Laws to obtain the second and third lines, respectively, we would have

$$
\begin{aligned}
8 \times 357 \ &= \ 8 \times (3 \times 100 \ + \ 5 \times 10 \ + \ 7 \times 1) \\
&= \ 8 \times (3 \times 100) \ + \ 8 \times (5 \times 10) \ + \ 8 \times (7 \times 1) \\
&= \ (8 \times 3) \times 100 \ + \ (8 \times 5) \times 10 \ + \ (8 \times 7) \times 1.
\end{aligned}
$$

The key thing to notice in this sequence is that as a consequence of the Distributive Law, each digit in the Arabic numeral gets multiplied by 8. Now there is nothing special about 8. Indeed, if we were multiplying by an arbitrary counting number, say m, we would just have:

$$
\begin{aligned}
m \times 357 \ &= \ m \times (3 \times 100 \ + \ 5 \times 10 \ + \ 7 \times 1) \\
&= \ m \times (3 \times 100) \ + \ m \times (5 \times 10) \ + \ m \times (7 \times 1) \\
&= \ (m \times 3) \times 100 \ + \ (m \times 5) \times 10 \ + \ (m \times 7) \times 1.
\end{aligned}
$$

As you can see, each digit is now multiplied by m. And there in nothing special about 357 either. If we take an arbitrary three-digit number in its usual form, multiplication

144

by 8 would yield:

$$\begin{aligned} 8 \times n_{100}n_{10}n_1 &= 8 \times (n_{100} \times 100 \ + \ n_{10} \times 10 \ + \ n_1 \times 1) \\ &= 8 \times (n_{100} \times 100) \ + \ 8 \times (n_{10} \times 10) \ + \ 8 \times (n_1 \times 1) \\ &= (8 \times n_{100}) \times 100 \ + \ (8 \times n_{10}) \times 10 \ + \ (8 \times n_1) \times 1. \end{aligned}$$

And of course, if we multiply $m \times n_{100}n_{10}n_1$, we would have:

$$\begin{aligned} m \times n_{100}n_{10}n_1 &= m \times (n_{100} \times 100 \ + \ n_{10} \times 10 \ + \ n_1 \times 1) \\ &= m \times (n_{100} \times 100) \ + \ m \times (n_{10} \times 10) \ + \ m \times (n_1 \times 1) \\ &= (m \times n_{100}) \times 100 \ + \ (m \times n_{10}) \times 10 \ + \ (m \times n_1) \times 1. \end{aligned}$$

These calculations are typical. We will see many instances of applications of the Distributive and Associative Law in what follows, so it is important you are comfortable with these manipulations. Most importantly, once you recognize that a particular manipulation is an application of one of these laws, you don't have to worry about why it is true, because, as you now know, it all comes back to counting and **CP**.

8.3.4 Multiplication by 10, 100, 1000, etc.

The rest of the multiplication procedure depends on the relationship of the Arabic System of numeration to the operation of multiplication. The basis of this relationship is the effect of multiplying by 10, 100, etc. Thus to go forward, you must completely understand how and why this works. The important theoretical facts we will apply are discussed above.

Multiplying a Two-digit Number by 10

To examine what happens when we multiply by 10, we start with a two-digit example, say, 42. Following the scheme developed in §8.3.3 with an extra step applying the Commutative Law to obtain line 4, when we multiply 42 by 10 we get the following:

$$\begin{aligned} 10 \times 42 &= 10 \times (4 \times 10 \ + \ 2 \times 1) \\ &= 10 \times (4 \times 10) \ + \ 10 \times (2 \times 1) \\ &= (10 \times 4) \times 10 \ + \ (10 \times 2) \times 1 \\ &= (4 \times 10) \times 10 \ + \ (2 \times 10) \times 1 \\ &= 4 \times (10 \times 10) \ + \ 2 \times (10 \times 1) \\ &= 4 \times 100 \ + \ 2 \times 10 \\ &= 4 \times 100 \ + \ 2 \times 10 \ + \ 0 \times 1 \\ &= 420. \end{aligned}$$

Observe, when we start, the 4 in 42 is a *tens* digit. When we finish, the 4 in 420 is a *hundreds* digit. **The 4 has been moved one place to the left**. Similarly, the 2 in 42 is a *ones* digit. In 420, the 2 is now a *tens* digit, so it also has been moved one place to the left.

Thus, after multiplying 42 by 10, we no longer have a *ones* digit, so we have to create one. We did that by putting in 0×1 on the next to last line. We can insert 0×1 exactly because the product is 0 and since 0 is the additive identity, it has no effect on the sum but it does fill the *ones* place. This place has to be filled if we are going to correctly interpret the digits 4 and 2 as part of a three-digit number.

The essence of multiplying by 10 is that

$$10 \times 42 = 42 \times 10 = 420.$$

Because we write language left-to-right, when we think about multiplying 10 times 42, we picture in our minds

$$10 \times 42.$$

However the Arabic number system reads right-to-left, and operations are performed right-to-left. Thus, although we think of 10 as the multiplier in 10×42, we write the product as 42×10 here because we want to emphasize the relation of multiplying by 10 on the right which merely introduces an extra zero on the right in the Arabic numeral as the equation

$$42 \times 10 = 420$$

shows.

We can emphasize the right-to-left aspect of the Arabic System if we apply the Commutative Law in the first step as in:

$$
\begin{aligned}
10 \times n_{10}n_1 &= n_{10}n_1 \times 10 \\
&= (n_{10} \times 10 \;+\; n_1 \times 1) \times 10 \\
&= (n_{10} \times 10) \times 10 \;+\; (n_1 \times 1) \times 10 \\
&= n_{10} \times (10 \times 10) \;+\; n_1 \times (1 \times 10) \\
&= n_{10} \times 100 \;+\; n_1 \times 10 \\
&= n_{10} \times 100 \;+\; n_1 \times 10 \;+\; 0 \times 1 \\
&= n_{10}n_1 0.
\end{aligned}
$$

The calculation is one step shorter, and the 10 is on the right, which is where we want it to be in the end. As in the numerical example, in the next to last line, the reader will notice the addition of the 0×1 so that the final result has a *ones* digit. The result of the computation is the numeral

$$n_{10}n_1 0,$$

which has three digits, the right-most of which is 0. The digit n_{10}, which originated as a *tens* digit, is now a *hundreds* digit, i.e., it has been moved one place to the left. Similarly, n_1, which originated as a *ones* digit, is moved one place to the left and is now a *tens* digit.

Before continuing, let's summarize:

> **When any two digit number is multiplied by 10, the result is the same two digits, in the same order, followed by a 0 on the right.**

The following are numerical examples:

$$25 \times 10 = 250, \quad 63 \times 10 = 630, \quad \text{and} \quad 71 \times 10 = 710.$$

A very important special case of this result is:

$$10 \times 10 = 100,$$

although we know this from the basic properties of the Arabic System.

Multiplying a Three-digit Number by 10

Consider the three-digit number 357. Using the Laws and the fact that $1000 = 100 \times 10$ on line 3 of what follows, multiplying 357 by 10 gives:

$$
\begin{aligned}
10 \times 357 = 357 \times 10 &= (3 \times 100 \ + \ 5 \times 10 \ + \ 7 \times 1) \times 10 \\
&= (3 \times 100) \times 10 \ + \ (5 \times 10) \times 10 \ + \ (7 \times 1) \times 10 \\
&= 3 \times (100 \times 10) \ + \ 5 \times (10 \times 10) \ + \ 7 \times (1 \times 10) \\
&= 3 \times 1000 \ + \ 5 \times 100 \ + \ 7 \times 10 \\
&= 3 \times 1000 \ + \ 5 \times 100 \ + \ 7 \times 10 \ + \ 0 \times 1 \\
&= 3570.
\end{aligned}
$$

The essential result is

$$357 \times 10 = 3570.$$

In other words, once again the result is the original digits in the same order with a 0 tacked on at the right.

You can start with any three-digit number whatsoever and put it into the calculation above for 357 and repeat the sequence of steps. The end result will be the three-digit number you started with, with a 0 added on as the right-most digit. Some additional examples are:

$$275 \times 10 = 2750, \quad 382 \times 10 = 3820, \quad \text{and} \quad 701 \times 10 = 7010.$$

We might now guess that to multiply any counting number in Arabic notation by 10, we can express the result by writing the digits in the same order, left-to-right followed by one additional 0 on the right. This indeed is correct, and the exact reasons why this is so simply follow the computational scheme laid out above.

Multiplying a Two-digit Number by 100

Consider now the problem of multiplying a two digit number by 100. For example, 56×100. Since we know $100 = 10 \times 10$ we can use the Associative Law to obtain,

$$\begin{aligned} 56 \times 100 &= 56 \times (10 \times 10) = (56 \times 10) \times 10 \\ &= 560 \times 10 = 5600. \end{aligned}$$

The net effect most simply stated is

$$56 \times 100 = 5600.$$

Observe that the 5, which starts as a *tens* digit in 56, is moved to the *thousands* place in the answer. Similarly, the 6, which starts in the *ones* place in 56, moves to the *hundreds* place in the answer. In other words, each original digit has been moved two places to the left, one place for each of the zeros in 100.

Once again the result above is true in general. using the Associative Law we have

$$\begin{aligned} n_{10}n_1 \times 100 &= n_{10}n_1 \times (10 \times 10) \\ &= (n_{10}n_1 \times 10) \times 10 \\ &= n_{10}n_10 \times 10 \\ &= n_{10}n_100. \end{aligned}$$

Thus, for clarity, we have:

$$n_{10}n_1 \times 100 = n_{10}n_100.$$

As the reader can see, each digit from the original two digit number, $n_{10}n_1$, has been moved two places to the left in the answer so that n_{10} is now in the *thousands* place and n_1 is now the *hundreds* place.

We can summarize the above as:

When any two digit number is multiplied by 100, the result is the same two digits, in the same order, followed by two 0's on the right.

The following are numerical examples:

$$38 \times 100 = 3800, \quad 65 \times 100 = 6500, \quad \text{and} \quad 92 \times 100 = 9200.$$

Multiplying a Two-digit Number by 1000

Consider now the problem of multiplying a two digit number by 1000. For example, 84×1000. Since we know $1000 = 100 \times 10$ using the Associative Law and the previous rules for multiplying by 100 and 10 gives,

$$
\begin{aligned}
84 \times 1000 &= 84 \times (100 \times 10) \\
&= (84 \times 100) \times 10 \\
&= 8400 \times 10 \\
&= 84000.
\end{aligned}
$$

To restate this result in its simplest form gives

$$84 \times 1000 = 84000.$$

Observe that the 8, which started as a *tens* digit in 84, is a *ten thousands* digit in the answer; in other words, it has been moved three places to the left, one place for each of the zeros in 1000. Similarly, the 4, which starts as a *ones* digit in 84, becomes a *thousands* digit in the answer. Again, it has shifted three places to the left, one place for each of the zeros in 1000.

Once again the computation does not depend on the particular number. Thus,

$$n_{10}n_1 \times 1000 = n_{10}n_1 000$$

by repeated application of the Associative Law and the rules for multiplying by 100 and 10. As the reader can see, each digit from the original two digit number, $n_{10}n_1$, has been moved three places to the left in the answer, so that n_{10} is now in the *ten thousands* place and n_1 is now the *thousands* place.

We can summarize the above as:

> **When any two digit number is multiplied by 1000, the result is the same two digits, in the same order, followed by three 0's on the right.**

The following are numerical examples:

$$45 \times 1000 = 45000, \quad 18 \times 1000 = 18000, \quad \text{and} \quad 30 \times 1000 = 30000.$$

The general rule for multiplying by a multiple of 10 is:

> **to multiply a two digit number by a 1 followed by some number of zeros, write the two digits of your original number, then add on the right, the same number of zeros as follow the 1.**

This is the rule. It's simple to use, and the above discussion explains why it works.

8.3.5 Multiplying an Arbitrary Number by a Single Digit

Recall, the purpose stated at the outset of this section, §8.3, was to develop methods for performing multiplication based on the Arabic System of numeration. At this point the reader may be wondering what all the details in the preceding subsections have to do with that problem. So let's consider exactly where we are. To keep things reasonably simple, we work with a three-digit number, 952.

Based on the reinterpretation of the Arabic System (see §8.3.1), we know that

$$952 = 9 \times 100 \ + \ 5 \times 10 \ + \ 2 \times 1.$$

Suppose we want to compute the product of 7 and 952. Using the Distributive and Associative Laws as in §8.3.3-4, we have

$$
\begin{aligned}
7 \times 952 \ &= \ 7 \times (9 \times 100 \ + \ 5 \times 10 \ + \ 2 \times 1) \\
&= \ 7 \times (9 \times 100) \ + \ 7 \times (5 \times 10) \ + \ 7 \times (2 \times 1) \\
&= \ (7 \times 9) \times 100 \ + \ (7 \times 5) \times 10 \ + \ (7 \times 2) \times 1.
\end{aligned}
$$

Notice that what is now required is to find three products, each of which consists of two single-digits, namely,

$$7 \times 9, \ \ 7 \times 5 \ \text{ and } \ 7 \times 2.$$

Each of these products can be found using the Multiplication Table in §8.3.2, and the results are

$$7 \times 9 = 63, \ \ 7 \times 5 = 35 \ \text{ and } \ 7 \times 2 = 14,$$

all of which are two-digit numbers. Incorporating these facts into the original computation and using the rules for multiplying by 10, etc., gives

$$
\begin{aligned}
7 \times 952 \ &= \ 63 \times 100 \ + \ 35 \times 10 \ + \ 14 \times 1 \\
&= \ 6300 + 350 + 14.
\end{aligned}
$$

To find the answer, we have to perform the indicated sums. Let's do this using the standard procedure. We write the numerals in columns as follows:

$$
\begin{array}{r}
14 \\
350 \\
+ \ 6300 \\
\hline
\end{array}
$$

As we know, each of the three numbers contributing to this sum arises as the product of two single-digits and a multiple of 10. One of the single-digits is 7 which is the single-digit multiplier. The remaining single-digits are the digits of 952 and the products are,

$$
\begin{aligned}
14 \times 1 &= 7 \times (2 \times 1) \\
35 \times 10 &= 7 \times (5 \times 10) \\
63 \times 100 &= 7 \times (9 \times 100).
\end{aligned}
$$

The multiple of 10 occurring in the product is the place multiplier of each single-digit that is a digit in 952. To be clear, the place multiplier on 2 is 1 which tells us that 2 is the *ones* digit in 952; the place multiplier on 5 is 10 which tells us that 5 is the *tens* digit in 952; and the place multiplier on 9 is 100 which tells us that 9 is the *hundreds* digit in 952. In each case, when we form the product of the single digit with 7 as shown above on the LHS, we get a two-digit number times the same place multiplier that is on the RHS. Thus, 4 is a *ones* digit in 14 and arises from the 2 in 952 which is also a *ones* digit; 5 is a *tens* digit in 350 and arises from the 5 in 952 which is also a *tens* digit; and 3 is a *hundreds* digit in 6300 and arises from the 9 in 952 which is also a *hundreds* digit.

Each of the single-digit products results in a two-digit answer, 14, 35, and 63. The second digit, considered as part of a two-digit numeral, will always be a *tens* digit, hence it will always contribute to a column one place to the left of the *ones* digit. This positioning is shown in the sum of $14 + 350 + 6300$ written in columns above.

We turn these ideas into a concise procedure as follows.

Example 1.

The setup amounts to writing 952 above 7 as shown:

$$
\begin{array}{r}
9 \quad 5 \quad 2 \\
\times \qquad 7 \\
\hline
\end{array}
$$

The 7 is a *ones* digit and its place in the setup is directly under the *ones* digit in 952. We have three single digit products that have to be computed,

$$
7 \times 2, \quad 7 \times 5, \quad 7 \times 9
$$

in the order shown (right-to-left). The remaining issue is how to record the results of these products. The first product is 14, so we record the 4 in the *ones* place below the line. The 1 in 14 is a *tens* digit and as such, has to be put in the *tens* column one place to the left of the 4. We do this by carrying the 1 to the *tens* place in a new top row (the **carry row**) above the line, as shown below.

$$
\begin{array}{ccc}
  & 1 & \\
9 & 5 & 2 \\
\times & & 7 \\
\hline
  & & 4 \\
\end{array}
$$

Now, the second product is $35 = 5 \times 7$ and since the 5 is in the *tens* place, we record the 5 in 35 in the *tens* place below the line and carry the 3 to the *hundreds* place in the carry row as shown:

$$
\begin{array}{ccc}
3 & 1 & \\
9 & 5 & 2 \\
\times & & 7 \\
\hline
  & 5 & 4 \\
\end{array}
$$

The last product is $63 = 9 \times 7$. Since the 9 is in the *hundreds* place, we record the 3 in 63 in the *hundreds* place below the line and carry the 6 to the *thousands* place in the carry row as shown:

$$
\begin{array}{cccc}
6 & 3 & 1 & \\
  & 9 & 5 & 2 \\
\times & & & 7 \\
\hline
  & 3 & 5 & 4 \\
\end{array}
$$

What remains is to sum the carried digits and the digits below the line, as shown below:

$$
\begin{array}{cccc}
6 & 3 & 1 & \\
  & 9 & 5 & 2 \\
\times & & & 7 \\
\hline
+ & 3 & 5 & 4 \\
\hline
6 & 6 & 6 & 4 \\
\end{array}
$$

The process above consists of two steps, multiplication/recording followed by summing the carry row and the row below the line. As we will see in the examples with multi-digit multipliers, we will need to combine these two steps. As well, our examples will reflect the fact that the CCSS emphasis is on multiplying two- and three-digit numbers by one- and two-digit numbers.

Before continuing to our next example, we note that multiplication of an arbitrary number by a single digit number requires us to use the various theoretical laws, the data in the multiplication table, together with our knowledge about multiplication by 10, our understanding of place, and what we learned about carrying when we studied addition. In short, multiplication applies all of our previous knowledge.

You can see from this sentence why learning arithmetic is like climbing a ladder. To get to the next rung, you have to stand on a lower rung. To learn the next item, you have to use what you learned previously. If previous knowledge is weak, or incomplete, it will make proceeding difficult. Eventually, progress becomes impossible. This is the central and inescapable fact. What has to be learned is a small amount, but it has to be learned perfectly. If you understand this and help your child achieve a solid foundation, your child will go far. Without a solid foundation, eventually your child starts to struggle, becomes needlessly frustrated, and ultimately is blocked from success.

Example 2.

For simplicity, our next example does not require carrying: 21×4. The setup places the *ones* digits in a single column.

$$
\begin{array}{cc}
2 & 1 \\
\times & 4 \\
\hline
\end{array}
$$

Two single digit products are required: $1 \times 4 = 4$ and $2 \times 4 = 8$. The first of these, namely 4 is recorded below the line in the *ones* column, the same column as the 1 in 21 as shown below.

$$
\begin{array}{cc}
2 & 1 \\
\times & 4 \\
\hline
 & 4 \\
\end{array}
$$

Because this product generated a single digit numeral, there is nothing to carry. The second product, namely 8, which is also a single digit numeral, is recorded in the *tens* column, which is the same column as the 2. Again, we show below:

$$
\begin{array}{cc}
2 & 1 \\
\times & 4 \\
\hline
8 & 4 \\
\end{array}
$$

The multiplication is now complete because there is nothing carried.

Example 3.

Let's find 84×9. The setup aligns the *ones* digits

$$
\begin{array}{cc}
8 & 4 \\
\times & 9 \\
\hline
\end{array}
$$

153

Two single digit products are required: $4 \times 9 = 36$ followed by $8 \times 9 = 72$. Following the outline above, the 6 from the 36 is recorded below the line in the *ones* column as shown below, and the 3 is recorded in the *tens* place in a new carry row, again as shown below:

$$
\begin{array}{ccc}
 & 3 & \\
 8 & & 4 \\
 \times & & 9 \\
\hline
 & & 6 \\
\end{array}
$$

The result of the second single digit product is 72. Since the 8 was in the *tens* column, the 2 will be recorded in the *tens* column. However, since there is a carried 3 in the *tens* column which we know will have to be added to the 72, we can do this now, as part of this step. When we perform this addition, $3 + 72 = 75$, we cross the 3 out as shown and record the 5 below the line. Finally since all single digit products are completed, we write the 7 below the line one place to the left of the 5 in the *hundreds* place as shown:

$$
\begin{array}{ccc}
 & \cancel{3} & \\
 8 & & 4 \\
 \times & & 9 \\
\hline
 7 & 5 & 6 \\
\end{array}
$$

Example 4.

The next example has three digits, 852×4 with the setup shown.

$$
\begin{array}{ccc}
 8 & 5 & 2 \\
 & \times & 4 \\
\hline
\end{array}
$$

There are three single digit products which, in order right to left, are

$$2 \times 4 = 8, \quad 5 \times 4 = 20 \quad \text{and} \quad 8 \times 4 = 32.$$

The first product, $8 = 2 \times 4$ is recorded below the line in the *ones* column since the 2 is in the *ones* column. There is nothing to carry.

$$
\begin{array}{ccc}
 8 & 5 & 2 \\
 & \times & 4 \\
\hline
 & & 8 \\
\end{array}
$$

The next single digit product is $5 \times 4 = 20$ where the 5 is in the *tens* column. So the 0 from 20 also gets recorded in the *tens* column below the line, and the 2 is carried and put in a new carry row in the *hundreds* place as shown:

$$\begin{array}{r} 2 \\ 8 \quad 5 \quad 2 \\ \times \quad 4 \\ \hline 0 \quad 8 \end{array}$$

The last single digit product is $8 \times 4 = 32$, where the 8 is in the hundreds place in 852. Thus, the 2 in 32 has to be recorded in the *hundreds* column below the line. But there is a already a 2 in the *hundreds* place in the carry row which must be added to the 32 at some point. We can do that now giving $2 + 32 = 34$ and record the 4 in the *hundreds* place below the line. The carried 2 is crossed out to show that it has been used. Since 8×4 is the last single digit product, the 3 from 34 is recorded one place to the left in the *thousands* place as shown.

$$\begin{array}{r} \not{2} \\ 8 \quad 5 \quad 2 \\ \times \quad 4 \\ \hline 3 \quad 4 \quad 0 \quad 8 \end{array}$$

Example 5.

We do a last example using a four digit number: 7568×9. The setup is:

$$\begin{array}{r} 7 \quad 5 \quad 6 \quad 8 \\ \times \qquad 9 \\ \hline \end{array}$$

Four single digit products are required which are, right to left:

$$8 \times 9 = 72, \quad 6 \times 9 = 54, \quad 5 \times 9 = 45 \text{ and } 7 \times 9 = 63.$$

Since the 8 in 8×9 is in the *ones* column, the 2 from 72 is recorded below the line in the *ones* column and the 7 is recorded in a new carry row in the *tens* place.

$$\begin{array}{r} 7 \\ 7 \quad 5 \quad 6 \quad 8 \\ \times \qquad 9 \\ \hline 2 \end{array}$$

The next single digit product is 6×9 where the 6 is in the *tens* place. So the 4 from 54 would be recorded in the *tens* place below the line. However, we have a 7 in the *tens* place in the carry row. When this 7 is added to 54, the result is $7 + 54 = 61$, so that 1 is recorded below the line in the *tens* place and 6 is carried to the *hundreds* place in the carry row. Again the 7 is crossed out to indicate the sum has been recorded:

```
        6   7̶
    7   5   6   8
        ×       9
    ─────────────
            1   2
```

The next single digit product is 5×9 where the 5 is in the *hundreds* place; so the 5 from from the product, 45, has to be recorded in the *hundreds* place below the line. However, we have a carried 6 also in the *hundreds* place. When this 6 is added to 45, it produces $6 + 45 = 51$ so that 1 is recorded below the line in the *hundreds* place and the 6 is crossed out. The 5 is carried to the *thousands* place in the carry row, giving:

```
    5   6̶   7̶
    7   5   6   8
        ×       9
    ─────────────
        1   1   2
```

The last single digit product is 7×9 where the 7 is in the *thousands* place along with the 5 in the carry row. So we have to add the product 63 and the carried 5 to get $5 + 63 = 68$; the 8 is recorded below the line in the *thousands* place. Since there are no more single digit multiplications to be performed, the 6 is recorded below the line in the *ten thousands* place.

```
        5̶   6̶   7̶
    7   5   6   8
        ×       9
    ─────────────
    6   8   1   1   2
```

As detailed at the end of this chapter, multiplication of whole numbers is introduced in Grade 2 and concludes in Grade 4 with the expectation that children can skillfully complete calculations like those presented above.

8.3.6   Multiplying by Multi-digit Numbers

Once your child masters multiplying a multi-digit number by a single digit number, multiplying multi-digit numbers by other multi-digit numbers is relatively straight forward. To see why, consider that we have a multi-digit number, say $k = 89747$ and we want to multiply it by 27. Using the theory, we know that this product is obtained as:

$$k \times 27 = k \times (20 + 7) = k \times 20 + k \times 7 = (k \times 2) \times 10 + k \times 7.$$

Using the same k and completing the calculation requires we sum the two numbers on the LHS of the following:

$$
\begin{aligned}
(89747 \times 7) \times 1 &= 89747 \times (7 \times 1) \\
(89747 \times 2) \times 10 &= 89747 \times (2 \times 10).
\end{aligned}
$$

In this case, the place multiplier is associated with the digits in 27. In all of this, the only new wrinkle is the existence of the place multiplier 10 on the second line. But we know what the effect of this multiplier is, it simply moves all digits one place to the left. Thus

$$
(89747 \times 2) \times 10 = 179494 \times 10 = 1794940.
$$

Suppose instead we want to multiply k by a three-digit number, say 975, we would have:

$$
k \times 975 = k \times (900 + 70 + 5) = (k \times 9) \times 100 + (k \times 7) \times 10 + (k \times 5) \times 1.
$$

Using the same value for k and completing this calculation would require we sum the three numbers on the LHS of the following:

$$
\begin{aligned}
(89747 \times 5) \times 1 &= 89747 \times (5 \times 1) \\
(89747 \times 7) \times 10 &= 89747 \times (7 \times 10) \\
(89747 \times 9) \times 100 &= 89747 \times (9 \times 100).
\end{aligned}
$$

Again, the new feature is the existence of the place multipliers, 10 and 100, in the number we are multiplying by. As discussed, the 10 moves the entire product one place to the left and the 100 moves the entire product two places to the left as the following shows:

$$
(89747 \times 9) \times 100 = 807723 \times 100 = 80772300
$$

The key to correctly performing this procedure is how we record and add the various products. Even this comes down to a couple of reasonably straightforward rules as our next examples will show.

Example 6.

We want to find 57×46. The setup puts the 57 above the 46 so that the *ones* are above the *ones* and the *tens* above the *tens*:

$$
\begin{array}{r}
5\ \ 7 \\
\times\ \ 4\ \ 6 \\
\hline
\end{array}
$$

157

The multiplication process starts by computing 57 times 6. The procedure exactly replicates what is done in Examples 3 and 4 and produces:

$$
\begin{array}{ccc}
 & \not{4} & \\
5 & 7 & \\
\times & \boxed{4} & 6 \\
\hline
3 & 4 & 2
\end{array}
$$

Reviewing the material previous to this example leads to the next step which is to find 57×40. Computing this product starts by finding 57×4 where we have written $\boxed{4}$ to distinguish it from the carried 4. To compute the product, we have to find the two single-digit products (written in the manner of Examples 3 and 4 above)

$$7 \times \boxed{4} = 28 \quad \text{and} \quad 5 \times \boxed{4} = 20,$$

in that order. The multiplier $\boxed{4}$ is actually $4 \times 10 = 40$ because it is in the *tens* column. This means that the 8 in 28 has to be recorded in the *tens* place, that is, **in the same column as its multiplier**, in this case $\boxed{4}$. Further, if we have to carry, which in this case we do, what is carried goes in the carry row in the *hundreds* place. These computations are shown as an intermediate step below:

$$
\begin{array}{cccc}
2 & \not{4} & & \\
 & 5 & 7 & \\
\times & & \boxed{4} & 6 \\
\hline
 & 3 & 4 & 2 \\
 & & 8 &
\end{array}
$$

Notice that by putting the 8 in the same column as the multiplier $\boxed{4}$, we have automatically accounted for the fact that we are multiplying by 40 as opposed to 4. This also ensures that the results of all other products with $\boxed{4}$ are correctly placed.

   To complete multiplication by $\boxed{4}$, we need to record the second single digit product, $5 \times \boxed{4} = 20$. In this case, since the 5 is in the *tens* place and the $\boxed{4}$ is in the *tens* place, the 0 from the 20 has to be in the *hundreds* place ($10 \times 10 = 100$), which puts it one column to the left of the previously placed 8. Since we have a 2 in the *hundreds* place on the carry line, we add it to the 0 and cross out the carried 2 as shown:

$$
\begin{array}{cccc}
\not{2} & \not{4} & & \\
 & 5 & 7 & \\
\times & & \boxed{4} & 6 \\
\hline
 & 3 & 4 & 2 \\
2 & 2 & 8 &
\end{array}
$$

As the reader can see, the 2 from 20 automatically ends up in the *thousands* place. What remains is to sum these two products using the standard procedure to obtain:

$$
\begin{array}{r}
\cancel{2} \quad \cancel{4} \qquad \qquad \\
5 \quad 7 \\
\times \quad 4 \quad 6 \\
\hline
{}^{1}3 \quad 4 \quad 2 \\
+ \quad 2 \quad 2 \quad 8 \qquad \\
\hline
2 \quad 6 \quad 2 \quad 2 \\
\end{array}
$$

Example 7.

Find the product of 683 and 37. Again we know that what is required is finding and summing the two products shown in the equation

$$683 \times 37 = 683 \times (30 + 7) = 683 \times 30 + 683 \times 7.$$

Aligning the *ones* the setup is:

$$
\begin{array}{r}
6 \quad 8 \quad 3 \\
\times \quad 3 \quad 7 \\
\hline
\end{array}
$$

The first required product is 683×7 which produces the following

$$
\begin{array}{r}
\cancel{5} \quad \cancel{2} \qquad \\
6 \quad 8 \quad 3 \\
\times \quad 3 \quad 7 \\
\hline
4 \quad 7 \quad 8 \quad 1 \\
\end{array}
$$

using the methods of §8.3.4.

The second required product is 683×30. For clarity, we repeat the last formulation with the 30 multiplier identified by a box.

$$
\begin{array}{r}
\cancel{5} \quad \cancel{2} \qquad \\
6 \quad 8 \quad 3 \\
\times \quad \boxed{3} \quad 7 \\
\hline
4 \quad 7 \quad 8 \quad 1 \\
\end{array}
$$

Finding this product will require computing and recording the single digit products

$$3 \times \boxed{3} = 9, \quad 8 \times \boxed{3} = 24, \quad \text{and} \quad 6 \times \boxed{3} = 18,$$

where the box is being used to remind the reader that the multiplier is in the *tens* place. As observed in the last example, the issue is how to record the results. Since the multiplier is in the *tens* place and $3 \times \boxed{3}$ yields a single digit, we record the 9 below the line in the *tens* column in a new row below the existing row

$$
\begin{array}{cccc}
 & \cancel{5} & \cancel{2} & \\
 6 & 8 & 3 & \\
 \times & \boxed{3} & 7 & \\
\hline
4 & 7 & 8 & 1 \\
 & & 9 &
\end{array}
$$

Notice that recording the 9 directly below the $\boxed{3}$ in the *tens* column automatically accounts for the fact that the actual product being performed is

$$3 \times \boxed{3}0 = 90$$

and by only recording the 9, we are suppressing the 0. Moreover, the remaining products will now be in their correct places.

The next product is $24 = 8 \times \boxed{3}$. The 4 is recorded in the new row below the line in the *hundreds* place to the left of the 9. As noted, the 4 is in the correct place, since the actual computation being performed is $80 \times 30 = 2400$. The 2 from 24 is a *thousands* digit and is recorded in the carry row in the *thousands* column. We can do that since the *thousands* place in the carry row is empty as shown

$$
\begin{array}{ccccc}
\boxed{2} & \cancel{5} & \cancel{2} & \\
 & 6 & 8 & 3 \\
 & \times & \boxed{3} & 7 \\
\hline
 4 & 7 & 8 & 1 \\
 & 4 & 9 &
\end{array}
$$

The last single digit product is $18 = 6 \times \boxed{3}$ where the 6 is a *hundreds* digit and the $\boxed{3}$ is a *tens* digit. So the 8 in the 18 has to be a *thousands* digit and the 1 has to be in the *ten thousands* place. Since we have a carried $\boxed{2}$ in the *thousands* column already, we add to obtain $\boxed{2} + 18 = 20$. The 0 is recorded below the line as $\boxed{0}$ in the *thousands* place as shown below. The 2, which must be carried, is in the *ten thousands* place. However, since there are no more products to compute, we can record this 2 directly below the line in the *ten thousands* column to the left as shown:

$$
\begin{array}{ccccc}
\cancel{2} & \cancel{5} & \cancel{2} & \\
 & 6 & 8 & 3 \\
 & \times & \boxed{3} & 7 \\
\hline
 4 & 7 & 8 & 1 \\
2 & \boxed{0} & 4 & 9
\end{array}
$$

What remains is to sum the two product lines. Note the utility of crossing out items on the carry row as they are used so that we know they have already been included in the numbers below the line. Summing gives:

$$
\begin{array}{ccccc}
\overset{2}{2} & \overset{3}{5} & \overset{2}{2} & & \\
 & 6 & 8 & 3 & \\
 & & \times & 3 & 7 \\
\hline
 & 4 & 7 & 8 & 1 \\
+ & 2 & 0 & 4 & 9 \\
\hline
2 & 5 & 2 & 7 & 1
\end{array}
$$

Example 8.

A last example finds the product of two three digit numbers. The setup is:

$$
\begin{array}{ccc}
3 & 0 & 5 \\
\times \quad 6 & 4 & 7 \\
\hline
\end{array}
$$

As the reader knows,

$$305 \times 647 = 305 \times (600 + 40 + 7)$$

so that we have to find the following three products

$$305 \times 7, \quad 305 \times 40 \ \text{ and } \ 305 \times 600$$

which then have to be summed.

The first computation required is 305×7 where the 7 is in the *ones* place and results in:

$$
\begin{array}{cccc}
 & & \overset{3}{3} & \\
3 & 0 & 5 & \\
\times \quad 6 & 4 & \boxed{7} & \\
\hline
2 & 1 & 3 & 5
\end{array}
$$

The second multiplier is 4 in the *tens* place and results in the following computation.

$$
\begin{array}{ccccc}
 & & \overset{2}{2} & & \\
 & 3 & 0 & 5 & \\
 & \times \quad 6 & \boxed{4} & 7 & \\
\hline
 & 2 & 1 & 3 & 5 \\
1 & 2 & 2 & \boxed{0} &
\end{array}
$$

Notice that by placing the 0 from $45 \times 4 = 20$ directly under the 4 in the *tens* column, the remaining digits in

$$305 \times 4 = 1220$$

161

are correctly positioned and reflect the actual calculation which is

$$305 \times 40 = 12200.$$

The third multiplier is a 6 in the *hundreds* place and results in the following computation:

```
              2
           3   0   5
    ×     [6]  4   7
       2   1   3   5
    1  2   2   0
  1  8  3  [0]
```

Notice that by placing the 0 from $5 \times 6 = 30$ directly under the multiplier 6, the product

$$305 \times 6 = 1830$$

is automatically correctly positioned and reflects the actual computation

$$305 \times 600 = 183000.$$

All that remains is to sum the three products

```
            3   2   3
            3   0   5
        ×   6   4   7
        2   1   3   5
      1 2   2   0
  +  1  8   3   0
     1  9   7   3   3   5
```

Using the Commutative Law to reverse the order of multiplication would produce a simpler, two-step, process as we can see below:

```
        1   2
        3   2   3
        6   4   7
    ×  [3]  0   5
    3   2   3   5
    0   0   0
  + 1  9   4   1
    1  9   7   3   3   5
```

162

Observe, we had to open a second carry row above the first because the position that we wanted to use for the 2 from $21 = 7 \times 3$ had already been used. While opening additional rows adds a bit of complexity, it is simply a matter of positioning the carried digit in the correct column and making sure it is crossed out as it is used. The overall computation is simplified by the second product being 0.

The procedure described in these examples is what was taught in schools when I was a child. Given its simplicity, the reader may well wonder why we spent so much time on place value and products with 10? The answer is that if you are going to be able to help your child, you need to know how things work at a fundamental level. If you understand the previous material, you know why this procedure works, and you know why it is really nothing more than a high-powered form of counting. Finally, this level of understanding will make helping your child with decimals easy.

8.4 What Your Child Needs to Know

I want you to think about how you found the process of reading the material in this chapter. Did you find it straightforward? Was it easy to understand? Did you have difficulty following the threads of arguments? Did you find computations like

$$70 \times 40 = (7 \times 10) \times (4 \times 10) = (7 \times 4) \times 100 = 28 \times 100 = 2800$$

intimidating? If so, why? If not, why not?

There are several comments to be made here. First, the multiplication algorithm (procedure) being discussed is complex, which may have something to do with why it took thousands of years of human learning to create. Second, as an adult, you bring to this process substantial mathematical training in the form of your previous education. Third, a child brings none of your sophistication and training to the learning of these things.

I would suggest that the degree to which the discussion above was transparent reflects the degree to which mathematical facts are available to you as recall. For example, if you have any question as to why, or if

$$70 \times 40 = (7 \times 10) \times (4 \times 10),$$

then at least some of your available mental abilities are being used to consider this question, and are not available for the larger question which had to do with *place*.

What I want to consider here is the importance of knowledge as recall. If you accept the premise that each of us has a fixed mental work space which we can use to solve problems, understand complex ideas, or in other words *think*, then at any given moment we want to minimize the percentage of that workspace that is being taken up with extraneous and/or subsidiary questions, for example, $7 \times 4 = ?$, which is needed in the above example. In other words, what we have not already learned at the level of recall, detracts from what we can learn in the future.

A central issue and area of conflict in school mathematics education during the last 30 years has been the role of **rote learning versus understanding**.

I would assert, that understanding why arithmetic works is straight forward, once you have figured out arithmetic is about modelling counting in the real world. The fundamental reason why most things are true, as we have already observed, is conservation (**CP**).

However, when we ask a child who knows that addition is counting, but has little knowledge of the addition table: Given 3, 6, 9, what comes next?, that child has next to no chance of arriving at the intended answer, namely, 12, because the child cannot *see* that each successive number is obtained from its predecessor by the addition of 3.

In my life experience I have worked with many outstanding mathematicians and physicists, generally some very smart folks. I have yet to meet a single one of them who did not have basic mathematical facts available to them as a matter of instantaneous recall. I cannot say how this factual knowledge was acquired; I can only observe it was there.[3] So, if you want to do one thing to help your child, make sure the addition and multiplication tables are known at the level of **instantaneous recall**. This is why we have included seggestions on how to help your child learn the tables (§6.4.3 and 8.3.2). The simplest way to complete the learning process is with flash cards containing questions like $5 \times 8 = \boxed{?}$ on them. But the reverse questions are also important, namely, $24 = \boxed{?} \times \boxed{?}$.

8.4.1 Multiplication Goals for Grade 2

Multiplication is formally introduced in Grade 2 according to the CCSS. The focus is on studying addition of equal groups of objects as a precursor to the formal definition of multiplication. In this respect, a child should be able to:

1. determine whether a group of objects having ≤ 20 members has an odd or an even number of members;

2. use addition to find the total number of objects in a rectangular array (grid) having ≤ 5 rows and/or columns.

---

[3]In my own case, I recall numerical experiences in the form of counting as part of children's' games. In the school curricula, there were requirements for rote learning in the form of being able to accurately reproduce addition and multiplication tables in timed tests.

8.4.2 Multiplication Goals for Grade 3

Multiplication is a key focus of Grade 3 in the CCSS. At the highest level, it is expected children will understand what multiplication is, when it is required to solve problems, and to use it in problem solving. We give a simple example:

> Bill is the team manager for a hockey team having 20 members. Each team member has to have 3 hockey sticks. How many sticks, total, are required to fulfill this requirement?

The relationship of multiplication and division should be known — division will be discussed in much greater detail in a later chapter. For our purposes, we think of division only in respect to whole numbers and only as solutions to equations of the form $\boxed{?} \times m = n$ where m and n are counting numbers. Thus, $24 \div 4 = 6$ because $6 \times 4 = 24$.

The CCSS mentions strategies. We refer the reader to our comments at the end of Chapter 6.

The following are specific expectations. On completion of Grade 3 your child should be able to:

1. explain that the meaning of a product, e.g., 5×3, is the sum of 3 added to itself 5 times, or the cardinal number of 5 groups of 3 objects combined in a single collection, or the total number of squares in a 5 by 3 grid;

2. recognize that a product can be partitioned by its factors, e.g., $24 = 3 \times 8$, so a group of 24 objects can be subdivided into 3 groups of 8 and also of 8 groups of 3 and that these groupings can be represented in the form of grids;

3. use multiplication within 100 to solve problems involving equal groups and arrays;

4. solve equations like: $3 \times 7 = \boxed{?}$, $\boxed{?} \times 6 = 24$. $5 = \boxed{?} \div 3$, and $5 \times \boxed{?} = 80$;

5. understand and use the Commutative, Associative and Distributive properties;

6. understand that by knowing $4 \times 3 = 12$ and $4 \times 5 = 20$, one can obtain $4 \times 8 = 32$ from the Distributive Law;

7. demonstrate fluent knowledge of the multiplication table sufficient to solve equations like those shown above;

8. be able to solve problems that require different operations, e.g., If Cindy's class has four girls and five boys and she has 27 pencils, how many will be left if she gives three pencils to each child?;

9. multiply single digit numbers by any single digit multiple of 10, e.g., $8 \times 30 = 240$;

10. recognize area as an attribute of plane figures and understand concepts of area measurement;

    (a) a square with side length 1 unit, called a unit square, is said to have one square unit of area, and can be used to measure area;

    (b) a plane figure which can be covered without gaps or overlaps by n unit squares is said to have an area of n square units;

    (c) measure areas by counting unit squares (square cm, square m, square in, square ft, and improvised units);

11. relate area to the operations of multiplication and addition;

    (a) find the area of a rectangle with whole-number side lengths by tiling it, and show that the area is the same as would be found by multiplying the side lengths;

    (b) multiply side lengths to find areas of rectangles with whole-number side lengths in the context of solving real world and mathematical problems, and represent whole-number products as rectangular areas in mathematical reasoning.

8.4.3 Multiplication Goals for Grade 4

By the end of Grade 4 your child should:

1. be able to interpret a multiplication equation as a **comparison**, e.g., $3 \times 7 = 21$, means 21 is three times the value of seven;

2. be able to solve multi-step word problems involving multiplication and division of whole numbers;

3. be able to find all factor pairs for whole numbers in the range 1 to 100;

4. understand the meaning of terms like: **multiple**, **prime**, **composite**, and **factor** as applied to whole numbers (discussed in a later chapter);

5. list multiples of one digit numbers as in: $4,\ 8,\ 12,\ 16,\ 20,\ 24,\ \ldots$;

6. understand that in the Arabic System the value of any digit in a numeral is ten times the value of that same digit when it occurs one place to the right, e.g., $800 = 80 \times 10$;

7. fluidly multiply four digit numbers by any of the ten single digit numbers;

8. use the standard place value method (see Example 5) to fluidly multiply pairs of two-digit numbers.

8.4.4 Multiplication Goals for Grade 5

By the end of Grade 5 your child should:

1. recognize **volume** as an attribute of solid figures and understand concepts of volume measurement;

   (a) a cube with side length 1 unit, called a unit cube, is said to have one cubic unit of volume, and can be used to measure volume;

   (b) a solid figure which can be packed without gaps or overlaps using n unit cubes is said to have a volume of n cubic units;

2. measure volumes by counting unit cubes, using cubic cm, cubic in, cubic ft, and improvised units;

3. relate volume to the operations of multiplication and addition and solve real world and mathematical problems involving volume;

4. find the volume of a right rectangular prism with whole-number side lengths by packing it with unit cubes, and show that the volume is the same as would be found by multiplying the edge lengths, equivalently by multiplying the height by the area of the base;

5. represent threefold whole-number products as volumes, e.g., to represent the associative property of multiplication;

6. apply the formulas $V = l \ w \ h$ and $V = b \ h$ for rectangular prisms to find volumes of right rectangular prisms with whole-number edge lengths in the context of solving real world and mathematical problems;

7. recognize volume as additive;

(a) find volumes of solid figures composed of two non-overlapping right rectangular prisms by adding the volumes of the non-overlapping parts, applying this technique to solve real world problems;

(b) recognize volume as additive;

(c) find volumes of solid figures composed of two non-overlapping right rectangular prisms by adding the volumes of the non-overlapping parts, applying this technique to solve real world problems;

Chapter 9

The Whole Numbers

Chapter Overview. In this chapter we define the set of **whole numbers**, or **integers** and study their arithmetic properties. Completing the whole numbers requires the introduction of negative numbers which are introduced by defining the notion of **additive inverse**. Addition and multiplication are then defined on the totality of integers. Eight essential rules of arithmetic are established based on the various laws that we know must be true in any model of the counting numbers. The chapter concludes with a discussion of progress goals for children.

A Brief Summary

In the beginning, there were the counting numbers, also known as the **Natural Numbers**.[1] These numbers were developed for the expressed purpose of providing a measure of how many items belonged to a real-world collection.

It was found that to effectively work with these numbers, an abstract notation was required. The system of notation that was found to be effective was the Arabic System, but its implementation required the recognition of a new symbol which was used as a place-holder in the notation system. Ultimately, it came to be realized that the place-holder representing **nothing** was also a number. The symbol given to this number was 0. Zero is assigned to what results when all elements are removed from a collection. Such a result may be thought of as an empty set, because collections with nothing in them do not exist in the real world. Empty collections can exist only as ideas in our minds which is why we use the abstract notion of set in this context. Thus, zero is the cardinal number of the empty set.

---

[1]A famous nineteenth century mathematician, Leopold Kronecker, once observed: God made the natural numbers; all else is the work of man.

The binary operation of addition of counting numbers was discussed and its major properties identified. This operation could be applied to any pair of known numbers, namely, any counting numbers or zero and effective procedures for performing this operation that connect to the Arabic System of notation were given.

The binary operation of subtraction of counting numbers was studied, where subtraction was thought of as *taking away* items from a real-world collection. Unlike addition, subtraction had certain restrictions in respect to how it was applied to pairs of counting numbers. Specifically, we could only compute $n - m$ in cases where $m \leq n$. An effective method for performing these computations was presented based on the Arabic System of notation.

Lastly, the binary operation of multiplication was studied and its important properties developed. As with addition, any pair of counting numbers or zero can be multiplied. As with addition and subtraction, an effective system for performing multiplication was given based on the Arabic System of notation.

The next question we pose is:

Given we can add and/or multiply any pair of counting numbers, why is there a restriction on subtraction?

It is clear that in the context of removing elements from a real-world collection, for example, buttons from a jar of buttons, we cannot remove more elements than are there in the first place. So this accounts for the original restriction. But numbers have no reality. They are abstract constructs. So we may ask again:

Why must $m \leq n$ in order to compute $n - m$?

9.1 Completing the Whole Numbers

Recall, in our discussion of subtraction of counting numbers we found that

$$n - m = p \text{ if and only if } n = p + m.$$

Thus, the operation of subtraction of $n - m$ is defined by an addition equation involving the unknown p. When these ideas are being introduced, the addition equation is likely to be written as

$$n = \boxed{?} + m.$$

In the case that $n < m$, it is a simple fact that there is no counting number, p, with the property that $n = m + p$. To take a numerical example, consider:

$$p + 20 = 4.$$

This equation was considered to be nonsensical by European mathematicians into the 17th century because there is no such counting number.[2] Moreover, so long as we only have counting numbers available, it tells us why $m \leq n$ is required.

But, as remarked, all numbers, including counting numbers, are abstract ideas. So creating new numbers, beyond the counting numbers and zero, which are the numbers we already have, is not a problem. All we have to do is think of them!

9.1.1 The Notion of Additive Inverse

To follow this path, let's start with what we know, namely, given any counting number n:

$$n - n = 0.$$

Here we are thinking of the process as being take away. Instead, let's think of the process as being addition, but with the same result:

$$n + (-n) = 0.$$

When we think this way, it is clear that the quantity symbolized by $(-n)$ has to be something new. Again, so we're clear, we give a numerical example:

$$6 + (-6) = 0.$$

It's clear the quantity symbolized by (-6) is neither a counting number nor 0, so this quantity has to be something new.

What we are really saying here is that given **any** counting number n, the equation

$$\boxed{?} + n = 0$$

should have a solution. The requirement, that for every counting number, n, the equation $\boxed{?} + n = 0$ has a solution, forces us to create new numbers, one for each counting number n. We call the solution to this equation the **additive inverse** of n and require the Commutative Law to hold so that

$$n + (-n) = (-n) + n = 0.$$

This equation is fundamental and so we will speak of this equation as the **Additive Inverse Equation** to emphasize its importance. Parents should note that when additive inverses are introduced in Grade 6, they may be referred to as **opposites**. Thus, every counting number has an opposite. The rationale for this language will be clear when we discuss the real line.

---

[2]See Wikipedia on Negative numbers.

9.1.2 Defining the Whole Numbers

We can now say precisely what we mean by the set of **whole numbers**, or **integers**. Something is a whole number exactly if it satisfies one of the following three conditions:

1. it is a counting number;

2. it is 0;

3. it is the additive inverse of a counting number.

For a number k to satisfy the last condition, means that we can produce a counting number, n such that the sum

$$n + k = 0.$$

It is important to understand what is happening here. We are identifying a number, namely the additive inverse of a counting number, by its behavior. Thus, when we speak of a particular number as being the additive inverse of some other number, in respect to behavior, we know with certainty that when we add these two numbers together, we must get 0 as the result. Another example of defining a number by its behavior occurred in respect to 0 which was required to satisfy the equation

$$0 + n = n + 0 = n.$$

This equation can, in fact, be taken as the defining property of 0.

We will use $\mathcal{I}$ to denote the set of whole numbers, integers. We note that the additive inverses of the counting numbers are generally referred to as **negative integers**, and the counting numbers are called **positive integers**. We will use integers and whole numbers interchangeably.

9.1.3 Historical Note

According to Wikipedia, Chinese mathematicians knew about negative numbers in 200 BCE, but not zero. Negative numbers were recognized as a measure of debt in commerce, and it was known how to calculate correctly with negative numbers. These ideas reached India by 400 AD and thence to the Middle East where they were being used by Islamic mathematicians in respect to debts by the 10th century. As we have already noted, European mathematicians resisted their use into the 17th century.

Fractions, as ratios of counting numbers, were known to the Pythagoreans, and so predate negative numbers in the historical development of ideas. So at this point our development is following a somewhat different path, both from history and from the standard school curriculum. Hopefully, it is a path of greater simplicity. We follow it because the properties of arithmetic summarized in §9.3.5 are used throughout the remainder of the text.

9.1.4 Notation for Additive Inverse

Recall, that in order to make counting numbers useful, we had to come up with a system of notation. At this point we have notations for numbers that are counting numbers, and a notation for zero, but no notation for new numbers that arise as additive inverses of counting numbers. We can deal with this problem generally by agreeing on a notation for all additive inverses.

Let n be any integer. We will use n preceded by a centered dash to denote the additive inverse of n. Thus, $-n$, is the additive inverse of n, and the proof of this fact is that

$$n + (-n) = 0$$

because satisfaction of the Additive Inverse Equation is what defines *additive inverse*. We will refer to $-n$ as **minus** n. Thus, minus n will be the name of the additive inverse of n where we stress that n can be any integer, even a positive integer.

Concrete examples of this notation are:

$$-10, \quad -501, \quad -1, \quad -(-25),$$

and so forth. The respective behavioral equations are:

$$10 + (-10) = 0, \quad 501 + (-501) = 0, \quad 1 + (-1) = 0,$$

and

$$(-25) + (-(-25)) = 0.$$

Other notations have been tried over the years for denoting the additive inverses of counting numbers. However, this notation, which uses parentheses and a centered dash, seems to have the fewest problems when it comes to rules of precedence which specify the order in which operations are to be performed.

9.2 Properties of Addition on the Integers

The next issue that must be addressed is extending the operation of addition to all integers. That is, given any two whole numbers, n and m, we have to be able to say what the sum of n and m is.

Our sole purpose here is that our arithmetic must model the real world. Thus, if our extended system of arithmetic, which now includes the negative integers, is to satisfy this condition, it must preserve everything that has been developed to date. To be clear, if n and m are counting numbers, then their sum as integers, $n + m$, must be the same as their sum as counting numbers, and we must be able to find this sum using existing procedures.[3]

In order to satisfy this requirement, the operation of addition as extended to **all** the integers must satisfy the following laws. Thus, let n, m and k be arbitrary integers (whole numbers). Then the following laws hold:

1. the **Commutative Law**:
$$n + m = m + n;$$

2. the **Associative Law**:
$$(n + m) + k = n + (m + k);$$

3. the **Identity Law**:
$$n + 0 = 0 + n = n.$$

As well, we require that for every integer n, we can find m which satisfies the Additive Inverse Equation:
$$n + m = m + n = 0.$$

> **To check that this assertion is true about a particular number, n, we simply find another number, m, with the property that $n + m = 0$.**

Let's check whether our three types of integers all have additive inverses.

For n a counting number (positive integer), take the negative integer $-n$ as the number we must find. Then $n + (-n) = 0$, as required.

For 0, we note $0 + 0 = 0$, so 0 is its own additive inverse.

Lastly, if n is a negative integer, then $n = (-p)$ for some counting number p. Substituting $-p$ for n as shown in the following equation gives
$$n + p = (-p) + p = 0,$$

whence we have found a number whose sum with n is 0. So every negative integer also has an additive inverse.

---

[3]For those comfortable with the jargon of software, as we extend addition from the counting numbers to the integers, we should have **forward compatibility**.

We emphasize that we did not have to create any new numbers in order for all the integers to have additive inverses. The reason is that additive inverses come in pairs and each member of the pair is the additive inverse of the other member of the pair. For example

$$(-7) + 7 = 0,$$

so 7 is the additive inverse of -7. Every child should be aware that additive inverses come in pairs and that each member of the pair is referred to as the additive inverse of the other member of the pair.

In previous chapters, we explained why these properties had to be true, based on the real-world behavior of collections and our intention that arithmetic capture this real-world behavior. Thus, for the integers under addition to be a satisfactory model, these four properties must be true. This is why we simply take these assertions about numbers to be Laws. What is important here is that you and your child have a sense of how these properties can be traced back to counting.

As noted above, our main purpose is to extend the addition operation to all the integers. In order to do that, we need the fact that each integer has **only one** additive inverse.

9.2.1 Additive Inverses Are Unique

Moving from a number to its additive inverse is essentially a unary operation. For this reason, we need to know that each integer has only one additive inverse. We provide a complete argument, including the reasons why we can make each step. Some readers may want to skip this. That's OK, but the fact that the additive inverse is unique is essential.

Let n be any integer. (Another way to say this, which we will also use is: let n be an arbitrary member of $\mathcal{I}$.) Suppose, n has more than one additive inverse. Let's call one of these $-n$ and the other k. (By giving the numbers names, we can speak about them.) Now since $-n$ is an additive inverse of n, we know

$$n + (-n) = 0,$$

and since k is an additive inverse of n, it must also be true that

$$n + k = 0.$$

Since *things equal to the same thing are equal to each other* (see E3 §6.2), we have

$$n + (-n) = n + k.$$

175

The Commutative Law tells us that:

$$(-n) + n = k + n.$$

Further, since $-n = -n$ and *equals added to equals are equal* (see E4 §6.2), we can add $-n$ to both sides of the equation to obtain

$$((-n) + n) + (-n) = (k + n) + (-n).$$

The Associative Law tells us

$$(-n) + (n + (-n)) = k + (n + (-n)).$$

Since $n + (-n) = 0$, carrying out the indicated computation inside the parenthesis gives:

$$(-n) + 0 = k + 0.$$

Finally, since $(-n) + 0 = -n$, and $k + 0 = k$,

$$(-n) = k,$$

as required. Thus, each integer n has only one additive inverse, for which the standard notation is $-n$.

On first reading, this argument may seem complicated. It is worth reviewing because it is typical of methods used to demonstrate uniqueness. In essence, one takes two of something with the required property and demonstrates the two things must be equal to each other, as in $(-n) = k$ above.

The key thing the reader should take from this is that

for each integer, there is only one other integer with which it will sum to 0.

9.2.2 Using Uniqueness to Demonstrate Equality

We will use uniqueness of additive inverses repetitively in what follows as a means of demonstrating equality. For this reason it is essential that the reader understand the process.

The first step in the process will be the identification of two, or more, quantities that we think might be equal. These quantities can be anything at all, but they have to have names that make them identifiable. For example, they could be $a + b$, $q \times r$, or simply m.

The next step is that we can find a single other number with which both, or all three, of the previous quantities sum to 0. This number will have to have a name.

176

For purposes of this discussion, let's call this other number n. But remember, this number could have any name at all, just so long as we have a means of identification.

Now the computation $n + \boxed{?}$ is performed, as in:

$$n + (a + b), \quad n + (q \times r), \quad \text{and} \quad n + m.$$

If the result of any one of these computations is 0, we know we have found an additive inverse of n. Thus, for example, if

$$n + (a + b) = 0,$$

then we know that $a + b$ is an additive inverse of n. But we now know more than this. Because of uniqueness, we know there is in fact only one integer having the property that its sum with n is 0. The standard notation for this number is $-n$. So, once we know $n + (a + b) = 0$, this enables us to write:

$$a + b = -n.$$

Further, if we also find that $n + (q \times r) = 0$, then

$$q \times r = -n,$$

and therefore that

$$a + b = q \times r,$$

since both quantities are equal to $-n$. Finally, if it is also the case that $n + m = 0$, then we can write

$$a + b = q \times r = m.$$

Example 1.

For illustration, we perform the numerical computation $(-5) + (-3)$ in detail. We know that

$$8 = 3 + 5.$$

Adding $(-5) + (-3)$ to both sides of the equation produces

$$8 + ((-5) + (-3)) = (3 + 5) + ((-5) + (-3)).$$

Applying the Associative Law to the RHS twice gives

$$\begin{aligned}
(3 + 5) + ((-5) + (-3)) &= ([3 + 5] + (-5)) + (-3) \\
&= (3 + [5 + (-5)]) + (-3).
\end{aligned}$$

Since (-5) is the additive inverse of 5, the RHS becomes

$$(3 + [5 + (-5)]) + (-3) = (3 + 0) + (-3).$$

Applying the Identity Law to $3 + 0$ followed by the fact that (-3) is the additive inverse of 3 gives

$$(3 + 0) + (-3) = 3 + (-3) = 0.$$

Thus,

$$8 + ((-5) + (-3)) = 0,$$

so that $(-5) + (-3)$ is an additive inverse of 8 and uniqueness gives

$$(-5) + (-3) = -8.$$

In addition, applying the Commutative Law tells us that

$$(-3) + (-5) = -8$$

as well.

9.2.3 Consequences of Uniqueness

Using the standard centered dash notation, the Additive Inverse Equation states that

$$n + (-n) = (-n) + n = 0.$$

If $n = 0$, since the Identity Law asserts

$$0 + 0 = 0 = 0 + (-0),$$

by uniqueness, we conclude $0 = -0$, whence 0 **is its own additive inverse**.

Applying the Additive Inverse Equation to $-n$, tells us that n is an additive inverse of $-n$, since we know

$$(-n) + n = n + (-n) = 0.$$

The standard notation for the additive inverse of $-n$ is $-(-n)$, so, by uniqueness,

$$-(-n) = n.$$

A simple numerical example is $-(-40) = 40$. In words, **minus, minus** 40 **is** 40.

9.2.4 Finding the Sum of Two Arbitrary Integers

In this section we establish the formula:

$$(-m) + (-n) = -(m+n),$$

which in words says

the sum of additive inverses is the additive inverse of the sum

and generalizes the fact expressed in Example 1.

This formula makes computations with negative integers trivial for any one who knows how to add counting numbers. For example

$$(-23) + (-61) = -(23 + 61) = -84.$$

As the reader can see, the computation comes down to adding a pair of counting numbers, namely, $23 + 61$.

Let's consider why

$$(-m) + (-n) = -(m+n)$$

might be true. On the RHS we have the additive inverse of $m + n$. If the LHS is also an additive inverse for $m + n$, then uniqueness of additive inverses will witness the truth of the equality.

We begin by observing that $m + n = n + m$, whence $-(m+n) = -(n+m)$, a fact we use below.

To show that $(-m) + (-n)$ is also an additive inverse for $n + m$, we start by applying the Associative Law twice as follows:

$$\begin{aligned}
((-m) + (-n)) + (n + m) &= ([(-m) + (-n)] + n) + m \\
&= ((-m) + [(-n) + n]) + m.
\end{aligned}$$

Since n and $-n$ are additive inverses, as are m and $-m$, we have

$$\begin{aligned}
((-m) + [(-n) + n]) + m &= ((-m) + 0)) + m \\
&= (-m) + m \\
&= 0.
\end{aligned}$$

Thus, $(-n) + (-m)$ is an additive inverse for $n + m = m + n$ and by uniqueness of additive inverses,

$$(-m) + (-n) = -(m+n)$$

is established. For readers put off by the use of letters, compare this with the computations in Example 1.

Using this equation, we are able to conclude that

$$(-55) + (-25) = -(55 + 25) = -80.$$

Thus, we now have a rule for computing the sum of any pair of negative integers. This is where theory really becomes helpful in practical computation and the same equation will also tell us how to compute the sum of an arbitrary positive integer with an arbitrary negative integer.

Let's see how to compute the sum of an arbitrary pair of integers, n and m. We already know how to compute the sum if both are positive integers (counting numbers) or negative integers, using the equation above. Thus, the only situation of interest is when one integer, say n, is positive and the other, m, is negative. That m is a negative integer means that $m = -k$ where k is a positive integer. After replacing m by $-k$, we see that we need to compute

$$n + (-k) = n + m.$$

There are two possibilities regarding k, namely, that $k \leq n$, or that $n < k$.

Case 1: $0 < k \leq n$. In this case, we know

$$n + m = n + (-k) = n - k,$$

where, $n - k$ is subtraction.[4]

Case 2: $0 < n < k$. Let's remind ourselves of what we need to do. We want to find $n + m = n + (-k)$ where $0 < n < k$. To perform this computation, we will twice use the fact that for any integer p, $-(-p) = p$ (see lines 2 and 4 of the computation). Also we will use the fact that the additive inverse of a sum is the sum of the additive inverses to obtain line 3.

Let's begin:

$$\begin{aligned} n + m &= n + (-k) \\ &= -(-[n + (-k)]) \\ &= -([-n] + [-(-k)]) \\ &= -([-n] + k) = -(k + [-n]). \end{aligned}$$

---

[4]Given this fact, the centered dash is used in two ways, to indicate subtraction, and to indicate finding the additive inverse. Some argue this is confusing. However, proper use of parentheses avoids any confusion.

Thus,

$$n + m = -(k + [-n]).$$

The expression inside parentheses on the RHS, $k + (-n)$, satisfies $0 < n < k$, so we now know how to perform the computation: simply use take away. Therefore, when $0 < n < k$, simply compute $k - n$ and the additive inverse of this number (see RHS of last equation) will be the required sum, $n + m$.

A couple of numerical examples would be helpful.

Example 2.

Find

$$18 + (-16) \quad \text{and} \quad 18 + (-25).$$

For $18 + (-16)$, since $16 < 18$, we proceed under Case 1 above:

$$18 + (-16) = 18 - 16 = 2.$$

To find $18 + (-25)$, since $18 < 25$, we proceed under Case 2 as follows:

$$18 + (-25) = -(25 - 18) = -7$$

To summarize, we now know that the sum of any two integers is another integer, and we have a procedure for finding that sum using methods already developed for addition and subtraction of counting numbers. As a result, we say the integers are **closed** under addition.

9.2.5 Subtraction as a Defined Operation

The rationale for negative numbers was founded in idea of take away. As a result, the reader may wonder whether it is possible to subtract arbitrary integers from one another. The answer to this question is: Yes. The reason is because we replace the old operation of take away by a new operation which is defined in terms of addition for any pair of integers.

Thus, let m and n be any two integers. Then $m - n$ is defined by

$$m - n \equiv m + (-n).$$

As shown in previous sections, the quantity on the RHS can always be computed and it is always another integer. Further, if m and n are counting numbers with $n \leq m$, then $m - n = m + (-n)$ is exactly the result which would be obtained by the take away process discussed in Chapter 7.

The use of the centered dash, $-$, as both the subtraction symbol and the additive inverse symbol has a built in level of ambiguity. However, this ambiguity is tremendously reduced by the intention that subtraction, as an operation, should be eliminated and replaced by addition of the additive inverse. Indeed, this intention is captured by the rule that was taught to children when I was going to school, namely, subtraction means: **change the sign and add**.

9.3   Multiplication of Integers

In Chapter 8, we studied multiplication of counting numbers. Multiplication was formulated as **repetitive addition**, and we found that as applied to counting numbers, multiplication satisfied the following essential properties:

1. the **Commutative Law**:

$$n \times m = m \times n;$$

2. the **Associative Law**:

$$(n \times m) \times k = n \times (m \times k);$$

3. the **Identity Law**:

$$1 \times n = n \times 1 = n;$$

4. the **Two-sided Distributive Law**:

$$n \times (m + k) = n \times m + n \times k;$$

and,

$$(m + k) \times n = m \times n + k \times n.$$

These properties were observed to hold in respect to real-world collections, and since arithmetic models the real world, these properties must also hold in respect to multiplication of integers. In previous sections, we found that addition on the integers extended the definition of addition on the counting numbers. By requiring these properties to hold in respect to multiplication, we ensure that multiplication on the integers extends the definition of multiplication on the counting numbers to all integers.

9.3.1 Multiplication is Still Repetitive Addition

Much of the remainder of this book will be concerned with what happens to arithmetic as we add new kinds of numbers. In the present case we have added negative integers. One of the features of our arithmetic as developed for **counting numbers** was that the operation of multiplication was repetitive addition. It is important to realize that this remains true and will remain true in the future. There is a simple reason why this must be so. It is a consequence of the Identity Law for multiplication and the Distributive Law. To see why, consider that

$$
\begin{aligned}
4 + 4 &= 1 \times 4 + 1 \times 4 \\
&= (1 + 1) \times 4 \\
&= 2 \times 4
\end{aligned}
$$

where the last step simply uses the fact that $2 = 1 + 1$. This is how these two laws ensure that 2×4 must be $4 + 4$.

The identical sequence of steps applies to any integer, n, thus,

$$
\begin{aligned}
n + n &= 1 \times n + 1 \times n \\
&= (1 + 1) \times n \\
&= 2 \times n.
\end{aligned}
$$

Moreover, in any system of arithmetic that has a multiplicative identity and a Distributive Law, we can use the same sequence to obtain

$$
\begin{aligned}
x + x &= 1 \times x + 1 \times x \\
&= (1 + 1) \times x \\
&= 2 \times x.
\end{aligned}
$$

where x represents an arbitrary number in the system being discussed. Thus, in any such system, the multiplication operation will be repetitive addition.

There is one other feature of this discussion worth noting. In each case above, the reasoning has not changed. The only difference between the first and last example is the change from 4 to x. The rest is identical. Once you realize this you will see that the same reasoning is repeated over and over again, not only here, but in many other situations as well. You should be on the look out for these kinds of repetitive reasoning because they make coming to terms with these ideas much easier.

9.3.2 Multiplication by 0

A key fact about multiplication found in Chapter 8 was

$$
0 \times n = n \times 0 = 0
$$

where n was any non-negative integer. Since multiplication is still repetitive addition, this equation continues to be true if n is a counting number or 0. We would like to know whether the equation is satisfied for all integers. To answer this question we have to extend the result to negative integers.

We show for an arbitrary negative integer, m,

$$0 \times m = 0.$$

For such an m, we know $m = -n$, where n is a positive integer. Now, since $0 \times 0 = 0$ and $n + (-n) = 0$, we have the first line of:

$$
\begin{aligned}
0 &= 0 \times 0 = 0 \times (n + (-n)) \\
&= 0 \times n + 0 \times (-n) \\
&= 0 + 0 \times (-n) \\
&= 0 \times (-n).
\end{aligned}
$$

To get the second line we have applied the left Distributive Law and to get the third line we use the fact that a positive integer, n, times 0 is 0. The last line follows from the fact that 0 is the additive identity. Thus, substituting $m = (-n)$ in the last line gives

$$0 = 0 \times (-n) = 0 \times m.$$

Since this extends the product result to negative integers, we now know that for any integer n,

$$0 \times n = n \times 0 = 0.$$

We will use this result below.

9.3.3 Computing Products of Integers

The results of the last two sections tell us how to compute the product of two positive integers, or any product with 0. Thus, let n and m be arbitrary counting numbers. We want to develop methods that will enable us to find the products:

$$n \times (-m), \quad (-n) \times m, \quad \text{and} \quad (-n) \times (-m).$$

Notice that if can compute products of these three forms, then we will be able to compute the product of any pair of integers.

Recall that in §9.2.4 we showed the **additive inverse of the sum was the sum of the additive inverses**. Since multiplication is repetitive addition, we can convert the product

$$n \times (-m)$$

to a sum of n integers, each of which is $(-m)$. Since each addend, $-m$, is an additive inverse, the product is the additive inverse of the sum. Thus,

$$n \times (-m) = -(n \times m)$$

where the product in parentheses on the RHS uses the methods of Chapter 8 to find $n \times m$.

Alternatively, we could obtain this equation by applying the Distributive Law, the Additive Inverse Equation, and the multiplication by 0 rule, to the LHS of:

$$\begin{aligned} n \times m + n \times (-m) &= n \times (m + (-m)) \\ &= n \times 0 = 0. \end{aligned}$$

Either way, we conclude $n \times (-m)$ is the additive inverse of $n \times m$.

Since the Commutative Law tells us that $(-n) \times m = m \times (-n)$, we have

$$(-n) \times m = -(n \times m),$$

so that $(-n) \times m$ is also the additive inverse of $n \times m$. Combining these results in one equation gives

$$(-n) \times m = n \times (-m) = -(n \times m).$$

Applying these results to a particular computation gives:

$$7 \times (-8) = (-7) \times 8 = -(7 \times 8) = -56.$$

Finally, to find the product of two negative numbers, $(-n) \times (-m)$, observe:

$$\begin{aligned} (-n) \times m + (-n) \times (-m) &= (-n) \times (m + (-m)) \\ &= (-n) \times 0 = 0, \end{aligned}$$

where line 1 uses the Distributive Law and line 2 is obtain from $m + (-m) = 0$. Since,

$$[(-n) \times (-m)] + [(-n) \times m] = 0$$

whence, $(-n) \times (-m)$ is an additive inverse of $(-n) \times m$, so uniqueness of additive inverses can be applied. We just established that $(-n) \times m = -(n \times m)$ and the additive inverse of $-(n \times m)$ is $n \times m$. Thus,

$$(-n) \times (-m) = n \times m$$

by uniqueness of additive inverses.

Using this equation in a numerical example gives

$$(-9) \times (-6) = 9 \times 6 = 54.$$

We are now able to find the product of any pair of integers because every non-zero integer is either a counting number or the additive inverse of a counting number. Moreover, we also know the product of any pair of integers is again an integer, whence the integers are **closed** under multiplication. Of course, since the integers are closed under addition and multiplication is repetitive addition, the integers must also be closed under multiplication.

9.3.4 Multiplication by -1

The results above suggest that products of the form $(-1) \times n$ might be special. In fact, they are and it is useful to know how.

First, if we apply

$$(-n) \times (-m) = n \times m$$

in the case where $n - m - 1$, we get

$$(-1) \times (-1) = 1 \times 1 = 1.$$

Next we consider $(-1) \times n$ where n is any integer. Since $n = 1 \times n$, observe[5]

$$
\begin{aligned}
n + (-1) \times n &= 1 \times n + (-1) \times n \\
&= (1 + (-1)) \times n \\
&= 0 \times n = 0.
\end{aligned}
$$

Thus, $(-1) \times n$ is an additive inverse for n, so by uniqueness of additive inverses it must be the additive inverse for n, that is,

$$(-1) \times n = -n,$$

or in words

> **the product of a number and minus 1 is the additive inverse of the number.**

From the fact that

$$(-n) \times m = n \times (-m) = -(n \times m),$$

---

[5]The second line is an application of the Distributive Law; the third uses $1 + (-1) = 0$.

and the above, we have

$$((-1) \times n) \times m = n \times ((-1) \times m) = (-1) \times (n \times m)$$

whence all the above are notations for the same number, namely, $-(n \times m)$, the additive inverse of $n \times m$.

9.3.5   Summary of Arithmetic Properties of Integers

The **integers** (whole numbers), $\mathcal{I}$, consist of exactly three types of numbers:

1. counting numbers, also referred to as **positive** integers;

2. the number 0, also known as the **additive identity**;

3. numbers n for which there is a specific counting number, m, with the property that
$$n + m = 0.$$
   Numbers of this type are called **negative** integers and are additive inverses of counting numbers.

   The operations of addition, $+$, and multiplication, $\times$ are defined on the integers and satisfy the following for n, m, and k arbitrary members of $\mathcal{I}$:

1. **Closure:** $m + n$ is an integer, and $m \times n$ is an integer;

2. the **Commutative Laws:**
$$m + n = n + m \ \text{ and } \ m \times n = n \times m;$$

3. the **Associative Laws:**
$$(m + n) + k = n + (m + k) \ \text{ and } \ (m \times n) \times k = n \times (m \times k);$$

4. the **Two-sided Distributive Law:**
$$(m + n) \times k = n \times k + m \times k \ \text{ and } \ k \times (n + m) = k \times n + k \times m;$$

5. the **Additive Identity Law:**
$$0 + n = n + 0 = n;$$

187

6. the **Additive Inverse Law**: for each n we can find m with the property,

$$m + n = n + m = 0;$$

7. the **Multiplicative Identity Law**:

$$1 \times n = n \times 1 = n.$$

The following arithmetic rules hold for all integers ($-n$ denotes the additive inverse of n). They are consequences of the Laws.

I1 $-0 = 0$;

I2 $-(n + m) = (-n) + (-m)$;

I3 $-(n \times m) = (-n) \times m = n \times (-m)$;

I4 $n \times m = (-n) \times (-m)$;

I5 $1 = (-1) \times (-1)$ and $n = -(-n)$;

I6 $-n = (-1) \times n$;

I7 $-(n \times m) = ((-1) \times n) \times m = n \times ((-1) \times m)$;

I8 $n \times 0 = 0 \times n = 0$.

We stress that these laws and rules are universal. As we will see, they apply to all the mathematical quantities that turn up in the modern world. They form the basis of how we think about, and manipulate, these quantities. It is their universal application that makes them so powerful and enabling.

9.4   What Your Child Needs to Know

This chapter appears heavily theoretical. The reader might well wonder what it is doing in a book that is intended to help parents support children in primary and elementary? The answer is that the concepts discussed are introduced early in a manner appropriate to children in a given Grade. For example in Grade 1 ideas are expressed using whole numbers in the range $0 \leq n \leq 20$; in Grade 2 this restriction is relaxed to $0 \leq n \leq 100$. By the time a student completes the arithmetic portion of the math curriculum, it is expected the student knows and can apply the rules given above in all situations and is comfortable with their formulation using standard mathematical notation such as found in this book.

9.4.1 Goals for Grade 1 ($n \leq 20$)

[6] By the end of Grade 1 it is expected your child will:

1. understand the meaning of the equal sign;

2. be able to correctly use and interpret equations like $2 + 5 = 5 + 2$, $9 = 11 - 2$;

3. apply Commutative and Associative Laws governing operations as strategies to add and subtract (Examples: If $8 + 3 = 11$ is known, then $3 + 8 = 11$ is also known by Commutative Law. To add $2 + 6 + 4$, the second two numbers can be added first to make a ten, so $2 + (6 + 4) = 2 + 10 = 12$ by the Associative Law.);

4. understand subtraction as an unknown-addend problem, e.g., $\boxed{?} + 3 = 8$ has solution $8 - 3 = 5$;

5. understand that any number subtracted from itself yields 0, for example, $8 - 8 = 0$; $17 - 17 = 0$, etc;

6. be able to solve equations like $2 + 5 = \boxed{?}$, $7 = 10 - \boxed{?}$ and $\boxed{?} + 4 = 9$.

9.4.2 Goals for Grade 2 ($n \leq 100$)

By the end of Grade 2 it is expected your child will:

1. be able to fluently work with equations using $+$ and $-$ with unknowns in any position, e.g., $\boxed{?} + 7 = 15$;

2. understands that $+$ is Commutative and Associative and can explain this through counting;

3. can correctly apply the Commutative Law to quantities like $5 + 4$;

4. can correctly apply the Associative Law to quantities like $5 + (7 + 4)$ and $(3 + 8) + 6$; notice that this means the child functionally has to understand the use of parentheses;

5. can explain how the Commutative and Associative Laws are applied in particular situations, e.g., since $12 + 28 = 40$, $12 + 28 + 5 = 45$;

6. understands that 0 acts as the additive identity in the sense that $0 + 12 = 12$.

---

[6]We remind our readers that the selected goals are based on the CCSS-M. In some cases, goals have been reproduced verbatim; in others, they have been slightly edited.

9.4.3 Goals for Grade 3 ($n \le 100$)

By the end of Grade 3 it is expected your child will:

1. understand the relationship between multiplication and addition, in particular the Distributive Law;

2. understand how the Distributive Law makes multiplication into repetitive addition via equations like $2 \times 7 = (1 + 1) \times 7$;

3. understand how the Distributive Laws turns any multiple of an even number into the sum of two equal addends, e.g., $17 \times 6 = 17 \times 3 + 17 \times 3$;

4. understand that the Distributive Law enables strategies like:
$$13 \times 6 = (10 + 3) \times 6 - 60 + 18 = 78;$$

5. can solve equations like $\boxed{?} \times 7 = 56$ and $9 \times \boxed{?} = 63$;

6. can solve two-step word problems involving the four operations, $=$, $\times$, $-$ and $\div$;

7. correctly represent two-step word problems using an equation with the unknown represented by a letter as in: $3n + 7 = 28$ to represent "the product of an unknown and 3 added to 7 is 28";

8. understand the Associative and Commutative laws for multiplication and correctly apply them in the context of equations, e.g., $8 \times 13 = 13 \times 8$;

9. know that 1 acts as the multiplicative identity, that is, that $1 \times n = n \times 1 = n$ for any n.

9.4.4 Goals for Grade 4

By the end of Grade 4 it is expected your child will:

1. solve multi-step word problems posed with whole numbers and having whole-number answers using the four operations, including problems in which remainders must be interpreted (see Chapter 11);

2. represent these problems using equations with a letter standing for the unknown quantity.

3. assess the reasonableness of answers using mental computation and estimation strategies including rounding.

9.4.5   Goals for Grade 5

By the end of Grade 5 it is expected your child will:

1. show a complete understanding of the role of parentheses, brackets and braces in numerical expressions and correctly evaluate expressions that use these symbols;

2. demonstrate a complete understanding of the place value system and its role in supporting arithmetic procedures;

3. be able to write and interpret arithmetic expressions in language, e.g., $9+(4\times3)$ means *first multiply* 4 *by* 3, *then add* 9; or *add* 8 *and* 5 *and then multiply by* 7 is written arithmetically as $(8+5)\times7$;

4. write simple expressions that record calculations with numbers, and interpret numerical expressions without performing the indicated operations;

5. recognize that $3\times(18932+921)$ is three times as large as $18932+921$, without having to calculate the indicated sum or product;

6. recognize volume as an attribute of solid figures and understand concepts of volume measurement;

   (a) a cube with side length 1 unit, called a unit cube, is said to have one cubic unit of volume, and can be used to measure volume;

   (b) a solid figure which can be packed without gaps or overlaps using n unit cubes is said to have a volume of n cubic units;

7. measure volumes by counting unit cubes, using cubic cm, cubic in, cubic ft, and improvised units;

8. relate volume to the operations of multiplication and addition and solve real world and mathematical problems involving volume;

   (a) find the volume of a right rectangular prism with whole-number side lengths by packing it with unit cubes, and show that the volume is the same as would be found by multiplying the edge lengths, equivalently by multiplying the height by the area of the base;

   (b) represent threefold whole-number products as volumes, e.g., to represent the associative property of multiplication;

   (c) apply the formulas $V = l \times w \times h$ and $V = b \times h$ for rectangular prisms to find volumes of right rectangular prisms with whole-number edge lengths in the context of solving real world and mathematical problems;

(d) recognize volume as additive, e.g., find volumes of solid figures composed of two non-overlapping right rectangular prisms by adding the volumes of the non-overlapping parts, applying this technique to solve real world problems;

9. use a pair of perpendicular number lines, called axes, to define a coordinate system, with the intersection of the lines (the origin) arranged to coincide with the 0 on each line and a given point in the plane located by using an ordered pair of numbers, called its coordinates;

10. understand that the first number indicates how far to travel from the origin in the direction of one axis, and the second number indicates how far to travel in the direction of the second axis, with the convention that the names of the two axes and the coordinates correspond (e.g., x-axis and x-coordinate, y-axis and y-coordinate);

11. represent real world and mathematical problems by graphing points in the first quadrant of the coordinate plane, and interpret coordinate values of points in the context of the situation.

9.4.6 Goals for Grade 6

By the end of Grade 6 it is expected your child will:

1. understand that positive and negative quantities are used together to describe real situations, for example, monetary assets and debt, temperature scales, altitudes in relation to sea level;

2. understand the role of 0 in relation to positive and negative quantities in real situations, e.g., debt, temperature, sea level, etc.;

3. know that the equation $-(-n) = n$ is valid for all integers and understand why it must be true in terms of the Additive Inverse Equation;

4. write and evaluate numerical expressions involving whole-number exponents;

5. write, read, and evaluate expressions in which letters stand for numbers;

   (a) write expressions that record operations with numbers and with letters standing for numbers, e.g, express the calculation "Subtract y from 5" as $5 - y$;

(b) identify parts of an expression using mathematical terms (sum, term, product, factor, quotient, coefficient);

(c) view one or more parts of an expression as a single entity, e.g., describe the expression $2(8+7)$ as a product of two factors and recognize the quantity $(8+7)$ as both a single entity and a sum of two terms;

(d) evaluate expressions at specific values of their variables;

(e) evaluate expressions that arise from formulas used in real-world problems such as volume formulae;

(f) perform arithmetic operations, including those involving whole-number exponents, in the conventional order when there are no parentheses to specify a particular order (Order of Operations/Precedence);

(g) for example, use the formulas $V = s^3$ and $A = 6s^2$ to find the volume and surface area of a cube with sides of length $s = 1/2$ (see Chapter 12);

6. apply the properties of operations to generate equivalent expressions, e.g, apply the Distributive Law to the expression $3(2 + x)$ to produce the equivalent expression $6 + 3x$, and to the expression $24x + 18y$ to produce the equivalent expression $6(4x + 3y)$; apply defining property of multiplication to $y + y + y$ to produce the equivalent expression $3y$.

7. identify when two expressions are equivalent (i.e., when the two expressions name the same number regardless of which value is substituted into them), e.g., the expressions $y + y + y$ and $3y$ are equivalent because they name the same number regardless of which number y stands for.

8. understand solving an equation or inequality as a process of answering a question: which values from a specified set, if any, make the equation or inequality true?

9. use substitution to determine whether a given number in a specified set makes an equation or inequality true;

10. use variables to represent numbers and write expressions when solving a real-world or mathematical problem;

11. understand that a variable can represent an unknown number, or, depending on the purpose at hand, any number in a specified set;

12. solve real-world and mathematical problems by writing and solving equations of the form $x + p = q$ and $px = q$ for cases in which p, q and x are all nonnegative rational numbers.

193

13. use variables to represent two quantities in a real-world problem that change in relationship to one another;

    (a) write an equation to express one quantity, thought of as the dependent variable, in terms of the other quantity, thought of as the independent variable;

    (b) analyze the relationship between the dependent and independent variables using graphs and tables, and relate these to the equation;

    (c) for example, in a problem involving motion at constant speed, list and graph ordered pairs of distances and times, and write an equation such as $d = 65t$ to represent the relationship between distance and time.

The reader should think about these goals in terms of the development of ideas. Consider the notion of equality which is introduced in Grade 1. There, the expectation is that children can deal with purely numerical expressions which might be supported by pictures of the type presented in the early chapters of this work. By Grade 6, there is an expectation that the child can deal with equations like

$$y + y + y = 3y,$$

which reflect a far higher level of abstraction and state of knowledge. Indeed, equations like this one are the foundation of algebraic computations that will be presented in later grades. It is critical that your child becomes comfortable with this level of abstraction by the end of Grade 6.

Chapter 10

Ordering the Integers

Chapter Overview. We develop the order properties of the integers, $\mathcal{I}$. These are based on the natural properties of order on the counting numbers as learned by children and expressed in Chapter 2. The order properties of $\mathcal{I}$ are used to develop a picture of the integers in terms of a line. Methods for comparing integers based on their Arabic notation are given. Learning goals are presented in respect to integers.

10.1   Order on the Counting Numbers

The counting numbers are ordered. In fact, our original ideas grew out of comparing the size of two collections. Key ideas in this development were notions like:

one collection having **more** than another, and the **next** counting number.

The process for determining order between counting numbers, m and n, had three steps and used pairing. Specifically, first construct collections having m and n members, respectively. Second, construct a pairing of the members of the collection having m members and the collection having n members. Third, if the pairing process leaves members of the first collection unpaired, the first collection has more members than the second and m is more than n; if the second collection has unpaired members, then n is more than m; if the pairing is exact then $n = m$.

An essential fact arising from the discussion of collections and order was that: if n was an arbitrary counting number, then

$$1 \leq n.$$

In words, 1 is the **smallest** counting number. This is the reason why there is a **next** counting number: if we have a collection, we can put in one more element and the augmented collection is the next largest. These ideas were fully developed in Chapter 3.

10.2   Extending the Order Relation $<$ to $\mathcal{I}$

The three-step approach to order through constructing and pairing collections is primitive. A more sophisticated approach makes use of the power of our arithmetic. Specifically, we can now express our ideas on order using addition.

To understand this approach, we consider the problem of determining the order relation between 8 and 5 using the three-step approach.[1] Step one requires the construction of two collections, as shown:

Collections with 5 and 8 members, respectively.

It is clear that any pairing of elements will leave members in the collection having 8 members unpaired. Now, recall that addition is based on counting the total number of members when two collections are combined as illustrated by the following diagram:

We have to **add** a counting number to 5 to get to 8. What we have to add is the amount left over when we try to construct an exact pairing between a collection having 5 members and one having 8.

10.2.1   Defining $<$ on $\mathcal{N}$ Using Addition

To apply these ideas to defining $<$ on the counting numbers, consider two counting numbers m and n. The three-step method tells us, m **is less than** n, in symbols, $m < n$, exactly if for any two collections A and B having sizes m and n, respectively, the process of pairing elements in A with elements in B leaves some elements in B unpaired. This is the real-world process illustrated in the first diagram. What it also tells us in respect to the counting numbers m and n and addition is:

---

[1] If you look at exercises on the Kindergarten level at the websites they have pairing problems like this.

196

$$m < n \text{ exactly if for some counting number, } k, \; m + k = n,$$

where k is the number of unpaired members of B (see second diagram).

This relation provides a suitable definition of $<$ on the positive integers (counting numbers). Specifically, for counting numbers n and m, we write

$$m < n \text{ \textbf{if and only if for some counting number} } k, \; m + k = n.$$

10.2.2   Defining $<$ on $\mathcal{I}$ Using Addition

The simplest way to extend our previous ideas to all integers would be to simply drop the requirement that n and m be counting numbers. Our definition of $<$ on $\mathcal{I}$ would then be: let n and m be integers, then

$$m < n \text{ \textbf{if and only if for some positive integer} } k, \; m + k = n.$$

Notice that we have replaced **counting number** by **positive integer**. This does not change the meaning, since the set of counting numbers and the set of positive integers are the same. But it does change the flavor, putting the emphasis on the notion of *positive*. We will explore the reasons for this as we proceed.

The following numerical examples illustrate the use of $<$ as as defined using addition:

1. $7 + 9 = 16$, so $7 < 16$;

2. $(-19) + 19 = 0$, so $-19 < 0$;

3. $(-5) + 1 = -4$, so $-5 < -4$;

and so forth.

The criterion given above suggests we have **to look** for a positive integer k, which might or might not exist. Looking might be a lengthy process, so we can ask: Is there a quick answer? There is and it is provided by the discussion in §7.2 and §9.2.5. For any pair of integers m and n, we have the following equivalence:

$$m + k = n \text{ if and only if } k = n + (-m).$$

For completeness, let's review why the two assertions are equivalent. If the assertion on the left holds, simply add $-m$ to both sides to obtain the assertion on the right, as in

$$(-m) + (m + k) = (-m) + n.$$

Applying the Associative Law and $(-m)+m=0$ to the LHS leaves k, while applying the Commutative Law to the RHS leave $n+(-m)$. On the other hand if we have $n+(-m)=k$, simply add m to both sides to obtain

$$[n+(-m)]+m = k+m.$$

Applying the same arithmetic rules gives $n=m+k$.

Now recall our criterion for $<$ based on addition:

$m<n$ **if and only if for some positive integer** k, $m+k=n$.

By the equivalence above, the k we have to find must be $n+(-m)$, so that to find out if $m<n$, we simply compute $n+(-m)$ and see if that is a positive integer. If it is, then $m<n$. If it is not, then $m\not<n$. So there is only one number to check and it is found by computing the sum of n and the additive inverse of m. Using the equivalence between the two equations, we can revise our definition of $<$ on $\mathcal{I}$ to read

$m<n$ **if and only if** $n+(-m)$ **is a positive integer.**

This second criterion is straightforward to use and requires only one computation. The two forms of the definition are completely equivalent and we will use either criterion depending only on ease of application.

We stress that we can make this definition precisely because we know how to perform the computation $n+(-m)$ for every pair of integers m and n. The following numerical examples illustrate its use for integers that are additive inverses of counting numbers (see also §10.3.1). To compare -23 and -25:

$$(-23)+(-(-25)) = 2 \ \text{we know} \ -25 < -23;$$

whereas to compare -28 and -25:

$$(-28)+(-(-25)) = -3 \ \text{so} \ -25 \not< -28.$$

10.3 The Nomenclature of Order

As we know, statements involving the symbol $<$, or the symbol $\leq$ are called **inequalities**.

Assertions involving the less than symbol, $<$, are referred to as **strict inequalities** because the possibility of equality is forbidden.

Another order symbol which is frequently used is $>$ which is read **greater than**. It is treated as being defined by the relationship

$$n > m, \text{ if and only if, } m < n.$$

Since $n > m$ is equivalent to $m < n$, an inequality of this type is also strict. The introduction of this symbol adds nothing to the theory, since any statement involving $>$ can be rewritten as a statement using $<$.

We also have **weak inequalities**. The symbols used to express weak inequalities are: $\leq$ and $\geq$. These are defined by:

$$m \leq n, \text{ if and only if, either, } m < n \text{ or, } m = n,$$

and,

$$n \geq m, \text{ if and only if, either, } m < n \text{ or, } m = n,$$

respectively. The statement $m \leq n$ is read: m is **less than or equal to** n. Similarly, the statement $n \geq m$ is read: n is **greater than or equal to** m.

To be clear, the statements

$$0 \leq 1 \quad \text{and} \quad 4 \leq 4$$

are both true. The former because $0 < 1$ and the latter because $4 = 4$.

This nomenclature is used throughout mathematics.

10.3.1   The Three Types of Integers

As set out in §9.1.2, there are three kinds of integers, counting numbers, additive inverses of counting numbers and zero. Counting numbers are the foundation. The recognition of the other two kinds of integer is based on arithmetic behavior. Here we relate the three types to order.

We start with the order relation between the counting numbers and 0. Since, $0 + n = n$ and for every counting number n, using the first criterion (§10.2.1) we can write

$$0 < n.$$

In words, 0 is less than every counting number, or, equivalently, every counting number is greater than 0. This is the underlying reason why we refer to the counting numbers as **positive** integers.

Next, given any counting number, n, we have

$$-n + n = 0,$$

whence again applying the first criterion we conclude

$$-n < 0.$$

199

So, the additive inverses of counting numbers are all less than 0 and for this reason are referred to as **negative** integers.

Thus, we have three essentially different kinds of integers based on arithmetic behavior and we divide them into groups based on order as follows:

1. counting numbers are said to be **positive** because they are all greater than zero;

2. additive inverses of counting numbers are said to be **negative** because they are all less than zero;

3. zero is neither, positive, nor negative, it is just zero.

Since our approach to $<$ is defined by an arithmetic computation, there is really nothing new here other than an explanation of our choice of positive and negative as descriptors.

Using this nomenclature, the **non-negative integers** consist of the counting numbers together with zero. Thus, to say n is a non-negative integer means we may immediately conclude:
$$0 \leq n.$$

Lastly, it is important in your mind to separate the notion of *minus* from the notion of *negative*. We use *minus* to take the additive inverse of a number as in minus 10 which we denote -10, or minus -15 which we denote $-(-15)$. Forming the additive inverse is an arithmetic concept. The result can be either *positive* or *negative* which are order concepts based on the relation of a number to 0. The following examples illustrate these facts:

1. minus 25 is the additive inverse of 25 and is negative because it is less than 0;

2. minus, -7 is the additive inverse of -7 and is positive because it is greater than 0.

10.4 A Graphical Interpretation of Order

The division of the integers into three groups based on order gives rise to a graphical description of the integers which is known as the **number line** and pictured below:

A graphical description of the integers. The positive integers extend indefinitely to the right of 0, the negative integers indefinitely to the left of 0. Numbers to the left of 0 use the names, minus 1, minus 2, minus 3, etc. This graph is referred to as the **number line**.

It is expected that every child will be able to use the number line to make inferences about the relationships between numbers. For example, if the position of m is to the left of n on the number line, the child is expected to infer that $m < n$. To achieve this ability, it is essential that children understand the details of the number line's construction.

Let's start with the simple idea of a line with 0 in the middle, positive numbers on the right and negative numbers on the left, as shown above. As we go through the construction process you may find it helpful to have paper, pencil and ruler.

Draw a horizontal line of, in principle, unlimited extent. Pick any place on the line and mark a point which is then labeled with 0, as shown in the figure. The position of 0 is called the **origin**. Pick any point to the right of 0, mark it and label it with 1. This choice determines one unit of distance on the line. The following is critical: **the distance between each positive integer and its successor must be the same as the distance between 0 and 1, that is, one unit**. So, starting at the point marked with a 1, move one unit of distance to the right and mark that point with 2. Move another unit to the right and mark that point with 3, and so forth. In this way, the entire RHS of the line is constructed. It consists of all the positive integers; so it is the **positive** side of the line. Notice, there will be exactly one unit of distance between any positive integer and its successor. The effect of this is that **each counting number n is exactly n units of distance from 0**.

The side of the line to the **left** of 0 is used to represent the negative integers. Each negative integer is the additive inverse of a positive integer. As such, this negative integer is placed the same distance to the left of 0 as its additive inverse, which is a positive integer, is to the right. Thus, for example, -5 is five units of distance to the left of 0 exactly because 5 is five units of distance to the right of 0.

The pairing of an integer and its additive inverse on opposite sides of the origin (zero) gives rise to the notion that 5 and -5 are **opposites** which is the common nomenclature in primary and elementary school used for additive inverse. Thus, the sum of any pair of opposites is 0 which is another way of stating the Additive Inverse Equation. The CCSS-M begin using the term additive inverse in Grade 6. Conceptually, this change in nomenclature reflects a change in focus from the position of an integer on the real line to an integer's behavior within arithmetic.

10.4.1   Length

The construction process described above ensures that between any pair of adjacent integers on the line, there is exactly one unit of distance (length). This fact is illustrated below:

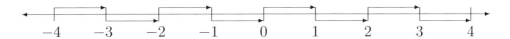

A graphical description of the integers showing the fact that each integer is exactly one unit of distance from its predecessor as illustrated by the horizontal arrows which all have the same length. This length is determined by the placement of 0 and 1 as described in the construction process (see previous text).

The relationship between adjacent integers on the line captures an important numerical fact. Suppose we pick an arbitrary integer, say, n. The adjacent integer to its right must then be $n+1$. We know this because moving to the right means moving in the positive direction and $n < n+1$ by the first criterion. We also have the following arithmetic relationship between n and $n+1$:

$$(n + 1) + (-n) = 1.$$

Our line construction captures this fact by ensuring that there is exactly one unit of distance between each pair of adjacent integers.

The relationship of **unit of length** and placement of integers on the line is something it is expected that every child will understand. These ideas are explored in respect to addition using sticks of unit length. For example, if we have 3 sticks of unit length, and 5 sticks of unit length, counting tells us we have a total of 8 sticks of unit length. Thus, the total length of the sticks placed end-to-end is 8 units, hence

$$3 + 5 = 8.$$

We illustrate this in the diagram below.

A graphical depiction of the addition of two positive integers. The horizontal arrows have lengths 3 and 5 units, respectively. Placing the arrows end-to-end starting at 0 and proceeding left-to-right takes us to 8.

10.4.2 Relating Addition to the Number Line

It is also expected that every child should understand the effect of adding a positive integer, versus adding a negative integer in respect to the real line. We explain this next.

For purposes of this discussion, we take n to be a fixed integer. For example, n might be 35 or -450. The key is that n is fixed. In the diagram below, we construct a portion of the number line centered on the number we chose, namely, n. In the diagram presented below, we show three units on either side of n (see below).

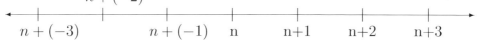

A portion of the number line is shown centered on a fixed integer, n. The numbers $n+1$, $n+2$ and $n+3$ are identified on the right. The numbers $n + (-1)$, $n + (-2)$ and $n + (-3)$ are identified on the left.

As shown in the diagram, $n + 1$ is positioned one unit of distance to the right of n. This positioning is fixed by the requirement that every pair of adjacent integers be positioned one unit of distance apart and by the fact that $n < n + 1$ so that $n + 1$ is to the right of n. Similarly, $n+2$ is one unit of distance to the right of $n+1$ and $n + 3$ is one unit of distance to the right of $n + 2$.

Since $n + (-1)$ is adjacent to n and $n + (-1) < n$, the position of $n + (-1)$ is one unit of distance to the **left of** n for the same reasons given above. Similarly, $n + (-2)$ is one unit of distance to the left of $n + (-1)$ and $n + (-3)$ is one unit of distance to left of $n + (-2)$.

As the diagram shows, adding 1 to n moves us one unit to the right. Using the fact that $2 = 1 + 1$, whence $n + 2 = (n + 1) + 1$, the same rule tells us that $n + 2$ must be two units to the right of n. Similarly, $n + 3 = (n + 2) + 1$, so $n + 3$ has to be positioned three units to the right of n. This discussion generalizes the remarks about addition and length. For example, adding 3 to n gives the following diagram.

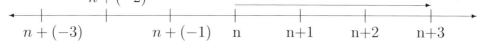

To add 3 to n find n and move three units of length to the right since 3 is **positive**.

There is nothing special about 3. To add a positive integer m, starting at n, move m units of distance to the right and you come to $n + m$.

The addition of negative integers to n moves us to the left. We can see why this must be the case by considering the equation

$$[n + (-1)] + 1 = n + [(-1) + 1] = n.$$

This equation tells us that n must be positioned one unit to the right of $n + (-1)$ as discussed above. Thus adding the integer -1 to n moves us one unit to the left. Once we understand this fact the only question is how many units to the left a given negative integer will move us. The simple way to think about this is to remember that a negative integer k satisfies:

$$k = (-1) \times p$$

where p is a positive integer. p tells us how many steps and this is illustrated below for

$$k = -2 = (-1) \times 2.$$

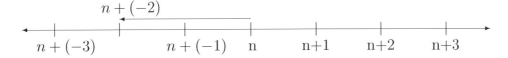

To add -2 to n find n and move two units of length to the left since -2 is **negative**. This process is captured by the left-pointing arrow two units in length that starts at n and finishes at $n + (-2)$.

10.5 Arithmetic Properties of $<$ on $\mathcal{I}$

As always, we let m, n and k stand for arbitrary integers. Then the following four assertions are satisfied:

1. the **Transitive Law**:

$$\text{if }\ m < n \ \text{ and }\ n < k, \ \text{ then }\ m < k;$$

2. the **Trichotomy Law**, given any n and m, exactly one of the following holds:

$$m < n \ \text{ or, }\ n < m \ \text{ or }\ n = m;$$

3. the **Addition Law**:

$$\text{if }\ m < n \ \text{ then }\ m + k < n + k;$$

4. the **Multiplication Law**:

$$\text{if }\ m < n \ \text{ and }\ 0 < k, \ \text{ then }\ m \times k < n \times k.$$

We consider why these four statements should be true by using the defining property of $<$.

For the transitive property, we assume we have $m < n$ and $n < k$. This means there are positive integers p and q, such that $m + p = n$ and $n + q = k$. Thus, after substituting for n in the last equation and then using the Associative Law,

$$k = n + q = (m + p) + q = m + (p + q).$$

Since the sum $p + q$ is again a positive integer applying the criterion in §10.2.1 gives

$$m < k,$$

as desired.

For Trichotomy, we simply compute $k = n + (-m)$. Since k is an integer, exactly one of the following must be true, namely, k is positive, negative, or zero. If k is a positive integer, $m < n$; if k is a negative integer, $n < m$; and if the $k = 0$, $m = n$.

For the Addition Law, suppose $m < n$, so that for some positive integer p, $m + p = n$. Adding k to both sides of the equality gives

$$(m + p) + k = n + k,$$

which quickly becomes

$$(m + k) + p = n + k.$$

Since p is a positive integer, by the first criterion we conclude that

$$m + k < n + k.$$

Finally, for the Multiplication Law, suppose $m < n$ and $0 < k$. To see that $m \times k < n \times k$, first note that for some positive integer p, $m + p = n$. Then multiply both sides of the equality by k to obtain

$$
\begin{aligned}
n \times k &= (m + p) \times k \\
&= m \times k + p \times k.
\end{aligned}
$$

Since the product of two positive integers, $p \times k$, is again a positive integer,

$$m \times k < n \times k,$$

as asserted.

The Multiplication Law requires that the **multiplier k be positive**. We can see why, because the product, $p \times k$, must also be positive to satisfy the first criterion for the inequality. This product will not be positive if the multiplier k is either less than zero or zero, in other words, **not positive**. A few numerical examples would be useful to help the reader understand how multiplication interacts with order.

We know $5 < 7$ and $0 < 4$. Multiplying through the inequality by the positive multiplier 4 gives

$$5 \times 4 = 20 < 28 = 7 \times 4.$$

Similarly, we also know $-3 < 8$ and $0 < 7$, so

$$(-3) \times 7 = -21 < 56 = 8 \times 7.$$

Suppose on the other hand we multiply through the same inequalities by -2, for which we note $0 \not< -2$. Multiplying through the first inequality by the non-positive multiplier -2 gives

$$5 \times (-2) = -10 \not< -14 = 7 \times (-2)$$

and the second gives,

$$(-3) \times (-2) = 6 \not< -16 = 8 \times (-2).$$

In both examples, multiplication by a negative number **reverses** the order relation; that is, $-14 < -10$ and $-16 < 6$.

In fact,

> **multiplication by a negative number will always reverse inequalities**.

Finally, we note that multiplication through any inequality by 0 produces an equality, because the result on both sides will be 0!

10.6 Arithmetic Properties of $\leq$ on $\mathcal{I}$

Recall that

$$m \leq n \text{ exactly if } m < n \text{ or } n = m.$$

We would therefore expect that the properties satisfied by $\leq$ would be some mixture of the properties of equality and less than. This is what we observe as the following shows.

Let n, m and k be arbitrary elements of $\mathcal{I}$. Then the following six statements are satisfied:

1. $\leq$ is Reflexive:
$$n \leq n;$$

2. $\leq$ is Anti-Symmetric:
$$\text{if } m \leq n \text{ and } n \leq m, \text{ then } n = m;$$

3. $\leq$ is Transitive:
$$\text{if } m \leq n \text{ and } n \leq k, \text{ then } m \leq k;$$

4. $\leq$ is a Linear ordering of $\mathcal{I}$:
$$\text{given } n \text{ and } m, \text{ either, } m \leq n \text{ or, } n \leq m;$$

5. $\leq$ satisfies the Addition Law:
$$\text{if } m \leq n, \text{ then } m + k \leq n + k;$$

6. $\leq$ satisfies the Weak Multiplication Law:
$$\text{if } m \leq n \text{ and } k \geq 0, \text{ then } m \times k \leq n \times k.$$

All of these properties follow directly from application of the properties of equality, or the properties of less than.

Note that the second property is referred to as **antisymmetry**. Recall that equality is symmetric, that is if $n = m$, then $m = n$. Note also that $<$ is **in no sense symmetric**, that is, if $n < m$, then it is **not true** that $m < n$.

But for $\leq$ we can have both $n \leq m$ and $m \leq n$. But the only way this can happen is if $n = m$.

10.7 Comparing Positive Integers

We give a three-step process for determining the larger of two counting numbers expressed in Arabic notation as follows:

Step 1 Count the digits in the numeral for each number. If they are not equal, the number requiring more digits is the larger.

Step 2 Otherwise, starting at the left-most place, compare the digits in each numeral place-by-place; at the point one is found to be larger than the other, the number represented by the numeral with the larger digit is the larger number.

Step 3 Lastly, if both numerals have the same number of the digits and these digits starting from the left are pairwise found to be equal, then the two numbers represented by these numerals are equal.

This process works for determining the relative size of counting numbers based on their numerals. Explaining in detail why it works is complicated and some readers may want to skim the rest of this section.

We have made much of how the Arabic System of numeration makes addition and multiplication easy. The three-step method also makes comparing two counting numbers straight forward. To understand the method, we need to recall how the Arabic System relates to multiplication as discussed in §8.3.1. Using the notation there, a general four-digit numeral was expressed by:

$$n_{1000}n_{100}n_{10}n_1 = n_{1000} \times 1000 \ + \ n_{100} \times 100 \ + \ n_{10} \times 10 \ + \ n_1 \times 1$$

where n_{1000}, n_{100}, n_{10} and n_1 are all single-digit numerals and the leading digit on the left is not 0.

There are two essential facts about this notation:

1. the single digits are ordered,

$$0 < 1 < 2 < 3 < 4 < 5 < 6 < 7 < 8 < 9;$$

2. the following order relations hold between places in multi-digit numbers:

$$n_{1000} \times 1000 \ > \ n_{100} \times 100 \ > \ n_{10} \times 10 \ > \ n_1 \times 1.$$

This line of inequalities between multi-digit numbers is true exactly because

a 1 followed by n zeros on the right is the successor of the largest number that can be expressed by a numeral using n digits.

Thus, 1000 is the successor of 999 which is the largest number that can be expressed using 3 digits. Similarly, 100 is the successor of 99 which is the largest two-digit number, and so forth. So in the line above, it does not matter what the digits n_{1000}, n_{100}, n_{10}, and n_1 are, the asserted ordering will hold.

Thus, the first effect of the inequalities in item 2 above when comparing positive integers expressed in Arabic Notation is that every five-digit number is larger than every four-digit number, which in turn is larger than every three-digit number, etc.

Thus, the first thing to check when comparing numerals for positive integers is whether they have the same number of digits. If not, the larger positive integer is the one requiring more digits.

For the underlying basis of Step 2, we need to recall the Multiplication and Addition Laws stated at the beginning of §10.5. According to the former, inequalities are preserved as long as the multiplier is positive. We apply these laws here in the following way. Given any two single digits, m and n, if $m < n$, then

$$m \times 10000 < n \times 10000, \quad m \times 1000 < n \times 1000,$$

$$m \times 100 < n \times 100, \quad m \times 10 < n \times 10,$$

and so forth. The effect of these inequalities is that m followed by some number of zeros is less than n followed by that same number of zeros.

To say $m < n$ for the single-digits m and n means there is a counting number $p > 0$ such that $m + p = n$ (see §10.2.1). Notice that $1 \leq p$, so that

$$n \times 1000 = (m + p) \times 1000 \geq (m + 1) \times 1000 = m \times 1000 + 1000$$

or more simply,

$$n \times 1000 \geq (m + 1) \times 1000 = m \times 1000 + 1000.$$

Since the largest three digit number 999 is less than 1000, the Addition Law tells us that

$$n \times 1000 \geq m \times 1000 + 1000 > m \times 1000 + 999.$$

In other words, if $m < n$, $n \times 1000$ exceeds every four-digit number having first digit m. Similarly for three-digit numbers we have, $n \times 100$ exceeds every three-digit number having first digit m and for two-digit numbers, $n \times 10$ exceeds every two-digit number having first digit m. Likewise analogous results apply to five-digit, six-digit, seven-digit numbers, and so forth.

Let's recall that our purpose here is to compare two four-digit numbers. The effect of the previous paragraph is that if the two numbers have different left-most digits, the one with the larger lead digit will be the larger. Thus we only need consider the case of two four-digit numbers where the lead digit in both is the same as in:

$$n_{1000}n_{100}n_{10}n_1 \quad \text{and} \quad n_{1000}p_{100}p_{10}p_1.$$

Using our knowledge of the Arabic System, the two numerals satisfy

$$n_{1000}n_{100}n_{10}n_1 = n_{1000} \times 1000 + n_{100}n_{10}n_1$$

and

$$n_{1000}p_{100}p_{10}p_1 = n_{1000} \times 1000 + p_{100}p_{10}p_1.$$

We can eliminate the leading digit, $n_{1000} \times 1000$, in both numerals leaving any order relations unaffected by adding $-n_{1000} \times 1000$ to both. (The reason is that a consequence of the Addition Law is that

$$m < n \quad \text{if and only if} \quad k + m < k + n.$$

The LHS can be obtained from the RHS by adding $-m$ to both sides of the inequality using the Addition Law.) Thus, the order relation between the four-digit numbers

$$n_{1000}n_{100}n_{10}n_1 \quad \text{and} \quad n_{1000}p_{100}p_{10}p_1$$

will be the same as the order relation between the two three-digit numbers

$$n_{100}n_{10}n_1 \quad \text{and} \quad p_{100}p_{10}p_1.$$

At this point we simply compare the left-most digits, n_{100} and p_{100}. If one is larger, it determines the larger number. If $n_{100} = p_{100}$, the Addition Law tells us the order relation between the two three-digit numbers will be the same as the order relation between

$$n_{10}n_1 \quad \text{and} \quad p_{10}p_1.$$

To determine the order relation between these two numbers we repeat the process, comparing the *tens* digits. If the *tens* digits are equal, we compare the *ones* digits. At each step we know the same reasoning applies that was used in the detailed considerations in respect to the four-digit case above.

Let's look at some concrete examples:

$$80301 \text{ and } 9983; \ 679428 \text{ and } 599983; \ 7124569 \text{ and } 7126059.$$

For the first pair, the number of digits in 80301 is greater than the number of digits in 9983 so by Step 1, $9983 < 80301$.

For the second pair, 679428 and 599983 both use six digits. But using Step 2, we note that the left-most digit in 679428 is a 6 and is greater than the 5, the left-most digit in 599983, so $599983 < 679428$.

For the third pair, 7124569 and 7126059, both require seven digits and for both, the digits working left-to-right are 7 followed by 1 followed by 2. In the next place, the fourth from the left, the digits differ, 6 versus 4, so we have $7124569 < 7126059$.

10.7.1 Comparing Arbitrary Integers

The situation for arbitrary integers is dealt with by noting that all positive numbers exceed all negative numbers, as we already know. So we only have to deal with situations in which both numbers are negative. Thus let $-n$ and $-m$ be two negative integers. From our previous considerations, we know n and m are positive integers, hence we may order m and n using the procedure given. Now observe that by adding $(-m) + (-n)$ to both sides of the inequality

$$m < n$$

we obtain:

$$m + ((-m) + (-n)) < n + ((-m) + (-n))$$

which, after performing the indicated sums, becomes

$$-n < -m.$$

Thus, if $m < n$, then $-n < -m$. That this relation must be so is also obvious from our graphical representation of the integers discussed in §10.4.

   In summary, given a pair of negative integers, we simply determine the order of their additive inverses. The original negative numbers then have the opposite order. For example, given -873 and -642, we compare 873 and 642, to find $643 < 873$, from which we conclude $-873 < -643$.

10.8 What Your Child Needs to Know

10.8.1 Goals for Kindergarten

By the end of Kindergarten it is expected your child will:

1. Understand the relationship between numbers and quantities;

2. connect counting to cardinality;

   (a) understand that each successive number name refers to a quantity that is one larger;

3. identify whether the number of objects in one group is greater than, less than, or equal to the number of objects in another group, e.g., by using matching and counting strategies;

4. compare two numbers between 1 and 10 presented as written numerals;

5. describe measurable attributes of objects, such as length or weight;

6. describe several measurable attributes of a single object;

7. directly compare two objects with a measurable attribute in common, to see which object has more of/less of the attribute, and describe the difference. For example, directly compare the heights of two children and describe one child as taller/shorter.

10.8.2 Goals for Grade 1

By the end of Grade 1 it is expected your child will:

1. order three objects by length; compare the lengths of two objects indirectly by using a third object;

2. express the length of an object as a whole number of length units, by laying multiple copies of a shorter object (the length unit) end to end;

3. understand that the length measurement of an object is the number of same-size length units that span it with no gaps or overlaps;

4. can perform measurements in contexts where the object being measured is spanned by a whole number of length units with no gaps or overlaps.

10.8.3 Goals for Grade 2

By the end of Grade 2 it is expected your child will:

1. be able to compare two three-digit numerals based on the place meaning of the digits;

2. be able to correctly use the symbols $\leq$, $<$, $\geq$, $>$ to express facts about numerals, for example, $45 < 53$, $92 \leq 95$, $88 \geq 41$, and so forth;

3. represent whole numbers as lengths from 0 on a number line diagram with equally spaced points corresponding to the numbers 0, 1, 2, ..., and represent whole-number sums and differences within 100 on a number line diagram.

The remaining goals associated with order relate to more advanced topics, for example decimals, and will be listed in the appropriate chapter.

Chapter 11

Division of Integers

Chapter Overview. Division of integers is developed based on solving the equation $m \times x = n$ for x. The concepts of quotient and remainder are defined and a mechanical procedure for finding them, the Division algorithm, is presented. Examples are worked out.

The last arithmetic operation on the integers that is introduced in primary and elementary grades is division. Although we have not done so, division is introduced simultaneously with multiplication. The reason for this simultaneous introduction is because of the fundamental relationship between the two operations. We will examine this relationship in what follows.

11.1   Division as a Concrete Process

One intention of the CCSS-M authors is that they want children to recognize division as a process for placing an **equal number** of objects into separate groups. For example:

> There are 18 candy bars in a box. If 6 children are coming to a birthday party, how many bars go in each loot bag so every child gets the same number?

It is expected that children will be able to turn this type of problem into the equation:

$$6 \times x = 18$$

which they then solve for x. They should understand that because 18 is a multiple (see §11.3 below) of 6, $x = 18 \div 6$ produces 6 groups of **equal size** and that 3 is the maximum number of items that can be placed in any group.

In addition, they are expected to understand the use of remainders in concrete settings:

> There are 30 candy bars in a box. If 8 children are coming to a birthday party, what is the maximum number of bars that can be put in each loot bag so every child gets the same number? How many bars, if any, are left for pop to have for desert?

11.2   The Idea of Division

The discussion above featured the equation $6 \times x = 18$ which needed to be solved for x. In order to understand where this equation comes from, let's recall how we looked at subtraction as applied to counting numbers. We started with the notion of take-away, which is concrete, a sound way to motivate subtraction, and a good way to suggest the idea of negative numbers.

Further, remember for a fixed pair of counting numbers, n and m, we saw that subtraction translates within arithmetic to the problem of finding a solution to the following equation for x

$$x + m = n.$$

For this equation to have a solution within the counting numbers required that $n \geq m$. However, once we had all the integers, we discovered we could drop the restriction that $n \geq m$, and the general solution to the equation was:

$$x = n + (-m).$$

Moreover, we could even drop the requirement that n and m were counting numbers and simply require that they be fixed integers. The solution was still the same, namely, $n + (-m)$. Using these ideas, we can motivate the idea of division.

The application of these ideas in the context of multiplication occurs when we replace the operation of $+$ by the operation of $\times$. So the equation $x + m = n$ becomes

$$x \times m = n$$

which is to be solved for an integer x. Here again n and m can be any fixed integers, but for the time being we assume they are positive. In the event that the equation $x \times m = n$ has a solution for x, it defines the process of division, and we write

$$x = n \div m$$

where $\div$ stands for the operation of division. We know division is an operation because it is defined in terms of multiplication. We also know that when the equation

$m \times x = n$ has a solution, the results will produce **equal groups** because as we recall, multiplication is **repetitive addition**.

To make these ideas concrete, and explore the possibilities, we consider numerical examples. The equation

$$x \times 4 = 20$$

has an integer solution ($x = 5$), since a check of the multiplication table shows

$$5 \times 4 = 20.$$

Thus, we would write $5 = 20 \div 4$ and understand this to mean we can divide any collection containing 20 members into 5 groups, each of which has 4 members.

As a second example, consider the equation $6 \times x = 42$. Since $6 \times 7 = 42$, we would write $7 = 42 \div 6$ and infer that any collection containing 42 members can be divided into 7 groups of 6 members each.

On the other hand, the equation

$$x \times 4 = 19$$

has no integer solution, since we cannot find any number in the multiplication table which when multiplied by 4 yields 19. Thus, the situation of division with respect to the integers is much like the situation of subtraction with respect to the counting numbers: sometimes the equation has a solution; sometimes it does not.

11.3   Multiples and Factors

In trying to find an integer solution to the equation

$$m \times x = n$$

we are asking two questions:

Is n **a multiple of** m?

Is m **a factor of** n?

For any two integers n and m, the answer to both questions is always the same, either both yes, or both no. But the focus of each question is quite different as we discuss below.

11.3.1 Multiples

The notion of **multiple of** is defined for any counting number, m. Multiples of m are the numbers:

$$m \times 1, \quad m \times 2, \quad m \times 3, \quad m \times 4, \quad m \times 5, \quad m \times 6, \quad m \times 7, \quad m \times 8, \quad \ldots$$

Thus, for example, the multiples of 6 are:

$$6, \quad 12, \quad 18 \quad 24, \quad 30 \quad 36, \quad 42 \quad 48, \quad \ldots$$

and so forth. In this context the phrase *and so forth* indicates the list of multiples is unlimited; it goes on forever. Moreover, each successive integer in the list can be obtained by adding the same number to its predecessor. In the above list we simply add 6 to get the next number in the list. This is referred to as **skip** counting.

Given a number on the list of multiples of 6, say 96, we know the equation

$$6 \times x = 96$$

has a positive integer as a solution. Picking a number not on the list, for example, 103, means the equation

$$6 \times x = 103$$

has no solution in the integers.

Thus, in our original numerical example $x \times 4 = 20$, we were able to find a solution precisely because 20 is a multiple of 4. On the other hand, 19 is not in the list of multiples of 4 and so the equation $x \times 4 = 19$ does not have a solution in the integers.

11.3.2 Factors and Divisors

The notion of **divisor of** is also defined for any counting number n. For example, consider $n = 24$. We know that

$$1, \quad 2, \quad 3, \quad 4, \quad 6, \quad 8, \quad 12, \quad \text{and} \quad 24$$

are all divisors, or **factors**, of 24. Notice that this list is short and includes **all** the factors of 24. Thus, when we are asking about factors, we are asking for a very short list. When we are asking about multiples, we are asking for an impossibly long list.

Consider again the equation

$$m \times x = n$$

where n and m are given.

When we focus on the product on the LHS of this equation, we are asking: Is n a **multiple of** m?

When we focus on the RHS of this equation, we are asking: Is m a **factor** or **divisor of** n?

We consider one last numerical example. Let's think of 5 as a divisor, and ask for which values of n, n a counting number, will the equation

$$5 \times x = n$$

have a solution? We know the list of values of n admitting solutions is:

$$5, \; 10, \; 15, \; 20, \; 25, \; 30, \; 35, \; 40, \; 45, \; 50, \; 55, \; 60, \; 65, \; 70, \; 75, \; \dots$$

This is exactly the list of multiples of 5 and it is exactly the list produced by **skip counting** by five starting at 5.

Instead of focussing on the exact multiples of 5, consider numbers not in the list. Specifically, between each successive pair of multiples, there are four numbers. For example, between 55 and 60 are the four numbers, 56, 57, 58, and 59. Let's write these four numbers in terms of 55, which is the largest multiple of 5 less than each of the four. Thus,

$$56 = 55 + 1, \;\; 57 = 55 + 2, \;\; 58 = 55 + 3, \;\; 59 = 55 + 4.$$

Or, consider the gap between 100 and 105. The numbers in this gap can be written as:

$$101 = 100 + 1, \;\; 102 = 100 + 2, \;\; 103 = 100 + 3, \;\; 104 = 100 + 4,$$

where $100 = 5 \times 20$ and $105 = 5 \times 21$. Again, notice that 100 is the largest multiple of 5 that is less than each of the four numbers in the gap. From these two examples we see the gaps between the successive multiples of 5 are all the same. Each such gap contains four counting numbers. Each of these counting numbers is the sum of the largest multiple of 5 less than itself and a **residual** that is either 1, 2, 3 or 4. What's important to realize here is that every counting number that is not a multiple of 5 can be found in such a gap. For example, neither 382 nor 868 is a multiple of 5, so each must sit in a gap between multiples, and in consequence, each can be written as the sum of a multiple of 5 and a residual that is 1, 2, 3 or 4 as shown:

$$382 = 380 + 2, \;\; \text{and} \;\; 868 = 865 + 3.$$

Recalling the idea of skip counting, note that if we start at 6 and skip count by five, every number on the list generated by this process will have the form

$$m \times 5 + 1.$$

Skip counting in this way produces the following list:

$$6, \ 11, \ 16, \ 21, \ 26, \ 31, \ 36, \ 41, \ 46, \ 51, \ 56, \ 61, \ 66, \ 71, \ 76, \ \ldots$$

You can try this yourself to see that any counting number, indeed any integer, that is not a multiple of 5 can be written as a multiple of 5 plus a residual counting number $k < 5$. Moreover, the number you pick will be found on a skip counting list that starts at $5 + k$ and proceeds by adding fives.

To summarize what we have learned about 5 as a divisor, we now know that given any counting number n, there are non-negative integers q and $r < 5$ such that

$$n = 5 \times q + r.$$

Notice that if $n < 5$ $q = 0$ and if n is an exact multiple of 5, $r = 0$.

11.4 Division of Positive Integers

Clearly, there should be nothing special about 5. So we reconsider these ideas using an arbitrary counting number, d, as the divisor. Fix in your mind any counting number you want to be the divisor. Now ask: Can we skip count by d? Of course we can. Doing so produces the following list:

$$0 \times d, \ 1 \times d, \ 2 \times d, \ 3 \times d, \ 4 \times d, \ 5 \times d, \ 6 \times d \ \ldots.$$

What we know is that every counting number that is a multiple of d appears in this list.

Now consider numbers in the gaps, that is the counting numbers, n, that occur between consecutive multiples of d:

$$q \times d < n < (q + 1) \times d = d \times q + d.$$

If we consider the first gap, namely when $q = 0$, then we know the list of counting numbers in the gap is:

$$1, \ 2, \ 3, \ 4, \ 5, \ldots, \ d - 1.$$

All of these counting numbers satisfy the inequality:

$$0 < 1, \ 2, \ 3, \ 4, \ 5, \ldots, \ d - 1 < d.,$$

that is, they are all less than d. Further, if we add $q \times d$ to each of the numbers in the gap we would have:

$$q \times d < q \times d + 1, \ q \times d + 2, \ \ldots, \ q \times d + (d - 1) < (q + 1) \times d$$

218

where this is a complete list of the counting numbers in the gap. Thus, n must be on this list, so n can be written as the sum of a multiple of d and a residual counting number that is less than d.

We can summarize this as follows:

> Given counting numbers n and d, we can find integers q and r such that $0 \leq r < d$ and $q \times d + r = n$.

The number d is called the **divisor**; it is what we are dividing by. The number n is called the **dividend**; it is the number being divided into. The integer q is called the **quotient** and r is called the **remainder**. The quotient q turns out to be the **largest integer** with the property that $q \times d \leq n$ so that $q \times d$ is the largest multiple of d that is less than or equal to n.

For example, suppose $n = 58$ and $d = 5$. Then q will be 11 and r will be 3, so that

$$5 \times 11 + 3 = 58$$

which is exactly what you would find by applying the algorithm you learned in school. We reiterate, 55 is the largest multiple of 5 that is ≤ 58.

Suppose we are given any pair of positive integers, d and n. Why should such a q and r exist? The reason is straight forward. The list of counting numbers produced by skip counting by d starting at 0 generates every counting number either as an exact multiple of d, or as a member of a gap. This is a direct consequence of the fundamental property that every counting number can be obtained using successor starting at 1, in other words, by the counting process.

The fact that quotients and remainders must exist does not help us find them. What is required is a method. We turn now to the algorithm for finding q and r that is taught to children.

11.4.1   The Division Algorithm: Long Division

The Division algorithm, commonly known as **long division** is a mechanical procedure that takes as input a dividend, n, and a divisor, d, and produces as output a quotient, q, and a remainder, r. We go over how this algorithm is implemented in practice. The reader should be aware that unlike our previous methods, **trial and error** will be a prominent feature of this algorithm. The reader should keep in mind the idea of skip counting through multiples of d to find the largest multiple of d that is less than n. We will continually stress this in what follows.

Example 1

Our first example has $n = 785$ and $d = 8$. This example might appear on a student's problem sheet as: Find $785 \div 8$.

Students should understand that they have to find the largest multiple of 8 that is ≤ 785. Finding this multiple involves trial and error using **educated guesses**. This algorithm works **left to right**, as we will explain below.

To find this multiple, the problem is set up as shown:

$$8 \;\Big|\; \overline{\begin{array}{ccc} 7 & 8 & 5 \end{array}}$$

Digits in the quotient will be written above the line, test multiplication data goes on lines underneath the 785.

The standard algorithm starts by asking: Does $d = 8$ divide 7 which is the leading digit in 785? The answer of course is: No. Equivalently, we ask: what is the largest multiple of 8 that is ≤ 7? Phrased this way, the answer is: 0.

But let's carefully analyze what this question means.

The lead digit in the divisor, 8, is a *ones* digit, whereas the lead digit in 785 is a *hundreds* digit. So, when we ask does 8 divide 7, we are really asking: Is there a positive multiple of 8 and 100 that is ≤ 785? Phrasing the question in this manner formulates the question as: Find the largest integer k such that:

$$8 \times (k \times 100) \leq 785.$$

The answer to this question is 0, which tells us that the quotient must have 0 in the *hundreds* place. Since this digit is 0, we proceed to the next step.

The next step in the standard algorithm is to ask: Is there a positive multiple of 8 that is ≤ 78? Here the answer is yes; it is 9. Again the question can be rephrased to: What is the largest multiple of 8 that is ≤ 78? This time the answer is the same, namely, 9.

Again we analyze the result. The question above is equivalent to asking: What is the largest single digit multiple of 8 and 10 that is ≤ 785? Notice that in asking this question, we have already determined that $785 < 8 \times (10 \times 10) = 800$. Again, we find this multiple by trial and error. Once it has been determined that the largest multiple is 9 as shown by

$$8 \times (\boxed{9} \times 10) = 720 \leq 785 < 800 = 8 \times (10 \times 10),$$

the issue is what to record where? Following the standard algorithm, the 9 is recorded above the 8 in 78 as shown below.

$$\begin{array}{r} \boxed{9} \\ 8 \;\Big|\; \overline{\begin{array}{ccc} 7 & 8 & 5 \end{array}} \end{array}$$

The analysis above tells us why the 9 is placed above the 8 in the dividend. The divisor is 8. The largest multiple ≤ 785 is

$$720 = 8 \times 90.$$

So the *tens* digit in the quotient is a 9. The quotient is recorded above the line and the *tens* place is determined by the *tens* place in the dividend as shown above.

The next step in the process is to record the 72, below the 78 in preparation for subtraction as shown:

$$
\begin{array}{r}
9 \\
8\ \ |\ \overline{7\ \ \ 8\ \ \ 5} \\
-\ \ 7\ \ \ \boxed{2}
\end{array}
$$

The 2 is placed in the *tens* column since, as discussed above, the 72 is actually 720.

The next step in the process is to subtract 72 from 78 using the standard method and record the answer below the line in the *tens* column. To see why we do this, recall our objective: find the largest multiple of 8 that is ≤ 785. What we have so far is the largest multiple of 8 and 10 that is ≤ 785. Only when we have performed the subtraction will we know whether this multiple also satisfies the criterion that it is the largest multiple of 8 that is ≤ 785.

$$
\begin{array}{r}
9 \\
8\ \ |\ \overline{7\ \ \ 8\ \ \ 5} \\
-\ \ 7\ \ \ 2 \\
\hline
\boxed{6}
\end{array}
$$

At this point, students must verify that the result of the subtraction, which we have put in a box for emphasis, is **strictly less than** the divisor, which is 8. If the divisor is less than or equal to the result of subtraction, the largest multiple of 8 has not been found.

Next, the student brings down the 5 directly in the *ones* column as shown:

$$
\begin{array}{r}
9 \\
8\ \ |\ \overline{7\ \ \ 8\ \ \ 5} \\
-\ \ 7\ \ \ 2 \\
\hline
6\ \ \ \boxed{5}
\end{array}
$$

Students should understand that the 65 results from

$$785 - 720 = 65$$

and also that because $8 < 65$ we have not yet found the largest multiple of 8 that is ≤ 785. This means the next step is to find the largest multiple of 8 that is ≤ 65. From recall we have:

$$8 \times \boxed{8} = 64 < 65 < 8 \times (8+1) = 72.$$

Since the 5 in 65 is in the *ones* place, the student records the $\boxed{8}$ from $8 \times \boxed{8}$ in the *ones* column on the top line (next to the 9) and records the product 64 below the 65 as shown:

$$
\begin{array}{r r r r}
 & & 9 & \boxed{8} \\
\hline
8 \mid & 7 & 8 & 5 \\
- & 7 & 2 & \\
\hline
 & & 6 & 5 \\
 & & 6 & 4 \\
\hline
\end{array}
$$

The last step is to compute the remainder $65 - 64$ and record the result as shown.

$$
\begin{array}{r r r r}
 & & 9 & 8 \\
\hline
8 \mid & 7 & 8 & 5 \\
- & 7 & 2 & \\
\hline
 & & 6 & 5 \\
- & & 6 & 4 \\
\hline
 & & 0 & \boxed{1} \\
\end{array}
$$

Again, children must check that the result of this subtraction is less than the divisor, that is, that

$$65 - 64 = 1 < 8.$$

Since digit $\boxed{8}$ is in the *ones* place, the process stops.

As students can now check, $98 \times 8 + 1 = 785$, and $r = 1 < 8$, so we have found $q = 98$ and r. Moreover, the steps just recreate the fact that

$$785 = 8 \times (9 \times 10) + 8 \times 8 + 1 = 8 \times (9 \times 10 + 8) + 1$$

where the RHS makes clear why each digit is in the place shown in the quotient.

This procedure will work for any pair of numbers. We perform one more computation, dividing a two digit number into a four-digit number, to illustrate the process. The instructions are intended to be suitable for students.

Example 2

Find q and r such that

$$27 \times q + r = 4316, \quad \text{and} \quad 0 \le r < 27.$$

The actual problem might well be stated as $4316 \div 27$. Children should understand that the request is to find the largest multiple of the divisor, 27, that is less than or equal to the dividend, 4316, with the residual on subtraction being the remainder.

The set up is shown below:

$$27 \mid \overline{ 4 \quad 3 \quad 1 \quad 6}$$

Since the divisor, 27, has two digits and the dividend has four digits, the process starts by asking whether there is a multiple of 27 that is ≤ 43? The answer is yes since, $27 \times 1 < 43$. Also, 1 is the largest such multiple, since $43 < 2 \times 27 = 54$. We rewrite this as

$$27 \times (1 \times 100) < 4385,$$

which tells us that 1 is the *hundreds* digit in the quotient, q. So the 1 is placed directly above the 3 in 4385. The 27 is recorded below 43 making sure to place the 7 in the *hundreds* column directly below the 3 because as we know from the above discussion, the 27 is actually 2700. We show this below:

$$
\begin{array}{r}
 1 \\
27 \mid \overline{4 \quad \boxed{3} \quad 1 \quad 6} \\
2 \quad 7
\end{array}
$$

To complete the first step, we subtract $43 - 27 = 16$ and verify the result is less than the divisor, 27. Since it is, we record the result as shown.

$$
\begin{array}{r}
 1 \\
27 \mid \overline{4 \quad 3 \quad 1 \quad 6} \\
- \quad 2 \quad 7 \\
\hline
1 \quad 6
\end{array}
$$

Our next step is to determine the *tens* digit in the quotient. To begin the process, we bring down the 1 from the *tens* place (marked with a box), as shown:

$$
\begin{array}{r}
 1 \\
27 \mid \overline{4 \quad 3 \quad \boxed{1} \quad 6} \\
- \quad 2 \quad 7 \\
\hline
1 \quad 6 \quad 1
\end{array}
$$

At this point we want the largest single digit multiple of 27 that is ≤ 161. We want to emphasize that asking the question this way is shorthand for the actual question: What is the largest single digit multiple of 27 and 10 that is $\leq 1616 = 4316 - 2700$? Finding the largest such multiple is trial and error. So, let's try 4. To do this, write 4 above the line in the the *tens* place which is the in same column as the 1 that was just brought down.

$$
\begin{array}{r|rr|c|r}
 & & 1 & 4 & \\
\hline
27\,| & 4 & 3 & \boxed{1} & 6 \\
- & 2 & 7 & & \\
\hline
 & 1 & 6 & 1 & \\
\end{array}
$$

Next we multiply 27×4 and record the result as shown. The setup actually facilitates the multiplication which is by a single digit, in this case 4.

$$
\begin{array}{r|rr|c|r}
 & & 1 & 4 & \\
\hline
27\,| & 4 & 3 & \boxed{1} & 6 \\
- & 2 & 7 & & \\
\hline
 & 1 & 6 & 1 & \\
 & 1 & 0 & 8 & \\
\hline
\end{array}
$$

The student performs the required subtraction as shown:

$$
\begin{array}{r|rr|c|r}
 & & 1 & 4 & \\
\hline
27\,| & 4 & 3 & \boxed{1} & 6 \\
- & 2 & 7 & & \\
\hline
 & 1 & 6 & 1 & \\
- & 1 & 0 & 8 & \\
\hline
 & & 5 & 3 & \\
\end{array}
$$

The result, 53, is greater than 27, so the student should know the largest multiple of 27 that is ≤ 161 has not been found and tries a larger multiple, in this case 5. The result of 5×27 is recorded below

$$
\begin{array}{r|rr|c|r}
 & & 1 & 5 & \\
\hline
27\,| & 4 & 3 & \boxed{1} & 6 \\
- & 2 & 7 & & \\
\hline
 & 1 & 6 & 1 & \\
 & 1 & 3 & 5 & \\
\hline
\end{array}
$$

This time the subtraction produces $161 - 135 = 26$ which is less than the divisor 27:

$$
\begin{array}{r}
1\ \ 5 \\
27\mid \overline{4\ \ 3\ \ 1\ \ 6} \\
-\ \ 2\ \ 7 \\
\hline
1\ \ 6\ \ 1 \\
-\ \ 1\ \ 3\ \ 5 \\
\hline
2\ \ 6
\end{array}
$$

Thus the *tens* digit in the quotient is a 5. Note that if the student had tried 6 as the multiplier, the result would have been 162 which is larger than 161 and fails the condition that $q \times d \le n$.

The last step is to find the *ones* digit in q. To begin this step, we bring down the 6 from the *ones* place as shown:

$$
\begin{array}{r}
1\ \ 5\phantom{\ \ \boxed{6}} \\
27\mid \overline{4\ \ 3\ \ 1\ \ \boxed{6}} \\
-\ \ 2\ \ 7\phantom{\ \ 1\ \ \boxed{6}} \\
\hline
1\ \ 6\ \ 1 \\
-\ \ 1\ \ 3\ \ 5 \\
\hline
2\ \ 6\ \ 6
\end{array}
$$

Again, the student is looking for the largest single digit multiple of 27 that is ≤ 266. Because the subtraction result in the previous step was so close to 27, a good guess for the required multiple would be 9. This results in:

$$
\begin{array}{r}
1\ \ 5\ \ 9 \\
27\mid \overline{4\ \ 3\ \ 1\ \ \boxed{6}} \\
-\ \ 2\ \ 7\phantom{\ \ 1\ \ \boxed{6}} \\
\hline
1\ \ 6\ \ 1 \\
-\ \ 1\ \ 3\ \ 5 \\
\hline
0\ \ 2\ \ 6\ \ 6 \\
-\ \ 2\ \ 4\ \ 3 \\
\hline
0\ \ 2\ \ 3
\end{array}
$$

The process stops because the 9 in the quotient, 159, is in the *ones* place directly above the 6. The remainder, 23, is less than the divisor, 27 and we now know that

$$27 \times 159 + 23 = 27 \times (1 \times 100 + 5 \times 10 + 9) + 23 = 4316.$$

11.5   What Your Child Needs to Know

The procedures for performing the arithmetic of whole numbers are complicated. Moreover the ability to carry them out accurately and with ease,[1] requires both

---

[1]In curriculum guides this capacity is referred to as **fluency.**

knowledge at the level of recall and practice. A natural question is:

> Why is fluency with the procedures of arithmetic necessary in an age of calculators?

We take as given, that the reader accepts the necessity for their child to successfully complete Algebra II as a gateway to future success. So the question posed really becomes:

> Can a calculator replace fluency with the procedures of arithmetic?

To answer this question, let's take the division procedure as an example. Performing computations with this algorithm requires a detailed understanding of the nature of the Arabic number system, and skill with previously developed algorithms for subtraction and multiplication. It is a complicated algorithm, and given the fact of cheap calculators, why should any one learn it?

To argue that one may need to divide and a calculator may not be available, seems to lack substance, given that the last time most of us had to divide anything was very likely in school.

The answer to this is that understanding and an ability to compute go hand-in-hand. Each time an additional skill is acquired, the capacity to acquire more skills is increased.

In addition, these algorithms are precursors to computations that are found useful in the future. For example, various computations in Calculus involve long division of polynomials. Such long division can be taught, on the spot, to students who are able to use the Division Algorithm discussed above. In the author's experience, attempting to teach long division of polynomials to students who have no knowledge of the long division is nearly impossible. Thus, the effect of choosing not to learn an item like the Division Algorithm starts the process of shutting down future options. Eventually, a door is closed and cannot be re-opened.

11.5.1   Goals for Grade 3

By the end of Grade 3 it is expected your child will:

1. understand division of whole numbers as a solution to equations like $\boxed{?} \times 8 = 48$, or equivalently, $48 \div 8 = \boxed{?}$;

2. understand division as a process for subdividing collections into equal groups, for example, $20 \div 5$ produces 5 groups of 4 from a collection of 20 objects;

3. understand that division by 0 is impossible, and why this is so in terms of division arising out of multiplication, specifically, $\boxed{?} \times 0 = 0$ for all counting numbers, so equations like $\boxed{?} \times 0 = 18$ can not have a solution;

4. fluently multiply and divide within 100, using strategies such as the relationship between multiplication and division (e.g., knowing that $8 \times 5 = 40$, one knows $40 \div 5 = 8$) or properties of operations.

In respect to the above, by the end of Grade 3, a child should immediately recall all the products of two one-digit numbers.

11.5.2 Goals for Grade 4

By the end of Grade 4 it is expected your child will:

1. be able to find all factor pairs for numbers ≤ 100, for example, 85 factors into 17 and 5;

2. be able to find integer quotients and remainders using the standard procedure for problems involving four digit dividends and single digit divisors;

3. multiply or divide to solve word problems involving multiplicative comparison, e.g., by using drawings and equations with a symbol for the unknown number to represent the problem, distinguishing multiplicative comparison from additive comparison;

4. solve multi-step word problems posed with whole numbers and having whole-number answers using the four operations, including problems in which remainders must be interpreted;

5. represent these problems using equations with a letter standing for the unknown quantity;

6. assess the reasonableness of answers using mental computation and estimation strategies including rounding;

7. find whole-number quotients and remainders with up to four-digit dividends and one-digit divisors, using strategies based on place value, the properties of operations, and/or the relationship between multiplication and division;

8. illustrate and explain multiplication and division calculations by using equations, rectangular arrays, and/or area models.

11.5.3   Goals for Grade 5

By the end of Grade 5 it is expected your child will:

1. be able to find integer quotients and remainders using the standard procedure for problems involving four digit dividends and two digit divisors;

2. understand the role of place in the division procedure and why it works;

3. find whole-number quotients of whole numbers with up to four-digit dividends and two-digit divisors, using strategies based on place value, the properties of operations, and/or the relationship between multiplication and division. (Illustrate and explain the calculation by using equations, rectangular arrays, and/or area models.)

11.5.4   Goals for Grade 6

By the end of Grade 6 it is expected your child will:

1. fluently divide multi-digit numbers using the standard algorithm.

Chapter 12

Arithmetic of Real Numbers

Chapter Overview. The concept of a unit fraction, $\frac{1}{n}$, is explained and defined in terms of its essential property. The concept of multiplicative inverse for non-zero numbers is defined and related to unit fractions. The real numbers are defined and their six fundamental arithmetic properties identified. These properties are used to demonstrate the fifteen rules of arithmetic which all numerical quantities must obey.

12.1   What Are Real Numbers?

All of us have had to make measurements of physical things in our lives. These might include measuring the size of a room before going to buy carpet, measuring the width of a space in the kitchen to figure out what size of new fridge would fit, or simply recording the height of a child as he/she grows up. What we know from these experiences is that there are lots of numbers wandering around out there that are not whole numbers.

In the school curricula, children are introduced to the idea that there are other kinds of numbers through the process of making measurements. These activities begin early in the Kindergarten curricula by having children identify numerical attributes of objects such as height and weight, and continues by having children make measurements using various standard units, for example, feet and meters for height, and pounds and kilograms for weight. One thing that children discover by making measurements is that in most instances the result of a measurement is not a counting number. So the notion that there are other numbers besides whole numbers becomes clear very early in a child's school experience.

Let's recall that in Chapter 10 we introduced the concept of the number line. We reproduce our diagram of the number line with some prominent additions. **We note** that much of the first section of this chapter requires an understanding of the construction of the number line and the other ideas developed in §10.4 and it is suggested the reader review this section before continuing.

A graphical description of the real numbers. Notice the vertical arrows that identify places on the line not associated with an integer. Each such place identifies a real number. This line is referred to as the **real line**.

We have added vertical arrows that point to places on the line. A fundamental question is whether each such arrow must also identify a number? The conclusion reached by thinkers in previous generations was that

> **wherever an arrow might point to on the line, there should be a unique number associated with that place.**

If one thinks of a measuring tape, it is mostly blank space. Does this mean that there are no numbers associated with the space in between consecutive marks on the tape? According to the above principle, there are numbers associated with every possible place that could be identified in the space in between the markings. But there is simply not room on the tape to show all these numbers and still have the tape be useful.

The set of numbers that can be found by identifying all possible places on the real line is called the **real numbers** and is denoted by $\mathcal{R}$. This set includes not only the integers which are labelled on our diagram above, but also fractions and numbers like the square root of 2. These numbers are the principal objects that serve as the basis for science and the technology in the modern world. This is why knowledge of their arithmetic is critically important.

Our task in this chapter is to develop the laws and rules of arithmetic as they apply to real numbers. In performing this task there are three things we need to recall based on the arithmetic of integers.

1. **The sum of any two real numbers must again be a real number**.

2. **The product of any two real numbers must also be a real number**.

3. **Multiplication of a real number by a counting number corresponds to repetitive addition**.

230

These are obvious facts about arithmetic, but they are essential to what follows.

We begin with a brief description of the kinds of numbers that have a place on the number line and are not integers. To do this we identify numbers that lie in the spaces between integers.

12.1.1   The Unit Fractions

Recall that the integers are also called the whole numbers, and the positive integers are the counting numbers. For the moment, we confine ourselves to speaking of the smallest counting number, namely, 1.

Let us suppose we have one of something. It might be one pie, one liter of water, or one stack of poker chips. Next, recall the lead into the idea of **division** as a process which subdivides something into an equal number of parts, as in $15 \div 5 = 3$ produces three equal parts. The notion of fraction is developed by dividing whatever we have one of into equal parts. The number of parts could, in principle, be equal to any counting number.

For example, if we take a circular pie and cut it once, along a diameter (right down the middle), we will end up with two equal parts. Similarly, if we had a square cake, we could divide it into two equal parts in various ways as shown below:

These types of examples and diagrams are typical of what appears in elementary school curriculum guides and textbooks for children.

Let's turn to a stack of poker chips. Suppose the stack has ten identical chips. Then passing out one chip to each of ten people would divide the stack into ten equal parts. There is nothing special about the stack having ten chips, it could have eleven, or nineteen, or eight hundred sixty one, and could be divided into eleven, nineteen, or eight hundred sixty one equal parts, respectively. The point is the original stack is taken to be the *whole*, the individual chips are taken to be the *equal parts*, and the number of chips comprising the stack could be any counting number whatsoever.

Thinking about a liter of water, we could, again in principle, divide it up into any counting number of equal parts.

We can interpret these ideas in the context of the real line. For example, we know there is one unit of distance between 0 and 1 on the number line (see §10.4). We could divide this interval, which we call the **unit interval**, into five equal parts as shown.

The vertical arrows subdivide the unit interval into five equal, non-overlapping parts. Each place of division identified by a vertical arrow is associated with a real number.

Using the principle that associated with every identifiable place on the line there is a unique number, we know that each of the points on the line identified by an arrow can be labelled with a unique real number.

Conceptually, the intrinsic idea is that, given one whole of something, we can divide that something into any counting number of equal parts. Since the parts all have the same size, each one of these parts can be described by the same number. We will explore this in respect to the attribute of length and the real line below. But first we need to make these new numbers useful.

Recall our discovery in Chapter 5 that for counting numbers to be useful, they had to have names. So let's consider the question:

What name should be assigned to a number that results from dividing one whole into five equal parts, as in the case of the line example?

Thinking about this we will quickly conclude that the name and notation we assign to the number representing the size of one equal part should reflect the total number of equal parts making up the whole, which in this case is 5, and the fact that the symbol represents a process of subdividing one whole, or 1 unit.

The notational answer that was arrived at was that if we have one whole, and we divide that whole into 5 equal parts, we will use the symbols

$$\frac{1}{5}$$

as the name for the number resulting from this division process. Thus, the 1 on top tells us we are dealing with one whole entity, and the 5 on the bottom tells us how many equal parts this whole is divided into. Further numerical examples of such numbers are:

$$\frac{1}{2}, \quad \frac{1}{3}, \quad \frac{1}{8}, \quad \frac{1}{17}, \quad \frac{1}{25}, \quad \frac{1}{132}, \quad \frac{1}{845}, \quad \frac{1}{9847}$$

and so forth. In general, for any **counting number** $n > 1$, as in the numerical cases above, we will write

$$\frac{1}{n}$$

to signify the number that corresponds to the result of dividing 1 into n equal parts and refer generally to numbers of this type as **unit fractions**.

Returning to the diagram subdividing the unit interval into five equal parts, we can see that the number that results is not a counting number, nor any whole number for that matter. We know this because as we see from the diagram (repeated below), none of the arrows identifying subdivisions on the line point to whole numbers. They are all between 0 and 1 so that our unit fraction must satisfy the following inequality:

$$0 < \frac{1}{5} < 1.$$

Because of this relation and the fact that the number named by $\frac{1}{5}$ is to the **right** of 0, we know that $\frac{1}{5}$ must be **positive**.

As mentioned above, the numbers identified in this way, by subdividing the unit interval into parts of equal length, are of a new type that we refer to as **unit fractions**. We use **unit** to stress that these numbers result from subdividing one whole into an equal number of parts. Alternatively, you can think of unit as referring to the 1 in the numerator. We use **fraction** to stress that the result is not a whole number, but the result of dividing something into equal parts.

The vertical arrows subdivide the unit interval into five equal, non-overlapping parts. The horizontal arrows each have identical length, namely, $\frac{1}{5}$ as discussed in the text. We label the place associated with the arrow that starts at 0 with the number we call $\frac{1}{5}$.

We have decided that the number that results from subdividing the unit interval into five equal, non-overlapping parts should be denoted $\frac{1}{5}$. This number is a size attribute which we know is length in the context of a line (see §10.4). The question we want to address is:

Where is $\frac{1}{5}$ in the unit interval?

The answer dictated by the geometry of the real line is that we should use the name $\frac{1}{5}$ for the place identified by the left-most vertical arrow in the diagram above. Indeed, in that diagram, the label has been applied. This choice is completely consistent with our method of placing integers on the line as described in §10.4. Moreover, as we shall see in what follows, it is in perfect agreement with our purpose of extending our arithmetic to include all real numbers.

12.1.2 Other Fractions

One observation about any number, x, that we identify to the right of 0 on the real line, is that $0 < x$. This observation is dictated by our construction of the line which placed all the positive integers to the right of 0 and our previous interpretation of order as it relates to the line (see §10.4). As already noted, this is the reason we were able to write $0 < \frac{1}{5}$ above. Now we have agreed to write the length (size) of any of the five intervals between adjacent markings in the diagram as $\frac{1}{5}$. Does this mean that all the places identified by arrows should be labelled $\frac{1}{5}$? The answer here must be no because each identifiable place on the line is associated with a unique number. So only one of these places can be identified by $\frac{1}{5}$.

Suppose we consider two adjacent intervals of length $\frac{1}{5}$ as shown in our diagram below.

0    $\frac{1}{5}$ 1

The vertical arrows subdivide the unit interval into five equal, non-overlapping parts. The horizontal arrow identifies two contiguous such intervals whose total length must be the sum of the lengths of the two individual intervals of which it is composed (see §10.4).

We ask:

What is the total length of these intervals?

These intervals are non-overlapping and have equal length, $\frac{1}{5}$. Because this new interval is composed of two parts which are non-overlapping and have the same length, the total length of this interval must be the sum of the lengths of the two intervals of which it is composed. Thus, the length of the horizontal arrow must be given by

$$\frac{1}{5} + \frac{1}{5} = 2 \times \frac{1}{5}$$

since repetitive addition corresponds to multiplication by a counting number. The reader will recall that equations like this one are enforced by the Distributive Law through computations like:

$$\begin{aligned} \frac{1}{5} + \frac{1}{5} &= 1 \times \frac{1}{5} + 1 \times \frac{1}{5} \\ &= (1+1) \times \frac{1}{5} \\ &= 2 \times \frac{1}{5} \end{aligned}$$

which we write out to stress the underlying principle and how it is used. Using the ideas developed in §10.4 concerning addition and the number line, we know that the place associated with $2 \times \frac{1}{5}$ must be as shown below.

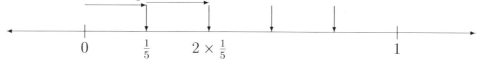

The horizontal arrows each have length $\frac{1}{5}$. As discussed in the text and §10.4, $2 \times \frac{1}{5}$ must be the number associated with the second vertical arrow to the right of 0.

In a similar manner, we conclude:

$$\frac{1}{5} + \frac{1}{5} + \frac{1}{5} = 3 \times \frac{1}{5},$$

and

$$\frac{1}{5} + \frac{1}{5} + \frac{1}{5} + \frac{1}{5} = 4 \times \frac{1}{5}.$$

Moreover each of these sums is associated with a place in the diagram since each represents the total length of three or four adjacent intervals. We show this in the next diagram.

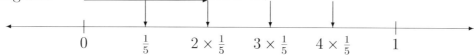

The horizontal arrows have lengths $2 \times \frac{1}{5}$ and $\frac{1}{5}$, respectively. Their sum has length $3 \times \frac{1}{5}$ and must be the number associated with the third vertical arrow to the right of 0. The place of $4 \times \frac{1}{5}$ is identified similarly.

There is nothing special about the unit fraction $\frac{1}{5}$. Any unit fraction can be placed on the line using the steps taken above. For example, for the counting number n, first divide the unit interval into n equal parts and identify the points of division. Second, label the left-most point of division with $\frac{1}{n}$. Third, label the remaining points of division using repetitive sums of $\frac{1}{n}$.

12.1.3 Notation and Nomenclature for Common Fractions

Numbers of the type discussed above are referred to as **common fractions**. As the reader may recall, $2 \times \frac{1}{5}$ is not the notation for common fractions used by school children. The notation taught in school and which appears on the RHS below is

$$2 \times \frac{1}{5} = \frac{2}{5}.$$

The notation $\frac{2}{5}$ is more compact and reflects our notation for unit fractions. But never forget, this notation is by **choice and convention**. A consequence of this convention is the equation

$$\frac{2}{5} = \frac{1}{5} + \frac{1}{5}.$$

The truth of this equation is not convention, it is a mathematical fact enforced by the Distributive Law. More generally, by convention we will agree that for any pair of counting numbers m and n the fractional form $\frac{m}{n}$ is the standard notation the for number on the RHS of

$$\frac{m}{n} \equiv m \times \frac{1}{n}$$

where $\equiv$ is read **by definition is**. In later discussion we will refer to this definition as the **Notation Equation**.

In a fraction, for example $\frac{m}{n}$, the number appearing below the line, in this case, n, is called the **denominator**. The number appearing above the line, in this case, a m, is called the **numerator**. Fractions of this form, $\frac{m}{n}$, with m, n positive integers, are referred to as **common fractions**. And if $m = 1$, they are called **unit fractions**. In speaking, we say m **over** n to denote the fraction having numerator m and denominator n.

Given specific numerical values for n, we have standard names for unit fractions. Thus, we write $1/2$, and say, one half; we write $1/3$, and say, one third; we write $1/4$, and say, one fourth; we write $1/5$, and say, one fifth; skipping on, we write $1/10$, and say, one tenth; we write $1/27$, and say, one twenty-seventh; and so forth.

More generally, we would write $\frac{2}{3}$, and say, two thirds; write $\frac{5}{8}$, and say, five eighths, and so forth for fractions generally. The names for common fractions may appear obvious, but they carry great meaning. The numerator tells how many of something. The denominator tell us what those somethings are. Thus, the 8 in $\frac{5}{8}$ tells us we are dealing with eighths ($\frac{1}{8}$), and the 5 tells us we have five of them. In this way the name and notation of each common fraction tells us a lot about its arithmetic properties as we shall see when we discuss their computations.

12.1.4 Other Real Numbers

There are still other numbers that turn up in the real world that are neither counting numbers nor fractions. Some such numbers arise as lengths. Others arise from geometric considerations having to do with circles. And still others arise from considerations having to do with the accumulation of interest on debts. Thus, the identification of these other numbers is based on the practical considerations of human activity, as the following example shows.

Long before the development of the Arabic system of numeration that has been critical to our development, it was shown that there are numbers arising as lengths that cannot be expressed as ratios of whole numbers (fractions). For example, consider the square shown in the figure.

Given the sides of the square have unit length, what is the length of the diagonal, d?

It was known to the followers of Pythagoras in 500 BC that the length of the diagonal of a square was incommensurate with the length of its side (which, in effect, means the length cannot be expressed as a fraction). From the diagram it is clear that the length of the diagonal is longer than the length of one side, but shorter than the length of two sides. So the number associated with the length of the diagonal is neither a fraction, nor a whole number.

What also seems clear is that the length of the diagonal can be used to identify a place on the real line, and hence a real number. So we take as a given that there is a number corresponding to this length. Using d to denote the length of the diagonal of the unit square and what readers may recall as the Pythagorean Theorem,[1] we have

$$d \times d = 1 \times 1 + 1 \times 1 = 1^2 + 1^2 = 2.$$

In other words, d is a solution to an arithmetical equation of the form

$$x^2 = x \times x = 2.$$

As most readers will know, the number d that solves this equation is called the square root of 2.[2] The equation $x^2 = n$, where n is a counting number always has a solution in $\mathcal{R}$. However, unless the counting number is a perfect square, like 1, 4, 9, or 16, the solution will not be a fraction or a whole number. Numbers of this type are called **irrational numbers**. It is a remarkable fact that most real numbers are neither fractions nor whole numbers. But such considerations are beyond the scope of this book and we confine ourselves to simply giving an example of a familiar number ($\sqrt{2}$) that is neither a fraction nor a whole number.

---

[1]The readers will may recall $a^2 + b^2 = c^2$ as the relation holding between the sides of a right triangle.

[2]Approximations to $\sqrt{2}$ date to 1000 years earlier. Wikipedia has an incredible picture of a Babylonian clay tablet showing how to approximate the square root of 2 to six decimal places.

12.2 Integer Based Arithmetic of The Real Numbers, $\mathcal{R}$

In what follows, we will develop arithmetic rules obeyed by all real numbers. We note that, for the most part, the quantities manipulated in sciences like physics and chemistry, by engineers and financial analysts in the course of their work, and by students taking calculus in college, are real numbers. Thus, the rules developed are exactly the rules followed by the quantities under discussion in these subjects. What is amazing is how few of these rules there are and how much power they provide those who master their use.

Since $\mathcal{R}$ consists of all numbers that can be identified on the real line, we know that all the integers belong to $\mathcal{R}$. For this reason we expect that the rules of arithmetic already identified that hold for integers must also hold for real numbers. In this respect, we are again observing that the rules are forward compatible. So that's where we begin.

12.2.1 Arithmetic Properties of $\mathcal{R}$ Derived from $\mathcal{I}$

We are about to set down the properties that are required of all real numbers. The reader may be somewhat intimidated by this prospect. To set your mind at ease, we emphasize that the properties we are about to list, for the most part, merely reiterate properties we have been studying and using right along. These properties for $+$ and $\times$ on the integers were summarized in §9.3.5. Comparison of the statements below and those in §9.3.5 reveal that the essential change is that integers are replaced by real numbers. For example, $m + n = n + m$ becomes $x + y = y + x$ in the Commutative Law so it now applies to all real numbers, not merely to integers.

We will use $x \in \mathcal{R}$, to assert that x is a member of $\mathcal{R}$ (see §2.3.1). Using this notation then, the following statements are true about $\mathcal{R}$:[3]

1. Closure: let $x,\ y \in \mathcal{R}$, then $x + y \in \mathcal{R}$, and $x \times y \in \mathcal{R}$;

2. let $x,\ y,\ z \in \mathcal{R}$, then the Commutative, Associative and Distributive properties hold as shown:

$$x + y = y + x \quad \text{and} \quad x \times y = y \times x$$

---

[3]My wife continually reminds me that many who are not mathematicians are intimidated by the use of variables like x and y. So if you feel that you would better understand any of the following equations by using a numerical example, simply write out the equation with the variables replaced by numbers. Any old numbers will do as long as they are not special, like 1 and 0. So for example, you could replace x by 5, y by 8 and z by 3 to obtain the following instance of the Distributive Law: $5 \times 3 + 8 \times 3 = (5 + 8) \times 3$. Performing the indicated computations will then give $39 = 39$.

$$(x + y) + z = x + (y + z) \quad \text{and} \quad (x \times y) \times z = x \times (y \times z)$$
$$(x + y) \times z = x \times z + y \times z \quad \text{and} \quad z \times x + z \times y = z \times (x + y);$$

3. $\mathcal{R}$ has an **additive identity**, denoted by 0, with the property that for any $x \in \mathcal{R}$

$$0 + x = x + 0 = x;$$

4. every element, x in $\mathcal{R}$ has an **additive inverse**, y, that satisfies

$$x + y = y + x = 0;$$

5. $\mathcal{R}$ has an **multiplicative identity**, denoted by 1, with the property that for any $x \in \mathcal{R}$

$$1 \times x = x \times 1 = x.$$

12.2.2 Additive Inverses and the Real Line

In §12.1 we identified the real numbers with places on the real line. Based on previous considerations with integers (see §10.4 and 12.1), we then said that numbers found on the RHS of 0 were positive. Let us consider an arbitrary real number x that has been identified as a place on the number line on the RHS of 0. Such a number is pictured below.

The number identified by x is found at a place to the right of 0 on the line. The additive inverse of x will be found x units to the left of 0 and is denoted by $-x$ as shown. See text for discussion.

Numbers having a place to the right of 0 on the line are said to be **positive**. For such numbers, like x, their name is the **length** of the interval between 0 and x. This interval is indicated in the diagram by an arrow starting at 0 and ending at x. The length of this arrow is x. Clearly there is a place to the left of 0 that is the exact same distance from 0 as x is to the right. This number, which we have denoted $-x$ in the diagram is x units to the left of 0. That is, the interval between $-x$ and 0 has the same length as the interval between 0 and x. The point of all this is that if we start at x and move x units to the **left** we end up at 0. We can summarize this property numerically as

$$x + (-x) = 0$$

239

and this is the essential reason we know our arithmetic must satisfy the additive inverse property for all real numbers. (The reader will recall that the additive inverse of a number is initially referred to as its **opposite**.)

As is already apparent, the **centered dash** notation will be used to denote the additive inverse of a real number. Thus, $-x$ is the additive inverse of x, and we know with certainty that

$$x + (-x) = (-x) + x = 0.$$

Notice that the second part of this equation, $(-x) + x$, captures the idea of starting at $-x$ and proceeding x units to the right puts us again at 0 (again see §10.4.)

Since x can be any real number, we immediately know that

$$\frac{4}{5} + \left(-\frac{4}{5}\right) = 0$$

to give a numerical example.

An essential fact about additive inverses of integers is that they are unique. It is also true that additive inverses of real numbers are unique for exactly the same reason. We could also note that additive inverses are unique because each place on the line denotes a unique real number. If the reader is not sure why additive inverse are unique, please review §9.2.1 and 9.2.2, because we will use uniqueness here in the same way.

12.2.3   Extending Rules Found for Integers

In addition to the five arithmetic laws (axioms) listed above, the following arithmetic rules derived from those laws hold for all real numbers. They are consequences of the axioms. As in the case of integers, their importance lies in the fact that they tell us how to perform computations, as for example finding $\frac{4}{5} + (-\frac{4}{5}) = 0$.

Let $x,\, y \in \mathcal{R}$. Then

I1 $-0 = 0$;

I2 $-(x + y) = (-x) + (-y)$;

I3 $-(x \times y) = (-x) \times y = x \times (-y)$;

I4 $x \times y = (-x) \times (-y)$;

I5 $1 = (-1) \times (-1)$ and $x = -(-x)$;

I6 $-x = (-1) \times x$;

I7 $-(x \times y) = ((-1) \times x) \times y = x \times ((-1) \times y)$;

I8 $x \times 0 = 0 \times x = 0$.

In Chapter 9 we demonstrated why each of these rules was true for integers. What we are doing here is building on that foundation by extending the rules developed for integers to all real numbers. Any rule which mentions only integers must automatically be true, for example I1, or the first equation in I5. We reiterate a few of the other arguments in the context of real numbers to remind readers of why these rules are true.

For I2 we show that both $-(x + y)$ and $(-x) + (-y)$ are additive inverses for $x + y$. Since $-(x + y)$ is the additive inverse of $x + y$, we only have to show the second quantity is also an additive inverse of $x + y$. Since $x + y = y + x$, we use $y + x$ and compute

$$
\begin{aligned}
(y + x) + [(-x) + (-y)] &= [(y + x) + (-x)] + (-y) \\
&= [y + (x + (-x))] + (-y) \\
&= [y + 0] + (-y) \\
&= y + (-y) = 0
\end{aligned}
$$

so that $(-x) + (-y)$ is also an additive inverse of $x + y$. Uniqueness of additive inverses means that $-(x+y)$ and $(-x)+(-y)$ are the same number, so the equality asserted in I2 holds.

Next consider the second equation in I5, namely, $x = -(-x)$. In words, this equation asserts that x is the additive inverse of $-x$. To see why this must be so, we test whether both $-(-x)$ and x are additive inverses for $-x$. By definition of the centered dash notation, $-(-x)$ is the additive inverse of $-x$. But we also know that $(-x) + x = 0$ because $-x$ is the additive inverse of x. Thus, x satisfies the additive inverse test in respect to $-x$. Thus, by the uniqueness of additive inverses, $x = -(-x)$.

While the reader can check that no new reasoning has been used in these arguments beyond what was presented in Chapter 9, they are important because they remind us of the utility of uniqueness. With these comments we take I1-I8 as being established in respect to real numbers.

12.3   The New Arithmetic Property of $\mathcal{R}$

12.3.1   Conservation Applied to Unit Fractions

A theme of this book has been that the computations of arithmetic model what we observe about the world. The most fundamental principle is conservation of number,

CP, and we return to it for the last time.

We began our discussion of numbers that are not integers by considering unit fractions. These fractions arise by dividing a physical whole into some number of identical parts. Associated with the whole was an attribute which had a numerical measure which could reasonably be thought of as 1. Of course in the real world there are units associated with any measure. In the examples previously given, the units measured were area, volume and length. Of a somewhat less directly numerical nature was 1 *whole pie* or 1 *stack of poker chips*.

For the moment, we confine our discussion to the collection of unit fractions that arises by subdividing a unit whole into n parts where n is a counting number. The key facts we want to draw on are that each of the resulting parts are identical, and each part has the numerical descriptor $\frac{1}{n}$ attached to it. As a particular example we may think of $1/10$ as the number representing the result of dividing something into 10 equal parts. The process of division results in 10 items which are identical in respect to being a part of the unit whole. And the number, $\frac{1}{10}$, is thought of as a **numerical measure** of the size of this part.

The question we want to consider is:

What happens when we bring all the divided parts back together?

To be concrete, consider a pie. We divide it into two identical parts. We bring the two parts back together, what do we have? We have a whole pie.

As another example, consider a square piece of land. We divide it into four equal parts by marking off dividing lines. Each of the four parcels contains one fourth of the original area. Now we remove the markers. What do we have? The original plot of land is restored (see below) and has the same area as when we started.

On the left: a square piece of land which has been subdivided into four equal parts by marking. On the right: the square piece of land restored to its original whole by removing the markings.

For larger subdivisions, say ten or more, we might start with a quantity of water, say a liter. Suppose we divide this into ten equal parts, in separate jars. When we pour the contents back into the original jar, we are back to having a liter of water, or whatever quantity of water you want to start with. The point is that when we bring all the equal parts back together, the whole is restored.

To summarize, in the ideal,

if we start with a whole something and we divide it into n **equal parts, when we bring those** n **equal parts back together, we reconstruct the whole.**

This fact about the real world is simply a consequence of conservation, **CP**. And as the reader will see in what follows, **this fact gives rise to an essential new property of numbers.**

12.3.2   The Fundamental Equation

As discussed above, when we divide something into equal parts, the result of bringing those parts back together is to restore the original whole. The question we want to consider is:

How should this aspect of conservation be captured by arithmetic?

Consider the diagram of the unit interval divided into fifths.

The vertical arrows subdivide the unit interval into five equal, non-overlapping parts.

Removal of the subdivisions produces the following diagram.

When the arrows marking subdivisions into fifths are removed, the unit interval is simply the unit interval and must have unit length as indicated by the horizontal arrow. Indeed, no markings identifying places on the line can change the length of the interval.

Now as discussed in §12.1, the number we attach to the length of each of the five equal subdivisions of the unit interval is $\frac{1}{5}$. Moreover, also as discussed in that section, summing five of these unit fractions must produce the number corresponding to the total length of the interval. Thus, we have the equation

$$1 = \frac{1}{5} + \frac{1}{5} + \frac{1}{5} + \frac{1}{5} + \frac{1}{5}$$

which is exactly what we must have if our arithmetic preserves conservation. Thus, it is clear that in our arithmetic model, which has to reflect the real world, the following must also be true:

$$1 = \frac{1}{2} + \frac{1}{2}$$
$$1 = \frac{1}{3} + \frac{1}{3} + \frac{1}{3}$$
$$1 = \frac{1}{4} + \frac{1}{4} + \frac{1}{4} + \frac{1}{4}$$

and so forth for any counting number. Since in our arithmetic, we know that repetitive addition is multiplication, the above equations can be reformulated as:

$$1 = 2 \times \frac{1}{2}$$
$$1 = 3 \times \frac{1}{3}$$
$$1 = 4 \times \frac{1}{4}$$
$$1 = 5 \times \frac{1}{5}.$$

Next, consider the following diagram of the portion of the number line from 0 to 5 as diagramed below.

The portion of the number line from 0 to 5. The distance from any integer to its successor is 1 unit.

Recall the construction of the number line in §10.4 required the successor of an integer to be placed one unit of distance to the right of its predecessor. Thus the markings 1, 2, 3 and 4 subdivide the interval from 0 to 5 into five equal parts. From our previous discussion, we know when we subdivide something into five equal parts, we refer to the result as being *one fifth* of whatever we started with. In this case, the unit whole we are starting with has length 5 as its numerical attribute. We are dividing it into 5 equal parts which correspond to the five intervals identified by the integer markings. Each has one unit of length. Thus, we can write

$$1 = \frac{1}{5} \times 5.$$

244

Obviously we know this must be a fact from the Commutative Law applied to the equation

$$1 = 5 \times \frac{1}{5}.$$

Alternatively, reflecting on the considerations in Chapter 11 on division of integers, we know $5 \div 5 = 1$, which ought to translate into $1 = \frac{1}{5} \times 5$. But these considerations are abstract. The former sits in the real world as a statement about the length that results from subdividing something five units long into five equal parts.

We can formulate this result generally as follows. For any counting number n, the **Fundamental Equation**

$$1 = \frac{1}{n} \times n = n \times \frac{1}{n}$$

must hold. We can think of the last as the Fundamental Equation because, as we discuss below, it determines arithmetic behavior of fractions. The Fundamental Equation is the essential fact that will play an important role in the development of fractions.

12.3.3   The Notion of Multiplicative Inverse

One of the arithmetic properties of $\mathcal{R}$ which is derived from integers is that every real number has an additive inverse; that is, for every real number x, there is a number $-x$ with the property that

$$x + (-x) = (-x) + x = 0.$$

Recall, 0 is the **additive identity**. We also have a **multiplicative identity**, namely, 1. Moreover, we just showed in the last section that every counting number n satisfies the Fundamental Equation:

$$n \times \frac{1}{n} = \frac{1}{n} \times n = 1.$$

This equation looks exactly like the additive inverse equation with the operation of addition replaced by multiplication and 0 replaced by 1. For this reason, it must be the case that at least some real numbers satisfy an equation that could reasonably be referred to as the **multiplicative inverse** equation.

If every real number satisfied a multiplicative inverse property in the same way as they satisfy the additive inverse property, the following would be true.

Given $x \in \mathcal{R}$, we can find $y \in \mathcal{R}$ with the property that

$$x \times y = y \times x = 1.$$

This equation is not universally satisfied because if we take $x = 0$, then no matter how y is chosen the product yields

$$0 \times y = y \times 0 = 0.$$

This last equation is a consequence of the fact that multiplication is repetitive addition. There is no escape!

Nevertheless, we know that all unit fractions and all counting numbers have multiplicative inverses. Therefore we might think that all real numbers **except** 0 have multiplicative inverses and this is indeed the case. We assert this as our last essential property of real numbers.

For every $x \in \mathcal{R}$, $x \neq 0$, there is a $y \in \mathcal{R}$ with the property that

$$x \times y = y \times x = 1.$$

The number y is referred to as the **multiplicative inverse** of x.

12.3.4 Summary of Arithmetic Properties of $\mathcal{R}$

The following are a list of the fundamental arithmetic properties of $\mathcal{R}$. In the jargon of mathematics they are known as **axioms**, which are statements which are obviously true. In preceding chapters and sections, we have discussed why these must be true if our arithmetic is to reflect the world around us.

Let x, y, $z \in \mathcal{R}$. Then the following statements are true about $+$ and $\times$ on $\mathcal{R}$:

1. Closure: $x + y \in \mathcal{R}$, and $x \times y \in \mathcal{R}$;

2. the Commutative, Associative and Distributive properties hold as shown:

$$x + y = y + x \quad \text{and} \quad x \times y = y \times x$$
$$(x + y) + z = x + (y + z) \quad \text{and} \quad (x \times y) \times z = x \times (y \times z)$$
$$(x + y) \times z = x \times z + y \times z \quad \text{and} \quad z \times x + z \times y = z \times (x + y);$$

3. $\mathcal{R}$ has an **additive identity**, denoted by 0, with the property that for every $x \in \mathcal{R}$

$$0 + x = x + 0 = x;$$

4. every element, x in $\mathcal{R}$ has an **additive inverse**, y, that satisfies

$$x + y = y + x = 0;$$

246

5. $\mathcal{R}$ has an **multiplicative identity**, denoted by 1, with the property that for every $x \in \mathcal{R}$

$$1 \times x = x \times 1 = x;$$

6. for every $x \neq 0$, we can find a **multiplicative inverse** $y \in \mathcal{R}$ such that

$$x \times y = y \times x = 1.$$

12.3.5 Nomenclature and Notation

Every subject has its own nomenclature and notation. Knowledge of nomenclature is essential for understanding any subject. Thus, **the content of this section is essential to what follows.**

We shall refer to the additive inverse of x as **minus** x, and use the symbols $-x$ to denote this number.

As in the case of integers §9.2.5, the operation of **subtraction** is then defined by:

$$x - y \equiv x + (-y).$$

Thus, there is no primary operation of subtraction, only a defined operation based on addition as shown. (Again, we use $\equiv$ which is read as: *is defined to be.*) The fact that subtraction is a form of addition gives rise to the computational rule for performing subtraction: *change the sign and add.*

Given $x \neq 0$, we shall refer to the **multiplicative inverse** of x as the **reciprocal** of x and denote it by the notational unit: x^{-1} where -1 is called an **exponent**. This means, for any $x \neq 0$:

$$x \times x^{-1} = x^{-1} \times x = 1.$$

The equation, $x \times x^{-1} = 1$, defines what is meant by x^{-1}. To give a numerical example,

$$5 \times 5^{-1} = 1.$$

There is an immediate ambiguity that arises out of the usage of the centered dash and the exponential notation, namely, what do we mean by:

$$-x^{-1} ?$$

Two different activities are specified. The centered dash tells us to find the additive inverse of x, while the exponent -1 tells us to find the multiplicative inverse of x. Which activity gets performed first? There is an order of precedence rule that removes the ambiguity. The rule says

all exponential operations are performed before addition and/or multiplication.

Because of these conventions, the expression on the LHS is not ambiguous and is equal to the RHS,

$$-x^{-1} = -(x^{-1}) = -\frac{1}{x},$$

which in words means the quantity being computed is the additive inverse of the multiplicative inverse of x. Alternatively, we have:

$$(-x)^{-1} = \frac{1}{-x}.$$

It turns out that for the case of $-x^{-1}$, the two expressions are equal by the rule R6 (see §12.6.1). However, if the exponent were divisible by 2, as in the expressions $-x^{-2}$ and $(-x)^{-2}$, the result would be a pair of additive inverses. We will discuss exponents in Chapter 17.

Recall the equation defining subtraction: $x - y \equiv x + (-y)$. By using this equation, subtraction becomes a defined operation subsidiary to addition. We do the same thing with division. Given $y \neq 0$, we define **the operation of division** by:

$$x \div y \equiv x \times y^{-1}.$$

In this context, x is the **dividend**, y is the **divisor** and $x \times y^{-1}$ is the **quotient**; that is, the result of multiplying x by the multiplicative inverse of y should be regarded as a quotient. To see why we should regard $x \times y^{-1}$ as a quotient, recall that for an integer quotient $q = n \div d$, dividend n, and one of its divisors d, we have:

$$q \times d = n.$$

In words, **the dividend equals the quotient times the divisor**. Now, for the reals x and $y \neq 0$, think of x as the dividend and y as the divisor. Then apply the Associative Law to the LHS of the following equation to obtain

$$(x \times y^{-1}) \times y = x \times (y^{-1} \times y).$$

Since $y^{-1} \times y = 1$, we conclude the RHS is:

$$(x \times y^{-1}) \times y = x.$$

So the quantity $x \times y^{-1}$ behaves like a quotient in respect to y acting as a divisor of the dividend x. We shall return to this idea in §15.2.

248

Recall the two key equations developed in respect to common fractions: the Notation and Fundamental Equations:

$$\frac{m}{n} \equiv m \times \frac{1}{n} \quad \text{and} \quad n \times \frac{1}{n} = 1.$$

The Notation Equation tells us that every common fraction is a product of a counting number and a unit fraction. The Fundamental Equation tells us $\frac{1}{n}$ is the multiplicative inverse of n. Together with the discussion above, this suggests making the following definition of fraction notation applicable to all real numbers. For x, $y \in \mathcal{R}$ and $y \neq 0$, we define the notation, $\frac{x}{y}$ by:

$$\frac{x}{y} \equiv x \times y^{-1}.$$

This notation merely extends to all real numbers, and all the expressions of algebra that represent real numbers, the notation developed for common fractions in §12.1.3. In respect to expressions of the form $\frac{x}{y}$, the standard nomenclature is retained. Thus, x is the **numerator** and y is the **denominator** of the fractional form: $\frac{x}{y}$.

Given any real number $y \neq 0$, an immediate consequence of this usage is:

$$y^{-1} = 1 \times y^{-1} = \frac{1}{y}$$

so that $\frac{1}{y}$ is simply another notation for the multiplicative inverse of y and

$$y \times \left(\frac{1}{y}\right) = y \times y^{-1} = 1.$$

The effect of this is that any fractional form can be expressed as a product:

$$\frac{x}{y} = x \times \frac{1}{y}.$$

Useful rules governing the behavior of these forms will be developed below.

12.4 Finding the Integers as a Subset of $\mathcal{R}$

The axioms for $\mathcal{R}$ and the operations of $+$ and $\times$ do not mention successor. However, they do mention both 0 and 1 and give them both very specific properties and we take advantage of these properties to identify all the numbers we have been discussing in previous chapters. The essential part of this is supplying names to specific members of $\mathcal{R}$ in the form of a system of numeration.

We define the numeration system on $\mathcal{R}$ by the following:

$$2 \equiv 1 + 1 \quad \text{and} \quad 3 \equiv 2 + 1 \quad \text{and} \quad 4 \equiv 3 + 1$$
$$5 \equiv 4 + 1 \quad \text{and} \quad 6 \equiv 5 + 1 \quad \text{and} \quad 7 \equiv 6 + 1$$
$$8 \equiv 7 + 1 \quad \text{and} \quad 9 \equiv 8 + 1 \quad \text{and} \quad 10 \equiv 9 + 1$$

and so forth. This process supplies the usual name from the Arabic system to a specific number in $\mathcal{R}$. Moreover, as the reader can see the names are given in such a way as to preserve the **successor** idea. Further, the naming process identifies a unique member of $\mathcal{R}$ for each member of what we had previously been referring to as the positive integers (counting numbers). Thus, through the naming process, we can think of every positive integer as being a member of $\mathcal{R}$. Once we have identified a particular positive integer n, we know its additive inverse must also be in $\mathcal{R}$, so very quickly, we have all integers belong to $\mathcal{R}$ since there are only three types of integers, counting numbers, zero and additive inverses of counting numbers. As mathematicians, we would say the positive integers are a **subset** of the reals, and write $\mathcal{I} \subseteq \mathcal{R}$.

There is an addition table in $\mathcal{R}$, attached to this system of naming. The reader may be quite happy to accept that this table is the same addition table we previously constructed, which it is. Nevertheless, we perform a calculation using the Commutative and Associative properties for addition, and the definitions of our symbols, to illustrate why the entries in the addition table are what we expect:

$$\begin{aligned} 4 + 3 &= 4 + (2 + 1) = 4 + (1 + 2) = (4 + 1) + 2 \\ &= 5 + 2 = 5 + (1 + 1) = (5 + 1) + 1 \\ &= 6 + 1 = 7 \end{aligned}$$

which is exactly the result in the addition table. We stress that the successor idea has been incorporated into the assignment of names and that fact is essential to the process of finding the sum. The computation on each line simply subtracts 1 from the current value on the right, and adds it to the number on the left. So $4 + 3$ on the first line becomes $5 + 2$ on the second. The process continues until a single integer remains, in this case, 7. In terms of our real-world button analogy, each line corresponds to moving one button from the jar on the right to the jar on the left and stops when the jar on the right is empty. It's as simple as that.

In the chapter on multiplication, we showed that multiplication was repetitive addition. The reader may wonder whether this is still the case. That multiplication in $\mathcal{R}$ must come down to repetitive addition on the integers is enforced by the Distributive Law and the definitions of the symbols, as the following calculations show: [4]

$$5 \times 2 = 5 \times (1 + 1) = 5 \times 1 + 5 \times 1 = 5 + 5,$$

$$5 \times 3 = 5 \times (2 + 1) = 5 \times 2 + 5 \times 1 = (5 + 5) + 5.$$

Equally well, since the Distributive Law is two-sided, we can write

$$\begin{aligned}
5 \times 2 &= (1 + 1 + 1 + 1 + 1) \times 2 \\
&= 1 \times 2 + 1 \times 2 + 1 \times 2 + 1 \times 2 + 1 \times 2 = 2 + 2 + 2 + 2 + 2
\end{aligned}$$

$$\begin{aligned}
5 \times 3 &= (1 + 1 + 1 + 1 + 1) \times 3 \\
&= 1 \times 3 + 1 \times 3 + 1 \times 3 + 1 \times 3 + 1 \times 3 = 3 + 3 + 3 + 3 + 3
\end{aligned}$$

so that in either direction, multiplication of counting numbers is repetitive addition.

It should seem credible to the reader that there is nothing special about 5 and 3, and that a similar set of calculations would establish that $m \times n$ is m added to itself n times for any pair of counting numbers m and n. It follows that the multiplication table for $\mathcal{I}$, as a subset of $\mathcal{R}$, is exactly as previously established, and that all procedures previously developed for performing arithmetic with integers calculations remain valid.

The Distributive Law also ensures that when an arbitrary real is multiplied by a counting number on either the right or the left, the result is repetitive addition. To see this, let a denote any real number. Then

$$2 \times a = (1 + 1) \times a = 1 \times a + 1 \times a = a + a$$

and

$$a \times 2 = a \times (1 + 1) = a \times 1 + a \times 1 = a + a.$$

This relation, $2 \times a = a + a = a \times 2$, shows how to verify that multiplication of an arbitrary real number by a counting number is repetitive addition of a added to itself the number of times determined by the counting number. This fact will play an essential role in the development of fractions in the next chapter.

12.5   Finding Common Fractions as a Subset of $\mathcal{R}$

As discussed above, **common fractions** are numbers that can be written in the form $\frac{m}{n}$ where m, $n \in \mathcal{I}$ and $n \neq 0$. In the last section we explained how the integers

---

[4]Remember, multiplication takes precedence over addition.

were identified within $\mathcal{R}$. In respect to the arithmetic of $\mathcal{R}$, we showed $\frac{1}{n} = n^{-1}$, whence

$$\frac{m}{n} = m \times \frac{1}{n} = m \times n^{-1} \equiv m \div n.$$

In other words, common fractions are **ratios** of whole numbers where the divisor is not zero. Thus every such ratio of whole numbers corresponds to a particular real number, and as discussed above, a place on the line. We therefore define the **rational numbers**, $\mathcal{Q}$, by:

$$\mathcal{Q} = \left\{ \frac{n}{m} : m, \ n \in \mathcal{I} \ \ and \ \ n \neq 0 \right\}.$$

For readers not totally comfortable with set notation, we provide an explanation: on the RHS, between the opening brace and the colon, is an expression, $\frac{m}{n}$. This expression specifies that every member of this set must have the form of a fraction. To the right of the colon and before the closing brace is a mathematical statement that puts requirements on things to the left of the colon. In this case, the statement asserts that m is an integer and n is a non-zero integer. Thus, in its entirety, we could read this line as:

> Q is the set of all fractions having an integer in the numerator and a non-zero integer in the denominator.

In still other words, the rational numbers are comprised of all numbers that can be obtained as a ratio of two integers, where the divisor must be non-zero. We stress that every common fraction is a real number. Thus the set $\mathcal{Q}$ is a **subset** of the set $\mathcal{R}$. Mathematicians use the symbol $\subseteq$ as a short hand for *is a subset of*. Thus we would write

$$\mathcal{Q} \subseteq \mathcal{R}.$$

12.6 Rules of Arithmetic for Real Numbers

Our purpose in this section is to develop the remaining facts about the arithmetic of real numbers not already listed in I1-I8. These facts end up as computational rules taught to children. What makes these rules so useful is their universal application; they apply to all the numerical quantities we use to describe the world.

These rules are not taught to children in Grade 2. But it is expected that by the end of the elementary curriculum that children will be comfortable applying these rules in their algebraic form and have a fair idea of how they are justified.

The arguments given as to why these rules are true are straight forward, and for the most part follow directly from the axioms and notation in §12.3.4-5. It will aid the reader to expend the effort to identify which property is being applied at each step. This will better enable you to help your child, as understanding is an essential goal of the CCSS-M. By the end of this section you will be thoroughly familiar with the various rules needed to successfully work with fractions, and where those rules come from. If you find the going really heavy, you might have a go at the next chapter which is aimed at developing simple explanations for understanding fractions which can be used with younger children. Then come back to this material. However, before doing that you should review the summary list of rules governing the arithmetic of real numbers that appears in §12.6.1.

We will identify each rule as it is discussed using the same identifier in §12.6.1. Throughout this section x, y, $z \in \mathcal{R}$ are arbitrary.[5]

Zero is the additive identity and I1 (§12.2.3) states that 0 is its own additive inverse. Since 1 is the multiplicative identity, the result for 0 suggests we check to see if 1 is its own multiplicative inverse and produces the other half of our first rule.

R1: $0 = -0$ and $1 = 1^{-1}$.

Observe:

$$1 = 1 \times 1^{-1} = 1^{-1}.$$

The first equality is due to the fact that 1^{-1} is the multiplicative inverse of 1, and so its product with 1 must be 1. The second equality is due to the fact that 1 is the multiplicative identity so its product with any real number returns that number. So 1 is also its own multiplicative inverse.

In respect to being its own additive inverse, 0 is unique; similarly, in respect to being its own multiplicative inverse, 1 is unique. All other inverses occur in pairs of two numbers.

R2: Each real number has a unique additive inverse. The reasons were given for integers (see §9.2.1) so we do not repeat those reasons here. Similarly, we might expect multiplicative inverses to be unique as well.

R3: Each non-zero real number has a unique multiplicative inverse.

We begin by fixing $x \neq 0$. Suppose that this x has two multiplicative inverses, y and x^{-1}, the latter being the one guaranteed to exist since $x \neq 0$. The following computation witnesses that y and x^{-1} are multiplicative inverses for x:

$$x \times y = 1 = x \times x^{-1}.$$

Transitivity of equality tells us that

$$x \times y = x \times x^{-1}.$$

Multiplication through this equality by x^{-1}, yields:

$$\left(x^{-1} \times x\right) \times y = \left(x^{-1} \times x\right) \times x^{-1}.$$

---

[5]Again, we remind the reader that $\in$ means *is a member of* (see §2.3.1).

Again, since $x^{-1} \times x = 1$, substitution gives

$$1 \times y = x^{-1} \times 1 = 1 \times x^{-1}.$$

Using the fact that 1 is the multiplicative identity gives $y = x^{-1}$, so there was really only one multiplicative inverse after all.

The uniqueness of multiplicative inverses is used in exactly the same manner as uniqueness of additive inverses. Readers can review §9.2.1, 9.2.2 and 9.2.3 for additional explanation and examples. To illustrate its use, we apply uniqueness to obtain the equations:

$$\left(x^{-1}\right)^{-1} = x \quad \text{and} \quad x = \left(x^{-1}\right)^{-1} = \frac{1}{\frac{1}{x}}.$$

Given $x \neq 0$, we have x^{-1} exists and

$$x \times x^{-1} = x^{-1} \times x = 1.$$

The fact that the expression on the LHS is equal to 1 tells us that x^{-1} is a multiplicative inverse for x. The fact that the expression in the middle is equal to 1 tells us that x is a multiplicative inverse for x^{-1}. Since multiplicative inverses are unique, using the standard notation for the multiplicative inverse of x^{-1}, we may write:

$$x = \left(x^{-1}\right)^{-1}.$$

The reader should compare this to $x = -(-x)$. Recall that $\frac{1}{x}$ is also a notation for the multiplicative inverse of x. In this notation we would write:

$$x = \left(\frac{1}{x}\right)^{-1} = \frac{1}{\frac{1}{x}}.$$

Recall our discussion of division in §12.3.5. There we found

$$y \div \frac{1}{x} \equiv y \times \left(\frac{1}{x}\right)^{-1} = y \times x$$

where the last step uses the previous equation. This is where the rule **invert and multiply** for division of fractions comes from. We will discuss this for common fractions in Chapter 15.

R4: $x \times 0 = 0 \times x = 0$ was previously discussed as I8 in §12.2.3.

A consequence of R4 is **R5**: if $x \times y = 0$, then either, $x = 0$ or, $y = 0$.

To see why this must be so, **assume** $x \times y = 0$ but that $y \neq 0$. Then, the reciprocal of y exists, and

$$x = x \times 1 = x \times (y \times y^{-1})$$

where the 1 in $x \times 1$ is replaced by $y \times y^{-1}$ in the first line. Applying the Associative Law to the last expression on the RHS and remembering that $x \times y = 0$ gives

$$x = (x \times y) \times y^{-1} = 0 \times y^{-1}.$$

Applying I8 (§12.2.3) to the RHS leaves

$$x = 0.$$

So one of x and y, in this case x, must be 0. We apply this fact to fractional forms by noting that $\frac{x}{y} = x \times y^{-1}$, where $y \neq 0$, whence

$$\frac{x}{y} = 0 \text{ exactly if } x = 0.$$

In short, a **fraction is zero only if the numerator is zero**.

Recall I6 (see §12.2.3):

$$(-1) \times x = -x$$

which tells us in words that we can obtain the additive inverse of any integer by multiplying it by the additive inverse of 1. For the reasons given in §9.3.4, I6 applies to all numbers. To reflect this fact, we will relabel it **R6**.

R6 has greatest application in products like I7 which asserts

$$-(x \times y) = (-1) \times (x \times y) = (-x) \times y = x \times (-y).$$

Specifically, it lets you move the centered dash anywhere in a product without changing the value. Again, I7 applies to all real numbers and is included in R6.

Recall, by the definition of the fractional notation in §12.3.5, $\frac{x}{y} \equiv x \times y^{-1}$ is a product. Applying R6 gives:

$$-\left(\frac{x}{y}\right) = \frac{-x}{y} = \frac{x}{-y}.$$

For the case when the numerator $x = 1$, we have

$$-\left(\frac{1}{y}\right) = \frac{-1}{y} = \frac{1}{-y}.$$

These formulae tell us that the additive inverse of a fractional form has the same value as the additive inverse of the entire numerator over the denominator, or the numerator over the additive inverse of the entire denominator.

R7: $(-1) \times (-1) = 1$ and $(-x) \times (-y) = x \times y$ is a consequence of R6 and was previously discussed in §9.3.3.

The next few rules again apply R3 which asserts the uniqueness of multiplicative inverses and the reader may wish to review §9.2.2 which discussed applying uniqueness in respect to additive inverses before continuing.

R8 states: suppose both x and y are not 0. Then

$$x^{-1} \times y^{-1} = (x \times y)^{-1} \quad \text{and} \quad \frac{1}{x} \times \frac{1}{y} = \frac{1}{x \times y}.$$

To obtain the first equation, since both x^{-1} and y^{-1} exist, we have

$$\begin{aligned}
(x \times y) \times (x^{-1} \times y^{-1}) &= (x \times y) \times (y^{-1} \times x^{-1}) = ((x \times y) \times y^{-1}) \times x^{-1} \\
&= (x \times (y \times y^{-1})) \times x^{-1} = (x \times 1) \times x^{-1} \\
&= x \times x^{-1} = 1.
\end{aligned}$$

Since multiplicative inverses are unique, we conclude:

$$(x \times y)^{-1} = x^{-1} \times y^{-1}.$$

In words we have the rule:

> **the multiplicative inverse of a product is the product of the multiplicative inverses.**

The second equation in R8 is obtained from the first by substitution using:

$$\frac{1}{x} \equiv x^{-1}, \quad \frac{1}{y} \equiv y^{-1}, \quad \text{and} \quad \frac{1}{x \times y} \equiv (x \times y)^{-1}.$$

Applying R8 to fractional notation gives **R9** which provides a simple rule for multiplying any fractional forms:

$$\frac{x}{y} \times \frac{w}{z} = \frac{x \times w}{y \times z}.$$

To see how, combining the second equation in R8 with the definition of fractional form followed by several applications of the Associative Law yields:

$$\begin{aligned}
\frac{x}{y} \times \frac{w}{z} &= (x \times y^{-1}) \times (w \times z^{-1}) = ((x \times y^{-1}) \times w) \times z^{-1}) \\
&= (x \times (y^{-1} \times w)) \times z^{-1}) = (x \times (w \times y^{-1})) \times z^{-1}) \\
&= ((x \times w) \times y^{-1}) \times z^{-1}) = (x \times w) \times (y^{-1} \times z^{-1}) \\
&= (x \times w) \times (y \times z)^{-1} = \frac{x \times w}{y \times z}.
\end{aligned}$$

Transitivity of equality now yields R9:

$$\frac{x}{y} \times \frac{w}{z} = \frac{x \times w}{y \times z}.$$

R10 is the special case of R9 when $w = z$: if y and z are both not 0, then

$$\frac{z}{z} = 1 \quad \text{and} \quad \frac{x \times z}{y \times z} = \frac{x}{y}.$$

To see this, simply apply the definition of the fractional notation as follows:

$$\begin{aligned}
\frac{x \times z}{y \times z} &= \frac{x}{y} \times \frac{z}{z} \\
&= (x \times y^{-1}) \times (z \times z^{-1}) = (x \times y^{-1}) \times 1 \\
&= x \times y^{-1} = \frac{x}{y}.
\end{aligned}$$

In essence we have

$$\frac{x \times z}{y \times z} = \frac{x \times \not{z}}{y \times \not{z}} = \frac{x}{y}$$

which is how fractional forms are simplified when the numerator and denominator contain a **common factor** (see Chapter 15).

We need some rules for addition of fractional forms.

R11: Given $y \neq 0$,

$$\frac{x}{y} + \frac{w}{y} = \frac{x + w}{y}.$$

Using the definition of fractional form and the Distributive Law, we have

$$\begin{aligned}
\frac{x}{y} + \frac{w}{y} &= x \times y^{-1} + w \times y^{-1} \\
&= (x + w) \times y^{-1} = \frac{x + w}{y}
\end{aligned}$$

for two forms having the same denominator. This equation provides an addition rule for fractional forms having the same denominator

$$\frac{x}{y} + \frac{w}{y} = \frac{x + w}{y}.$$

R12 provides a rule for adding fractional forms, $\frac{x}{y}$ and $\frac{w}{z}$, where y, $z \neq 0$ and are not equal:

$$\frac{x}{y} + \frac{w}{z} = \frac{x \times z + w \times y}{y \times z}.$$

To see why this rule is valid, we produce equivalent fractional forms that have the same denominator, after which we can apply R11. The computation is as follows:

$$
\begin{aligned}
\frac{x}{y} + \frac{w}{z} &= \frac{x}{y} \times \frac{z}{z} + \frac{w}{z} \times \frac{y}{y} \\
&= \frac{x \times z}{y \times z} + \frac{w \times y}{z \times y} \\
&= \frac{x \times z + w \times y}{z \times y}.
\end{aligned}
$$

Notice, the first step is to multiply both fractional forms by $1 = \frac{y}{y} = \frac{z}{z}$ so that both fractional forms end up with the same denominator, namely $y \times z$. At this point, we can just add the numerators to complete the calculation. Written in a single line the rule is

$$
\frac{x}{y} + \frac{w}{z} = \frac{x \times z + w \times y}{z \times y}.
$$

And finally **R13**, the two Cancellation laws are satisfied, namely:

1. if $x + y = x + z$, then $y = z$;

2. if $x \times y = x \times z$ and $x \neq 0$, then $y = z$.

The first is obtained by adding minus x on both sides of $x + y = x + z$, thus eliminating x. The second is obtained by multiplying both sides of $x \times y = x \times z$ by x^{-1}, again eliminating x.

12.6.1 Summary of Arithmetic Rules for Real Numbers

Together with the six axioms set out in §12.3.4, the following rules of arithmetic apply to all real numbers. (As shown above, all denominators are non-zero.)

R1 $0 = -0$ and $1 = 1^{-1}$;

R2 the additive inverse (see §12.3.5), $-x$, of any number is unique, and $-(-x) = x$;

R3 the multiplicative inverse, also referred to as the reciprocal (see §12.3.5), x^{-1}, of any non-zero number is unique, and satisfies the equations

$$
x^{-1} = \frac{1}{x}, \quad \text{and} \quad (x^{-1})^{-1} = x;
$$

R4 $0 \times x = 0$, hence 0 has no reciprocal and we cannot divide by 0;

R5 $x \times y = 0$ exactly if one of x and y is 0;

R6 $(-1) \times x = -x$, whence

$$-(x \times y) = (-x) \times y = x \times (-y)$$

and,

$$-\frac{x}{y} = \frac{-x}{y} = \frac{x}{-y};$$

R7 $(-1) \times (-1) = 1$ and $(-x) \times (-y) = x \times y$;

R8 for x and $y \neq 0$

$$x^{-1} \times y^{-1} = (x \times y)^{-1} \quad \text{and} \quad \frac{1}{x} \times \frac{1}{y} = \frac{1}{x \times y}.$$

R9 for y and $z \neq 0$

$$\frac{x}{y} \times \frac{w}{z} = \frac{x \times w}{y \times z};$$

R10 for y and $z \neq 0$

$$\frac{z}{z} = 1 \quad \text{and} \quad \frac{x \times z}{y \times z} = \frac{x}{y};$$

R11 for $y \neq 0$

$$\frac{x}{y} + \frac{w}{y} = \frac{x + w}{y};$$

R12 for y and $z \neq 0$

$$\frac{x}{y} + \frac{w}{z} = \frac{x \times z + w \times y}{y \times z};$$

R13 the Cancellation laws:

$$\text{if} \quad x + z = y + z \quad \text{then} \quad x = y,$$

and,

$$\text{if} \quad x \times z = y \times z \quad \text{and} \quad z \neq 0 \quad \text{then} \quad x = y;$$

R14 subtraction is defined by:

$$x - y \equiv x + (-y),$$

R15 if $y \neq 0$, division of x by y is defined by:

$$x \div y \equiv x \times y^{-1} \equiv \frac{x}{y}.$$

Consistent with this notation, the fractional form $\frac{x}{y}$ is defined by

$$\frac{x}{y} \equiv x \times y^{-1}.$$

12.7 How to Use the Facts of Arithmetic

In this chapter we have listed the properties, nomenclature and rules that govern the arithmetic of the real numbers. The reader should treat this information as **factual content**. The various facts fall into categories: **Axioms**, **Definitions**, **Theorems**, and **Conventions** which we describe below.

12.7.1 Axioms

Axioms are statements which are taken to be so patently obvious that they do not require justification, or what mathematicians call **proof**. The six properties listed in Section 12.3.4 are in the category of axioms. In the chapters leading up to this one, we discussed how these statements reflect what we know about the behavior of collections in the real world.[6] In this respect, arithmetic is the original experimental science capturing truths about the behavior of the world.

12.7.2 Definitions

In Section 12.3.5, we list various notations that will be used. These are, in essence, definitions. All that is happening in a definition is that a group of symbols, for example, x^{-1}, or a new word, for example, **reciprocal**, is being given an exact arithmetical meaning. Thus, in respect to x^{-1}, we are told, if $x \neq 0$, then a number denoted by x^{-1} must exist and that number satisfies:

$$x \times x^{-1} = x^{-1} \times x = 1.$$

---

[6]It is sometimes asserted by physicists that either, the Special Relativity theory developed by Einstein, or Quantum Mechanics, is the best tested and most accurate theory we have that applies to the physical world. I would argue that this description properly applies to arithmetic, since every commercial computation, scientific computation, or engineering computation, verifies the truth of arithmetic as applied to real-world activities.

The arithmetical meaning, or property, is that for a non-zero quantity, the product of that quantity and its reciprocal will be 1. No computation has to be performed to know this fact about x and its reciprocal.

We have also used the symbol $\equiv$ to define new operations. For example, the operation of division, $\div$, is given meaning with: for x, $y \in \mathcal{R}$, $y \neq 0$,

$$x \div y \equiv x \times y^{-1}.$$

12.7.3 Theorems

Theorems are arithmetic facts that are derived from the axioms. One can think of a theorem as a statement that will be satisfied in any system that satisfies the axioms. For example, the equation

$$-(x \times y) = (-x) \times y = x \times (-y),$$

which is Rule 6, is a theorem. Sometimes theorems assert that in prescribed circumstances we can draw certain conclusions. The Cancellation Law for addition (Rule 13) is like this:

$$\text{if} \quad x + z = y + z \quad \text{then} \quad x = y.$$

The prescribed condition is: if $x + z = y + z$. The conclusion is: $x = y$.

12.7.4 Conventions

Conventions involve the **application of choice**. The most important conventions affecting arithmetic as taught to children are the **order of precedence** rules. These rules tell us in what order computations are to be performed. They exist to remove ambiguity from ambiguous expressions. The simplest example is the rule that, unless otherwise indicated by parentheses, all multiplications should be performed before any additions. This means that

$$4 \times 5 + 3 = 23,$$

as opposed to computing $5 + 3$ first which leads to the incorrect calculation:

$$4 \times 5 + 3 \neq 32.$$

If we intend the latter, we have to introduce parentheses:

$$4 \times (5 + 3) = 32,$$

which require that $5 + 3$ be calculated first.

There are two rules of precedence not yet mentioned. The first states that the centered dash, $-$, applies to the **smallest quantity to its right in any expression**. Thus, in the expression

$$x - y + z,$$

the centered dash applies only to the y so that this expression properly interpreted is

$$x - y + z = x + (-y) + z.$$

Thus, in the following numerical example,

$$6 - 3 + 2 = 5, \quad \text{not} \quad 1.$$

To obtain 1, we have to use parentheses as follows

$$6 - (3 + 2) = 1.$$

The second precedent rule states that the exponent -1 apples to the smallest quantity occurring to its left in any expression. To see what we mean consider $x \times y^{-1}$. This expression produces

$$x \times y^{-1} = \frac{x}{y}$$

from which we see that the exponent -1 applies only to the y and not to the entire product which would yield an incorrect calculation

$$x \times y^{-1} \neq \frac{1}{x \times y}.$$

To force the exponent to apply to the entire product, we must use parentheses as follows:

$$(x \times y)^{-1} = \frac{1}{x \times y}.$$

Another convention is the choice of base for the numeration system. We have chosen base 10. However, computers choose to operate in base 2. One effect of this choice for computers is that $1 + 1 = 10$ which changes the name of the successor of 1, but not the fact that this number is the successor of 1.

What is critical to note is that in making these choices, the value of $1 + 1$ has not changed; it is still the successor of 1. Only the symbols we use to identify this number have changed. This is why these are conventions, not fundamental truths. That said, these **conventions must be carefully followed in order to correctly perform computations**.

12.7.5   Using Particular Rules

In the remainder of this chapter, we will use the various rules to develop procedures for working with fractions. And in Chapter 17, the rules will be used to develop procedures for dealing with decimals. Before doing so we want to go over how these rules are to be used.

Each of the rules is an instruction. Let's consider **Rule 10** which asserts: given y and $z \neq 0$,

$$\frac{z}{z} = 1 \quad \text{and} \quad \frac{x \times z}{y \times z} = \frac{x}{y}.$$

Each equation has occurrences of various letters, as in

$$\frac{z}{z} = 1$$

which is simple because it only has one letter, namely z. To use this equation, we simply replace each occurrence of z by the same quantity, so long as it is not 0. The quantity being substituted for z can be a number, a combination of numbers, or an algebraic expression. Valid forms to be substituted would be:

$$5, \ -2.3, \ 5 \times w, \ \sqrt{w-7},$$

and so forth. There is one constraint in respect to z; we must know that the quantity is not zero. There may also be other constraints. For example in the present case, we need that $w - 7 \geq 0$, otherwise $\sqrt{w-7}$ will not be a real number. Given the constraints are satisfied, we know with surety that

$$\frac{5}{5} = \frac{-2.3}{-2.3} = \frac{5 \times w}{5 \times w} = \frac{\sqrt{w-7}}{\sqrt{w-7}} = 1.$$

No computation is necessary to replace anything on the LHS by 1.

Consider the second equation in Rule 10. Once again each letter can be replaced by any quantity. Here the first restriction is that if we put a quantity in for x, we must put that same quantity in for all other x's in the equation. The same is true for y and z. So the result of replacing both x's by $a + 2$ would be:

$$\frac{(a+2) \times z}{y \times z} = \frac{a+2}{y}.$$

Another, correct substitution, which sets $x = \sqrt{w-7}$, $y = 5$ and $z = (s+4)^{-1}$ where $s + 4 \neq 0$, is:

$$\frac{\sqrt{w-7} \times (s+4)^{-1}}{5 \times (s+4)^{-1}} = \frac{\sqrt{w-7}}{5}.$$

Again, no calculation is required to get from the LHS to the RHS. This is why these rules are so powerful. But, in applying them one must be meticulous in verifying that the form you have for a LHS really can be obtained by substitution from the generic form in the rule and really does satisfy any restrictions, for example, in substituting for z in R10 $s + 4 \neq 0$.

12.8 What Your Child Needs to Know

This chapter has presented the basic rules of arithmetic. Every child should be fully able to apply these rules with facility by the end of Grade 7. If your child succeeds in this, he or she will have no difficulty with the mathematics component of any course of study or career that she or he might choose to pursue. The next chapter deals with numerical computations with fractions and I will make some further observations teaching these to children there. Reading through the grade-by-grade goals should provide guidance in respect to the rate at which you should expect your child to come to terms with these ideas.

12.8.1 Goals for Grade 1

The child should be able to

1. partition circles and rectangles into two and/or four equal shares;

2. correctly use the words *half, quarter, fourths* to describe the portions;

3. correctly use the phrases *half of, quarter of, fourth of* to describe the portions in respect to the whole;

4. understand that combining two halves (four quarters) physically restores the whole.

12.8.2 Goals for Grade 2

The child should be able to

1. measure the length of an object;

2. understand that the result depends on the units, e.g., centimeters vs inches vs feet vs meters;

3. measure to determine how much longer one object is than another;

4. use addition and subtraction to relate (compare) lengths of objects in common units;

5. partition rectangles into three equal parts and correctly use *thirds* and *third of*.

12.8.3 Goals for Grade 3

The child should be able to

1. understand the fraction $\frac{1}{n}$ partitions a whole into n equal parts, e.g., $\frac{1}{15}$ partitions a whole into fifteen equal parts called *fifteenths*;

2. understand the fraction $\frac{m}{n}$ as being the quantity obtained by taking m parts of size $\frac{1}{n}$, e.g., $\frac{3}{n} = 3 \times \frac{1}{n} = \frac{1}{n} + \frac{1}{n} + \frac{1}{n}$;

3. represent the fraction $\frac{1}{n}$ by dividing a unit length into n equal parts;

4. explain the equivalence of fractions in special cases, e.g., $\frac{1}{2} = \frac{2}{4}$ and $\frac{2}{3} = \frac{4}{6}$, both computationally and using a visual model;

5. express counting numbers as fractions, e.g., $5 = \frac{5}{1}$ or $8 = \frac{8}{1}$;

6. understand the concept of a **unit of area** and partition geometric shapes into equal parts based on equal areas.

12.8.4 Goals for Grade 4

The child should be able to

1. explain computationally why $\frac{n}{m} = \frac{n \times p}{m \times p}$ (See §12.5.1 R10);

2. apply previous understanding of multiplication, i.e., $\frac{m}{n} = m \times \frac{1}{n}$, to multiply fractions by whole numbers;

   (a) fully understand $p \times \frac{m}{n} = \frac{p \times m}{n}$, computationally;

   (b) fully understand $p \times \frac{m}{n} = \frac{p \times m}{n}$, visually, e.g., $3 \times \frac{2}{5}$ is equivalent to $(3 \times 2) \times \frac{1}{5}$;

   (c) solve word problems involving multiplication of fractions by whole numbers using equations and visual models. For example, if each child at a party drinks $\frac{2}{3}$ of liter of lemonade and five children are attending the party, how much lemonade must be made?;

12.8.5   Goals for Grade 5

The child should be able to

1. understand that a correct interpretation of a fraction is as division of the numerator by the denominator, i.e., $\frac{m}{n} = m \div n$.

12.8.6   Goals for Grade 6

The child should be able to

1. write, read, evaluate and interpret algebraic expressions in which letters stand for numbers;

   (a) write expressions that record numerical operations. e.g., subtract 5 from x as $x - 5$;

   (b) identify parts of a mathematical expression using descriptors like sum, product, term, factor, quotient. e.g., in the expression

   $$4(7 + 3),$$

   4 is a factor, $7 + 3$ is a sum composed of two terms, the entire expression is a product;[7]

   (c) evaluate expressions for specific values of the variables, e.g., find the value of A given $A = (s + 2) \times t$ and $s = 3$ $t = 7$;

2. apply the essential properties to generate equivalent algebraic expressions, e.g., $5(2 + 3x)$ becomes $10 + 15x$ by applying the Distributive Law, $2y + 4$ becomes $(y + 2)2$ again using the Distributive Law;

3. identify when two expressions are equivalent, e.g., $x + 2y$ and $(x + y) + y$ can be identified as equivalent using Associative and Distributive Laws;

4. understand that solving an equation for an unknown is a process of determining whether a particular value from a specified set substituted for the unknown will make the equation true;

---

[7]This type of material introduces your child to algebra which is not really covered in this book. However, the quantities manipulated in algebra are real numbers, so they must behave using the rules we have developed. For this reason, to help your child with this type of material, using the grounding provided by this book, all you have to do is to go through the associated material in your child's text book and you should have no trouble figuring out what is required.

5. solve equations of the form $x + p = q$ and $px = q$ for cases in which p, q and x are non-negative rational numbers;

6. understand ratio and proportion in relation to fractions, e.g., if the ratio of chairs to tables is four to one and we have 80 chairs, what fraction should we multiply by 80 to determine the number of tables?;

7. understand rates as a result of a division process, e.g., speed is distance divided by time;

8. use two variables to represent quantities in real world problems, e.g., d and t to represent time and distance in $d = 4t$ where 4 is speed in appropriate units.

12.8.7   Goals for Grade 7

The child should be able to

1. understand that for any pair of rational numbers p and q, $p - q = p + (-q)$ where $-q$ is the additive inverse of q;

2. understand that the addition and multiplication properties of the rationals arise by extending the Commutative, Associative and Distributive Laws to all numbers;

3. know and be able to use the fact that $(-1)(-1) = (-1) \times (-1) = 1$;

4. know that $(-p) \times q = p \times -q$ and more generally

$$-\left(\frac{p}{q}\right) = \frac{-p}{q} = \frac{p}{-q};$$

5. use the rules of arithmetic to generate equivalent linear algebraic expressions $\frac{2x}{3} = \frac{2}{3} \times x$;

   (a) add linear algebraic expressions with rational coefficients, e.g., $2x + 5x = 7x$;

   (b) subtract algebraic expressions with rational coefficients;

   (c) factor algebraic expressions with rational coefficients, e.g., $\frac{2x}{3} + 3x = \left(\frac{2}{3} + 3\right)x$;

   (d) multiply algebraic expressions with rational coefficients.

Chapter 13

Fractions for Kids: I

Chapter Overview. The concept of **unit fraction** is discussed. It is used to develop the more general concept of common fraction. The notation for fractions is carefully explained and related to the Notation Equation. The Fundamental Equation is discussed. These two equations will become the basis for computations with fractions. It is shown how to position a fraction in the unit interval. CCSS-M goals for Grades 1–3 are given.

The last chapter developed the essential rules of arithmetic that govern the use of all numbers. In Chapters 13-15 we develop the arithmetic of **common fractions** in a manner that is accessible to kids in elementary school. Our focus in this chapter is on the portion of that work that occurs in Kindergarten to Grade 3.

For those who don't recall, the **common fractions** are fractional forms having an integer in the numerator and a non-zero integer in the denominator. Numbers of this form considered as a set are referred to as the **rational numbers** and denoted by $\mathcal{Q}$ as discussed in §12.5. In primary and elementary, the discussion of these fractions is confined to common fractions where both the numerator and the denominator are counting numbers.

The topic of fractions is considered the hardest in the arithmetic curriculum. It consists of two separate parts, namely,

1. What are fractions as numbers?

2. How do we perform calculations with fractions?

School mathematics curricula generally do a reasonable job explaining the former, and a terrible job, in the author's opinion, with the latter. This assertion is based on many years of teaching students having weak computational backgrounds as a result of inadequate instruction and curricula in the early grades.

In this chapter, we concentrate on the first question, emphasizing the essential properties of fractions whose understanding is demanded by the CCSS-M. Beginning computational procedures are discussed in Chapter 14 and advanced procedures in Chapter 15. The material is presented at a level that will enable you to help your child succeed with fractions, even if fractions were a total mystery to you at the end of your school years.

13.1 Numbers That Are Not Whole

The first important fact is that every identifiable place on the number line, first introduced in §10.4, is associated with a number. This fact is illustrated in the following diagram.

> A graphical description of the real numbers. Notice the vertical arrows that identify places on the line not associated with an integer. Each such place identifies a real number. This line is referred to as the **real line**.

That **every identifiable place on the line is associated with a unique number** is a fact that your child must know. A great way to study this idea is to practice making measurements of the length of real objects. This activity begins in Kindergarten and the child must come to terms with the fact that not every object has a length which is a whole number. The website *softschools* has worksheets organized by grade level to help your child with this activity.

13.2 Unit Fractions

Unit fractions are numbers that result by the process of dividing 1 into n equal parts, where n is a counting number. They are the foundation on which fraction arithmetic is based, and are introduced in Kindergarten or Grade 1 for small values of n, for example 2 and 4. To make this idea concrete, we take the **unit interval**, that is, the portion of the number line having 0 at the left and 1 at the right and divide it into some counting number of equal parts. For example, in the following diagram we take $n = 5$.

The vertical arrows subdivide the unit interval into five equal, non-overlapping parts. Each of the parts is identical to every other in respect to **length**.

The essential fact that each child must understand about this diagram is that the number associated with the length of each interval between the vertical arrows is the same. In respect to the attribute of length, **each of the five subintervals is identical to every other**.

We illustrate this idea again by subdividing the unit interval into $n = 8$ identical parts.

The vertical arrows subdivide the unit interval into eight equal, non-overlapping parts. Each place of division identified by a vertical arrow is associated with a unique real number. Each horizontal arrow has the same length.

The intrinsic idea expressed in these diagrams is simple: given one whole of something, we can divide that something into any counting number of equal parts. Since the parts all have the same size, **each one** of these parts can be described by the same number. We stress that the process of dividing results in parts that are **indistinguishable**.

We illustrate this concept again using the numerical attribute of area. The box in the diagram has unit area and we divide it into 16 equal parts, as shown.

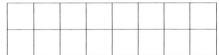

A rectangle of unit area subdivided into 16 equal parts.

We stress that in respect to the attribute of area, each part is identical to every other. They are indistinguishable.

13.2.1 Naming Unit Fractions

Recall our discovery in Chapter 5 that for counting numbers to be useful, they had to have names. Thus for unit fractions to be useful, they must have names as well.

Unit fractions arise by subdividing a unit whole into some counting number of equal parts. Our first two examples involved subdividing the unit interval into 5 and 8 parts, respectively. Consider the following diagram.

The down-pointing arrows subdivide the unit interval into eight equal, non-overlapping parts. The up-pointing arrows subdivide the unit interval into five equal, non-overlapping parts.

Notice that each of the 4 up-pointing arrows subdividing the unit interval into five equal parts identifies a different place on the line from the 7 down-pointing arrows. Using the principle stated at the beginning of this section that associated with every identifiable place on the line there is a unique number, we know that each of the points on the line identified by an arrow can be labelled with a unique real number. Moreover, because the numbers must be unique, the names for the numbers associated with places identified in the diagram by down-pointing arrows for $n = 8$ must be different from those identified in the diagram by up-pointing arrows for $n = 5$.

The notational answer that was arrived at was that if we have one whole, and we divide that whole into 5 equal parts, we will use the combined symbols

$$\frac{1}{5}$$

as the notation for the number associated with each of the identical parts resulting from this division process. Thus, the 1 on top tells us we are dealing with one whole entity, and the 5 on the bottom tells us how many equal parts this whole is divided into. Similarly, if we divide 1 whole into 8 equal parts, we would use the combined symbols

$$\frac{1}{8}$$

where the 1 tells us that 1 is being subdivided and the 8 tells us that the 1 is divided into 8 equal parts. And in the case of the area diagram where $n = 16$, we would use

$$\frac{1}{16}.$$

Further examples of the notation for such numbers are:

$$\frac{1}{2}, \ \frac{1}{3}, \ \frac{1}{7}, \ \frac{1}{17}, \ \frac{1}{25}, \ \frac{1}{132}, \ \frac{1}{845}, \ \frac{1}{9847}$$

and so forth. In general, for any **counting number** $n > 1$, as in the numerical cases above, we will write

$$\frac{1}{n}$$

271

to signify the number that corresponds to the result of dividing 1 into n equal parts and refer generally to numbers of this type as **unit fractions**. Their distinguishing feature is a 1 in the numerator. It is expected that children in Grade 3 will be fully conversant with the meaning and notation for unit fractions and be able to correctly use their names as in: *one fourth*, *one fifteenth* and *one one-hundred and twenty-fifth* to speak of $\frac{1}{4}$, $\frac{1}{15}$ and $\frac{1}{125}$, respectively.

Most importantly, children will recognize that when a whole is divided into some counting number of equal parts, the resulting parts are **indistinguishable** in respect to the numerical attribute targeted by the division. For example, that attribute might be length, area, volume, weight, etc.

13.3   The Effect of Indistinguishable

Let's explore why it is important to recognize that the process of subdividing a whole into equal parts results in parts that are indistinguishable with respect to a target numerical attribute. Consider the following problem which is typical of many beginning work sheets on fractions:

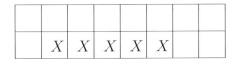

Write the fraction that represents the portion of the total area of the rectangle that is associated with squares marked with an X.

First, it is expected the child will recognize that the underlying attribute being considered is **area**. To solve the problem the child must determine the number of equal parts into which the whole is being divided. This number is found by counting and, as the reader can check, is 16. Counting also informs us that a total of 5 parts are marked with an X. At this point the child can mechanically write an answer, namely, $\frac{5}{16}$, which is the standard notation for this common fraction.

However, the child needs to understand that because each of the squares (parts) has the same area, the total area associated with 5 squares will be the same however these squares are chosen. So in the following diagram the squares marked with Y have the same total area as the squares marked with X in the diagram above.

			Y		Y		Y
						Y	Y

Write the fraction that represents the portion of the total area of the rectangle that is associated with squares marked with an Y. Understand that this fraction must be the same as the fraction identified by squares marked with an X in the diagram above.

Thus, the numerical measure of the portion of the area associated with Y—squares is the same as for X—squares. Indeed, any way we choose 5 squares from the 16 will result in this same portion of the area, albeit from a different physical part. This is the essential consequence of indistinguishability.

But there is another step that is even more critical, namely that counting corresponds to addition. In this case we are counting unit fractions. So the total has to be given by the following sum:

$$\frac{1}{16} + \frac{1}{16} + \frac{1}{16} + \frac{1}{16} + \frac{1}{16}.$$

Let's consider a second example using a different numerical attribute, length. In the physical example of unit fractions given in the last section, we asserted that when we divide the unit interval into equal parts, each part will have the same length. In the diagram that follows the unit interval is divided into six equal parts, 3 of which are marked with horizontal arrows. We know that each of these horizontal arrows that starts at a vertical arrow and ends at the next vertical arrow to the right will have length $\frac{1}{6}$ when measured in the same units as the unit interval. Using counting as above, we also know that the fractional part of the length of the unit interval associated with the 3 horizontal arrows is $\frac{3}{6}$ in standard notation. This time we frame a different question:

Is there any way to choose 3 of the subintervals that will produce a different total portion of the length?

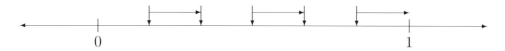

The vertical arrows subdivide the unit interval into six equal, non-overlapping parts. Each horizontal arrow has length $\frac{1}{6}$ (see text for discussion).

The answer is: No. And the reason is that each of the subintervals is indistinguishable in respect to its length. Further, because counting and addition correspond, the resulting fractional part of the total length must be the sum of the unit fractions measuring length of each subinterval identified by a horizontal arrow, namely,

$$\frac{1}{6} + \frac{1}{6} + \frac{1}{6}.$$

We will explore this idea in the next several sections.

13.4 The Fundamental Equation

Forming a unit fraction in the context of the real world is intended to divide a whole into some number of indistinguishable parts. When these parts are brought back together it should restore the whole. For example, if we divide a pie into two indistinguishable parts, when we bring those parts back together we should end up with a whole pie.

In respect to subdividing the interval into eight parts of equal length, we have the following diagram.

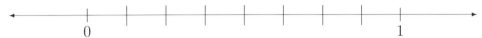

The vertical lines subdivide the unit interval into eight equal, non-overlapping parts.

Bringing the equal parts together in this context should amount to nothing more than removing the lines of subdivision. When we do this we end up with a diagram of the unit interval.

The vertical lines subdividing the unit interval into eight equal, non-overlapping parts have been removed restoring the unit interval.

We want to think about how to capture this idea with our arithmetic model. What we know is that when we bring all the subdivided parts of the whole back together, it restores the whole. This is an instance of **conservation**.

We can express this idea with an addition equation. For example if we subdivide a pie into 2 equal parts, each described by the unit fraction $\frac{1}{2}$, the equation

$$1 = \frac{1}{2} + \frac{1}{2}$$

274

expresses the result of bringing those two parts back together.

If we think instead of the unit interval divided into 8 parts, the relevant equation would be

$$1 = \frac{1}{8} + \frac{1}{8} + \frac{1}{8} + \frac{1}{8} + \frac{1}{8} + \frac{1}{8} + \frac{1}{8} + \frac{1}{8}.$$

This equation is visually represented by the following diagram.

The vertical lines subdivide the unit interval into eight equal, non-overlapping parts. Each horizontal arrow has length $\frac{1}{8}$. A total of 8 arrows placed end-to-end and directed left-to-right beginning at 0 will terminate at 1. In the context of the last section, we could ask: What fraction of the length of the unit interval is associated with all subintervals identified by a horizontal arrow?

The last diagram is completely consistent with the interpretation of addition and the number line presented in §10.4. In answer to the question posed, we can say when all the subunits are identified as in the diagram above, we must have the whole and this should be evident to children who have experience with the sorts of worksheets on fractions based on counting that are typical of K-Grade 2.

The deeper point comes by realizing that counting unit fractions having the same denominator corresponds to addition. Thus, given unit fractions, like $\frac{1}{2}$, $\frac{1}{3}$, $\frac{1}{4}$, $\frac{1}{5}$, and so forth, each satisfies an equation of the form

$$
\begin{aligned}
1 &= \frac{1}{2} + \frac{1}{2} \\
1 &= \frac{1}{3} + \frac{1}{3} + \frac{1}{3} \\
1 &= \frac{1}{4} + \frac{1}{4} + \frac{1}{4} + \frac{1}{4} \\
1 &= \frac{1}{5} + \frac{1}{5} + \frac{1}{5} + \frac{1}{5} + \frac{1}{5}
\end{aligned}
$$

where the number of unit fractions contributing to the sum on the RHS is exactly the number given in the denominator of the fraction. **Every child should know and understand why these equations are true by the end of Grade 3**. This understanding should be evidenced by an ability to produce a diagram, like the one above for *eighths*, that corresponds to each equation.

13.5 Repetitive Addition is Multiplication

In §8.1 we defined multiplication as repetitive addition. In §9.3.1 and 12.5 we discussed how this definition is enforced by the Distributive Law. It is critical that our readers remember that when we multiply any real number by a counting number the effect is adding that real number to itself however many times are specified by the counting number. Thus,

$$4 \times x = x + x + x + x$$

where x stands for any real number whatsoever. But this equation can be read the other way. It also says that

$$x + x + x + x = 4 \times x,$$

that is, whenever we add the same real number to itself four times, we get $4 \times x$.

The sum on the RHS of an equation like

$$1 = \frac{1}{7} + \frac{1}{7} + \frac{1}{7} + \frac{1}{7} + \frac{1}{7} + \frac{1}{7} + \frac{1}{7}$$

is repetitive addition. The same number, in this case $\frac{1}{7}$ is being added to itself 7 times. So this equation translates to

$$1 = \frac{1}{7} + \frac{1}{7} + \frac{1}{7} + \frac{1}{7} + \frac{1}{7} + \frac{1}{7} + \frac{1}{7} = 7 \times \frac{1}{7}.$$

We have made the point previously that what enforces the fact that multiplication is repetitive addition is the Distributive Law.

The CCSS-M expect every child in Grade 3 to understand the computations in their numerical form, so we write it out for the equation involving $\frac{1}{3}$. Recall the $3 \equiv 1 + 1 + 1$ and $1 \times \frac{1}{3} = \frac{1}{3}$. Using these facts we have,

$$
\begin{aligned}
3 \times \frac{1}{3} &= (1 + 1 + 1) \times \frac{1}{3} \\
&= 1 \times \frac{1}{3} + 1 \times \frac{1}{3} + 1 \times \frac{1}{3} \\
&= \frac{1}{3} + \frac{1}{3} + \frac{1}{3} = 1
\end{aligned}
$$

The second line results from the Distributive Law, and it is essential that children understand that we do not need parentheses in this line because multiplication must be performed before addition. The last line is obtained because 1 is the multiplicative identity, and all children should know this fact. That the last sum must be 1 is conservation.

In Chapter 12, we have expressed this fact as the Fundamental Equation (see 12.3.2) which is stated using multiplication as

$$1 = n \times \frac{1}{n} = \frac{1}{n} \times n,$$

where n is any counting number. In that context, the number $\frac{1}{n}$ was referred to as a **multiplicative inverse** of n. The product at the far right is also 1 due to the Commutative Law and is included to emphasize that the product of a number and its multiplicative inverse computed in either order results in 1.

13.5.1 Unit Fractions and Repetitive Addition in General

Let us fix our attention on a single unit fraction. Rather than take a specific numerical value for this fraction, we simply identify it by its denominator which we take to be the counting number n. Suppose we have some number of these unit fractions and we want to add them up. As discussed in the previous section in respect to *sevenths* and *thirds*, adding the fraction $\frac{1}{n}$ to itself repetitively comes down to multiplication by another counting number. The number that we multiply by is simply the number of times $\frac{1}{n}$ is added to itself and could be any counting number whatsoever. It does not have to be n as in the discussion in respect to the Fundamental Equation. The fact that repetitive addition is multiplication applies in this situation equally well. Specifically, for counting numbers from 2 to 7, using the fact that repetitive addition is multiplication gives us the following equations

$$
\begin{aligned}
2 \times \frac{1}{n} &= \frac{1}{n} + \frac{1}{n} \\
3 \times \frac{1}{n} &= \frac{1}{n} + \frac{1}{n} + \frac{1}{n} \\
4 \times \frac{1}{n} &= \frac{1}{n} + \frac{1}{n} + \frac{1}{n} + \frac{1}{n} \\
5 \times \frac{1}{n} &= \frac{1}{n} + \frac{1}{n} + \frac{1}{n} + \frac{1}{n} + \frac{1}{n} \\
6 \times \frac{1}{n} &= \frac{1}{n} + \frac{1}{n} + \frac{1}{n} + \frac{1}{n} + \frac{1}{n} + \frac{1}{n} \\
7 \times \frac{1}{n} &= \frac{1}{n} + \frac{1}{n} + \frac{1}{n} + \frac{1}{n} + \frac{1}{n} + \frac{1}{n} + \frac{1}{n}.
\end{aligned}
$$

What is critical to absorb from this sequence of equations is that every sum of unit fractions **having the same denominator** can be written as a product and determining the multiplier is simply a matter of counting!

13.6   The Equations and the Notation for Fractions

To really understand the effect of the equations in the last section, let's consider again the typical worksheet problem discussed in §13.1.3.

> Write the fraction that represents the portion of the total area of the rectangle that is associated with squares marked with an X.

As discussed above, the solution is obtained by counting the squares having an X to determine the numerator of the fraction, counting the total number of squares to determine the denominator, and writing the result in standard form. So the child would write the result as $\frac{5}{16}$ as a direct result of counting, just like counting identical buttons. This process is **mechanical**. It assigns a piece of notation to a process, but tells us nothing about the arithmetic properties of the number the notation represents, any more than the symbol 2 tells us anything about the number it identifies. In the case of 2 it is only when we say that 2 names the successor of 1, or write the equation $2 = 1 + 1$ that we know what number is being named by 2.

It is only when we connect counting to addition at the next level of abstraction that the fraction $\frac{5}{16}$ gains arithmetic meaning. The equation that provides this meaning is:

$$\frac{5}{16} = \frac{1}{16} + \frac{1}{16} + \frac{1}{16} + \frac{1}{16} + \frac{1}{16}$$

and informs us we should think about $\frac{5}{16}$ as a sum of unit fractions all of which have the same denominator, namely, 16. This idea is reinforced by saying the name: *five sixteenths* and thinking to oneself that we have 5 unit fractions each of which is $\frac{1}{16}$.

But there is an even higher level of abstraction that arises when we recognize repetitive addition as multiplication. With this recognition we can write

$$\frac{5}{16} = 5 \times \frac{1}{16}$$

and we are told we should really think of $\frac{5}{16}$ as a product.

The CCSS-M demand that every child should understand these two equations and the relation between them because, as we shall see, comfort with the numerical form of the two equations as in

$$\frac{5}{16} = \frac{1}{16} + \frac{1}{16} + \frac{1}{16} + \frac{1}{16} + \frac{1}{16},$$

and

$$\frac{5}{16} = 5 \times \frac{1}{16}$$

is critical to the child's success with common fractions. To come to terms with equations like these, the child must completely understand that

1. unit fractions represent distinct numbers determined by their denominator;

2. combining unit fractions having the same denominator is like counting identical buttons;

3. counting unit fractions having the same denominator amounts to repetitive addition of the unit fraction;

4. that because repetitive addition corresponds to multiplication, adding the unit fraction $\frac{1}{n}$ to itself m times is the same as multiplying the unit fraction by m.

One way to teach these ideas to young children is to get them to think of the notation $\frac{7}{8}$ in terms of a jar which contains seven items, each of which is the unit fraction $\frac{1}{8}$.

13.7   Notation for Common Fractions

Common fractions can have any integer in the numerator. Since negative numbers are not introduced until Grade 6, most fractions discussed in elementary school are composed of two counting numbers as in

$$\frac{4}{7}, \quad \frac{3}{5}, \quad \frac{2}{8}, \quad \frac{6}{15}, \quad \frac{8}{10}, \quad \frac{5}{24},$$

and so forth. It is expected that children in Grade 3 will know that equations like

$$\frac{4}{5} = 4 \times \frac{1}{5}$$

are valid. This may result from treating this equation as a definition, as we explained in §12.1.3, or it may result in viewing $\frac{4}{5}$ as resulting from

$$\frac{4}{5} = \frac{1}{5} + \frac{1}{5} + \frac{1}{5} + \frac{1}{5},$$

but either way the first equation is satisfied. What is essential is that by Grade 3 children should generally come to recognize every fractional form as a product as in

$$\frac{m}{n} = m \times \frac{1}{n}$$

because this recognition is critical to correctly performing and understanding computations. This recognition is so important that we give this equation a special name: **Notation Equation**. The reason we stress this equation's importance is that it is only when fractions are written as products that we can apply the full power of the Associative, Commutative and Distributive Laws to make computations with common fractions easy. So this idea needs to be carefully explained to children.

Referring to $\frac{1}{n}$ as the **reciprocal** of n is appropriate in Grade 3, but the introduction of negative exponents in the form n^{-1} has to wait until Grade 6.

In a fraction, for example $\frac{m}{n}$, the number appearing below the line, in this case, n, is called the **denominator**. The number appearing above the line, in this case, a m, is called the **numerator**. The terms denominator and numerator apply to all fractional forms, not merely common fractions.

As already noted, given specific numerical values for n, we have standard names for unit fractions. Thus, we write $1/2$, and say, one half; we write $1/3$, and say, one third; we write $1/4$, and say, one fourth; we write $1/5$, and say, one fifth; skipping on, we write $1/10$, and say, one tenth; we write $1/27$, and say, one twenty-seventh; and so forth. These names are introduced starting in Grade 1.

13.8   Placing Proper Fractions in the Unit Interval

Let's see how to apply these relations to placing fractions on the line.

To make things concrete, we work with the unit interval subdivided into five parts of equal length. We know that each of the subintervals has a length of $\frac{1}{5}$ and also that

$$1 = \frac{1}{5} + \frac{1}{5} + \frac{1}{5} + \frac{1}{5} + \frac{1}{5}$$

which physically corresponds to placing five horizontal arrows end-to-end starting at 0 and terminating at 1. This last relation, between the arrows and the equation, forces us to identify as $\frac{1}{5}$ the number corresponding to a horizontal arrow starting at 0 and terminating at the first vertical arrow as shown in the following diagram.

The vertical arrows subdivide the unit interval into five equal, non-overlapping parts. The horizontal arrow identifies the position of $\frac{1}{5}$ as shown.

Any other choice for the name of the number identified by the left-most vertical arrow will be inconsistent with the Fundamental Equation.

If we take two horizontal arrows, each of length $\frac{1}{5}$, and place them end-to-end, the total length is given by the sum of the lengths and we can write this as a product

$$\frac{1}{5} + \frac{1}{5} = 2 \times \frac{1}{5} = \frac{2}{5}.$$

Since addition can be represented on the line by placing arrows end-to-end starting at 0, we represent this sum as follows.

The horizontal arrows each have length $\frac{1}{5}$. As discussed in the text and §10.4.1, $\frac{2}{5} = 2 \times \frac{1}{5} = \frac{1}{5} + \frac{1}{5}$ must be the number associated with the second vertical arrow to the right of 0.

In a similar manner, we conclude that the remaining places on the line must identify the following numbers as shown below.

We have now found a number to go with each place identified by vertical arrows dividing the unit interval into 5 equal parts.

Most importantly, we understand that the determination of which number is associated with a particular place on the line is determined by the arithmetic relationships:

$$\frac{k}{5} = k \times \frac{1}{5} \quad \text{and} \quad 1 = 5 \times \frac{1}{5}.$$

There is nothing special about the unit fraction $\frac{1}{5}$. Any unit fraction can be placed on the line using the steps taken above. For example, for the counting number 6, first divide the unit interval into 6 equal parts and identify the points of division. Second, label the left-most point of division with $\frac{1}{6}$. Label the remaining points of division using repetitive sums of $\frac{1}{6}$.

13.9   What Your Child Needs to Know

The material in this chapter is covered in Kindergarten to Grade 3. For this reason, the goals for Grade 4 and beyond are left to the next chapter.

13.9.1   Goals for Grade 1

The child should be able to

1. partition circles and rectangles into two and/or four equal shares;[1]

2. correctly use the words *half, quarter, fourths* to describe the portions;

3. correctly use the phrases *half of, quarter of, fourth of* to describe the portions in respect to the whole;

4. understand that combining two halves (four quarters) physically restores the whole;

5. understand that dividing a whole into fourths, as opposed to halves, produces smaller shares.

13.9.2   Goals for Grade 2

The child should be able to

1. measure the length of an object;[2]

    (a) understand that the result depends on the units, e.g., centimeters vs inches vs feet vs meters;

    (b) measure to determine how much longer one object is than another;

    (c) use addition and subtraction to relate (compare) lengths of objects in common units;

2. partition rectangles into rows and columns of equal size squares and count to find the total numbers;

3. partition rectangles into three equal parts and correctly use *thirds, third of*;

4. recognize that equal portions of identical wholes need not have the same shape; for example, in a 3 by 2 partition of a rectangle into squares, any three adjacent squares have the same share of the area, although not necessarily the same shape.

---

[1]The **math-aids** website has visual material to help children conceptually. The **softschools** website has interactive fraction games. The **superkids** website has worksheets at all levels.

[2]The **softschools** website has measurement worksheets by grade level. These can be helpful in learning about the real line, fractions and decimals.

13.9.3   Goals for Grade 3

The child should be able to

1. understand the fraction $\frac{1}{n}$ partitions a whole into n equal parts, e.g., $\frac{1}{15}$ partitions a whole into fifteen equal parts called *fifteenths*;

2. understand the fraction $\frac{m}{n}$ as being the quantity obtained by taking m parts of size $\frac{1}{n}$, e.g., $\frac{3}{n} = 3 \times \frac{1}{n} = \frac{1}{n} + \frac{1}{n} + \frac{1}{n}$;

3. understand a fraction as a number on the number line;

4. represent the fraction $\frac{1}{n}$ by dividing a unit length into n equal parts;

5. represent fractions on a number line diagram;

   (a) represent a fraction $\frac{1}{n}$ on a number line diagram by defining the interval from 0 to 1 as the whole and partitioning it into n equal parts and recognize that each part has size $\frac{1}{n}$, and that the endpoint of the part based at 0 locates the number $\frac{1}{n}$ on the number line;

   (b) represent a fraction $\frac{m}{n}$ on a number line diagram by marking off m lengths $\frac{1}{n}$ from 0;

   (c) recognize that the resulting interval has size $\frac{m}{n}$ and that its endpoint locates the number $\frac{m}{n}$ on the number line;

6. understand two fractions as equivalent (equal) if they are the same size, or locate the same point on a number line;

7. explain the equivalence of fractions in special cases, e.g., $\frac{1}{2} = \frac{2}{4}$ and $\frac{2}{3} = \frac{4}{6}$, both computationally and using a visual model;

8. express counting numbers as fractions, e.g., $5 = \frac{5}{1}$ or $8 = \frac{8}{1}$;

9. understand that equivalent expressions such as, $5 = \frac{5}{1}$ or $8 = \frac{8}{1}$, identify the same place on the real line;

10. understand the concept of a **unit of area** and partition geometric shapes into equal parts based on equal areas.

Chapter 14

Fractions for Kids: II

The last chapter focussed on the question:

What are fractions?

In this chapter our focus will be on how to compute with common fractions. §14.2 and 14.3 present straightforward methods, easily taught to children, for performing computations that **always work**. The CCSS-M expects facility with these rules to be achieved during Grades 4 and 5. Complete knowledge of why these methods work is founded in the material in Chapter 12. Complete understanding as to why these rules work is not expected before Grade 7, but is a continuing process that starts when fractions are introduced in Grades 1-3. For this reason, we will provide an explanation why the procedures work as they are developed.

14.1 The Key Equations

In Chapters 12 and 13 two essential equations were developed, the **Fundamental Equation**:

$$n \times \frac{1}{n} = \frac{1}{n} \times n = 1$$

and the **Notation Equation**:

$$\frac{m}{n} = m \times \frac{1}{n}.$$

The Fundamental Equation asserts that the unit fraction with denominator n is the **multiplicative inverse** (see §12.3.3) of the counting number n which means their product must be 1. Why the Fundamental Equation must be a fact about common fractions was explained in §13.4 in a manner that can be taught to children.

The **Notation Equation** relates the general notation for fractions to the arithmetic operations of multiplication and addition. Explanatory material for the Notation Equation suitable for children is given in §13.7.

These two equations connect unit fractions to the abstract operations of arithmetic. If they are understood, they make computations with common fractions simple! Our development of the computational procedures will also use the Commutative, Associative and Distributive Laws which we, and the CCSS-M, expect that the reader and every child that has completed Grade 3 is comfortable with the use of these laws in computations. To illustrate the use of these ideas, we begin with the problem of multiplying an integer times a fraction.

14.1.1 Applying the Notation Equation: Multiplying a Whole Number by a Common Fraction

We start with the simplest type of numerical example, a counting number, say 6, times a common fraction, say $\frac{2}{9}$.

The Notation Equation asserts that $\frac{2}{9} = 2 \times \frac{1}{9}$, so that

$$6 \times \frac{2}{9} = 6 \times \left(2 \times \frac{1}{9}\right).$$

Applying the Associative Law for multiplication to the RHS and performing the indicated multiplication gives:

$$6 \times \frac{2}{9} = (6 \times 2) \times \frac{1}{9} = 12 \times \frac{1}{9}.$$

The computation is completed by again applying the Notation Equation to the expression at the far right:

$$6 \times \frac{2}{9} = \frac{12}{9}.$$

This computation illustrates the importance of the Notation Equation as a tool for computation with fractions. While we have given a numerical example, it is clear that the computation results from the general rule:

$$m \times \frac{p}{q} = \frac{m \times p}{q}.$$

It is expected that children will be completely comfortable with the use of this rule starting in Grade 4. It is also expected that children will understand each step in the process (see Goals for Grade 4).

14.2 Multiplication of Common Fractions

Suppose we want to multiply two common fractions $\frac{m}{n}$ and $\frac{p}{q}$, or, in numerical form $\frac{2}{5}$ and $\frac{3}{7}$. The Notation Equation tells us that each of these fractions is actually the product of a counting number and a unit fraction so that

$$\frac{m}{n} \times \frac{p}{q} = \left(m \times \frac{1}{n}\right) \times \left(p \times \frac{1}{q}\right).$$

The numerical example illustrates this fact:

$$\frac{2}{5} \times \frac{3}{7} = \left(2 \times \frac{1}{5}\right) \times \left(3 \times \frac{1}{7}\right).$$

The Commutative and Associative Laws for multiplication tell us we can write the four products on the RHS in any order we choose, so that:

$$\frac{m}{n} \times \frac{p}{q} = (m \times p) \times \left(\frac{1}{n} \times \frac{1}{q}\right)$$

or numerically,

$$\frac{2}{5} \times \frac{3}{7} = (2 \times 3) \times \left(\frac{1}{5} \times \frac{1}{7}\right).$$

Since m and p are integers, we know how to form their product, in this case $2 \times 3 = 6$. What we **need to know** is how to compute the product of the two unit fractions which we explore next.

14.2.1 Multiplying Unit Fractions

For the counting numbers n and q, R8 (see §12.6.1) asserts

$$\frac{1}{n} \times \frac{1}{q} = \frac{1}{n \times q} \quad \text{or} \quad \frac{1}{5} \times \frac{1}{7} = \frac{1}{5 \times 7}.$$

So R8 tells us what we want to know, but not why it is so. To answer that question, we turn to the Fundamental Equation.

Consider the unit fractions $\frac{1}{n}$ and $\frac{1}{q}$. The Fundamental Equation asserts that

$$n \times \frac{1}{n} = 1 \quad \text{and} \quad q \times \frac{1}{q} = 1$$

which means that these fractions are the multiplicative inverses for n and q, respectively. In terms of our numerical example, the corresponding equations are:

$$5 \times \frac{1}{5} = 1 \quad \text{and} \quad 7 \times \frac{1}{7} = 1.$$

The Fundamental Equation also tells us that

$$(n \times q) \times \left(\frac{1}{n \times q} \right) = 1$$

or numerically,

$$(5 \times 7) \times \left(\frac{1}{5 \times 7} \right) = 1.$$

Now multiplying the first two instances of the Fundamental equation together produces

$$\left(n \times \frac{1}{n} \right) \times \left(q \times \frac{1}{q} \right) = 1.$$

Applying the Associative and Commutative Laws to the LHS we can obtain:

$$(n \times q) \times \left(\frac{1}{n} \times \frac{1}{q} \right) = 1$$

whence by Transitivity of Equality

$$(n \times q) \times \left(\frac{1}{n \times q} \right) = 1 = (n \times q) \times \left(\frac{1}{n} \times \frac{1}{q} \right).$$

Since $n \times q$ is a positive integer, hence not 0, we can apply the Cancellation Law, R13, to obtain:

$$\frac{1}{n \times q} = \frac{1}{n} \times \frac{1}{q}.$$

The numerical essence of this is that

$$\cancel{35} \times \frac{1}{35} = \cancel{35} \times \left(\frac{1}{5} \times \frac{1}{7} \right)$$

whence

$$\frac{1}{35} = \frac{1}{5} \times \frac{1}{7}$$

by an application of the Cancellation Law for multiplication, R13 of §12.6.1.

The computation can be repeated for any pair of unit fractions, and your child should become fluent in using this formula,

$$\frac{1}{n} \times \frac{1}{q} = \frac{1}{n \times q},$$

in Grade 5.

287

14.2.2 Multiplying Common Fractions

Now that we have R8 for unit fractions let's return to our original problem of multiplying the two common fractions $\frac{m}{n}$ and $\frac{p}{q}$. The Notation Equation tells us that

$$\frac{m}{n} \times \frac{p}{q} = \left(m \times \frac{1}{n}\right) \times \left(p \times \frac{1}{q}\right).$$

and the Commutative and Associative Laws for multiplication turn the RHS into:

$$\frac{m}{n} \times \frac{p}{q} = (m \times p) \times \left(\frac{1}{n} \times \frac{1}{q}\right).$$

Applying R8 to the RHS takes us a further step:

$$\frac{m}{n} \times \frac{p}{q} = (m \times p) \times \frac{1}{n \times q},$$

and one more application of the Notation Equation to the RHS gives:

$$\frac{m}{n} \times \frac{p}{q} = \frac{m \times p}{n \times q}.$$

The last equation, R9 in §12.6.1, can be stated succinctly in words as:

the product of two common fractions is the product of the numerators over the product of the denominators.

Since m, n, p and q are all integers, a child knows how to do the required multiplications on the RHS and that the results of these multiplications are integers. Thus, finding the product of two fractions comes down to finding the product of two pairs of integers (the product of the numerators and the product of the denominators). You may have been wondering why the CCSS-M demands that your child should know and understand the Notation Equation:

$$\frac{m}{n} = m \times \frac{1}{n}.$$

As you can see, the Notation Equation is an essential relation that enables computations with common fractions.

Let's review these steps for our numerical example: $\frac{2}{5} \times \frac{3}{7}$. The Notation Equation tells us:

$$\frac{2}{5} \times \frac{3}{7} = \left(2 \times \frac{1}{5}\right) \times \left(3 \times \frac{1}{7}\right).$$

The RHS involves only multiplications which can be manipulated using only the **Commutative and Associative Laws** to obtain

$$\left(2 \times \frac{1}{5}\right) \times \left(3 \times \frac{1}{7}\right) = (2 \times 3) \times \left(\frac{1}{5} \times \frac{1}{7}\right).$$

Applying R8 we have $\frac{1}{5} \times \frac{1}{7} = \frac{1}{5 \times 7} = \frac{1}{35}$ followed by the Notation Equation gives

$$\frac{2}{5} \times \frac{3}{7} = \frac{2 \times 3}{5 \times 7} = \frac{6}{35}$$

It is expected that children will understand all aspects of multiplication of fractions in Grade 5.

14.2.3   Concrete Meaning for Products of Unit Fractions

We have made much of the idea that multiplication was repetitive addition. Indeed, this was a key idea used in the development of the integers. And in the next section we will use this idea again as a means for developing procedures for adding common fractions. However, the reader may be wondering, how can I interpret the the computation

$$\frac{1}{4} \times \frac{1}{3} = \frac{1}{12}$$

as repetitive addition. If you have been thinking about this and find yourself stumped, there is a reason. There is no sensible way to interpret this product as repetitive addition. This is because of the requirement that one factor in the product has to be an integer for this interpretation to be valid.

There is, however, an alternative interpretation, namely, **scaling**. Consider the following list of numbers

$$2, 10, 50.$$

Speaking in comparative terms, we would say that 10 is five times as big as two, but only one fifth as big as 50. We would also say that 2 is one fifth as big as 10 and one twenty-fifth as big as 50. The factor that we multiply one number by to obtain another number is often referred to as a **scale factor**. Thus, in the equation

$$2 = \frac{1}{25} \times 50,$$

we would say *the scale factor is one twenty-fifth*.

To get the idea of scaling in respect to products, think of a unit square. If we reduce the width by a scale factor of one third, we obtain a rectangle having an area that is $\frac{1}{3}$ as big as the original. On the other hand, if we

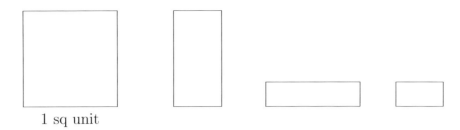

1 sq unit

Start with unit square. Scale the width by one third. Scale the height by one quarter. Scale both to get the rectangle on right. Its area is scaled by one twelth.

reduce the height by a factor of one fourth, the resulting rectangle has an area $\frac{1}{4}$ as large as the original. Scaling the width by one third and the height by one fourth results in the rectangle on the far right which has its area scaled by $\frac{1}{12}$. This is the notion of scaling.

14.3   Adding Common Fractions

14.3.1   Adding Fractions with the Same Denominator

The Notation Equation,

$$\frac{m}{n} = m \times \frac{1}{n},$$

combined with the **Distributive Law** makes adding fractions with the same denominator easy. For example, suppose we want to add $\frac{9}{17}$ and $\frac{4}{17}$. The computation, which uses the Distributive Law to factor out $\frac{1}{17}$, in full detail looks like

$$
\begin{aligned}
\frac{9}{17} + \frac{4}{17} &= 9 \times \frac{1}{17} + 4 \times \frac{1}{17} = (9+4) \times \frac{1}{17} \\
&= \frac{9+4}{17} = \frac{13}{17}.
\end{aligned}
$$

We remind the reader that in the middle sum of the first line, the requirement to perform multiplication before addition avoids the use of parentheses.

The computation is an instance of R11 of §12.6.1 which asserts for common fractions, $\frac{m}{n}$ and $\frac{k}{n}$ that

$$\frac{m}{n} + \frac{k}{n} = \frac{m+k}{n}.$$

To review where R11 §12.6.1 comes from, let's see how this computation is worked out in general. Consider the two integer numerators m and k each over the denominator

n. Then the Notation Equation tells us that

$$\frac{m}{n} + \frac{k}{n} = m \times \frac{1}{n} + k \times \frac{1}{n}.$$

Applying the Distributive Law on the RHS to pull out a factor of $\frac{1}{n}$ gives:

$$\frac{m}{n} + \frac{k}{n} = (m + k) \times \frac{1}{n}.$$

The Notation Equation applied to the RHS now produces R11:

$$\frac{m}{n} + \frac{k}{n} = \frac{m + k}{n}.$$

It is expected that children in Grade 4 will be able to understand and apply these ideas with facility. Understanding why this works comes down to understanding why the Notation Equation is true. Achieving this understanding depends on the child coming to realize that unit fractions are like identical buttons as discussed in Chapter 13 and this is the central idea that underlies the Notation Equation.

We can state this equation in the form of a rule:

To add fractions having the same denominator, put the sum of the numerators over the common denominator.

There is a companion rule:

Do not try to add fractions this way, unless they have the same denominator!

It is always helpful to have easy ways to remember things. In my own mind, I think of these two rules as examples of the rule:

Always add apples to apples.

To see what I mean by *adding apples to apples*, if we want to add

$$\frac{4}{15} + \frac{7}{15},$$

we would think of adding 4 *fifteenths* to 7 *fifteenths*. Here we are regarding each *fifteenth* as an **apple**. So we are adding four apples to seven apples, and we end up with eleven apples or more precisely 11 *fifteenths*:

$$\frac{4}{15} + \frac{7}{15} = \frac{4 + 7}{15} = \frac{11}{15}.$$

Thought of in this way, we realize the computation merely amounts to counting up the total number of *fifteenths*. Thus, adding the numerators and leaving the denominator alone really is the correct procedure. Moreover, the actual computation, in this case $4 + 7$, is mere recall, provided the child knows the addition table.

On the other hand, if we want to add

$$\frac{3}{7} + \frac{2}{9},$$

we don't satisfy the *apples to apples* rule, since when we put this in words, we have to add 3 *sevenths* and 2 *ninths*. If the *sevenths* are apples, then the *ninths* have to be oranges, or fish, or whatever, but not apples.

14.3.2   Adding Fractions with Different Denominators

Suppose we want to add fractions having different denominators, as in:

$$\frac{3}{7} + \frac{2}{9}.$$

The companion rule tells us not to try to perform the computation unless the two fractions have the same denominator. The obvious question is:

Can we make the denominators the same?

When we ask this question what we really want to know is: Can we make the denominators the same without changing the problem in any arithmetic way?

There are two ways to change an arithmetic problem without affecting the answer: add 0, or multiply by 1. In this case, we multiply by 1, as follows:

$$\frac{3}{7} \times 1 + \frac{2}{9} \times 1.$$

As the reader can see, the problem is unchanged, but we are still no better off. This is where someone way back when got clever. Since $\frac{7}{7} = \frac{9}{9} = 1$, we have

$$\frac{3}{7} \times 1 = \frac{3}{7} \times \frac{9}{9} = \frac{27}{63} \quad \text{and} \quad \frac{2}{9} \times 1 = \frac{2}{9} \times \frac{7}{7} = \frac{14}{63}$$

where the resulting fractions now have the same denominator, called a **common denominator**. The **common denominator**, 63, is the product of the two starting denominators

$$9 \times 7 = 63$$

292

and can **always** be used in problems where the starting denominators are different. Using the product, 63, we have

$$\frac{3}{7} + \frac{2}{9} = \frac{27}{63} + \frac{14}{63}$$

where the RHS now satisfies the **apples to apples** rule, where **apples** are *sixty thirds*. The computation is now straight forward using the rule in the last section:

$$\frac{3}{7} + \frac{2}{9} = \frac{27}{63} + \frac{14}{63} = \frac{27 + 14}{63} = \frac{41}{63}.$$

The example just given merely implements R12 of §12.6.1 for finding the sum of two fractions.

To generate R12, we first multiply each of the fractions by a form of 1 to obtain:

$$\frac{m}{n} + \frac{p}{k} = \frac{m}{n} \times \frac{k}{k} + \frac{p}{k} \times \frac{n}{n}.$$

Remember, on the RHS above, multiplication takes precedence over addition, so no parentheses are needed. By applying the rule for multiplying fractions to the RHS we obtain:

$$\frac{m}{n} + \frac{p}{k} = \frac{m \times k}{n \times k} + \frac{p \times n}{k \times n}.$$

The Commutative Law for multiplication applied to the second denominator on the RHS produces:

$$\frac{m}{n} + \frac{p}{k} = \frac{m \times k}{n \times k} + \frac{p \times n}{n \times k}$$

so that both summands on the RHS have the same denominator. We now use the rule for adding fractions having the same denominator to obtain:

$$\frac{m}{n} + \frac{p}{k} = \frac{m \times k + p \times n}{n \times k}.$$

14.3.3   Adding Fractions: A General Rule for Children that Always Works

We could teach the content of these examples to children as a step-by-step process for adding fractions. Given the problem:

$$\frac{m}{n} + \frac{p}{k} = ?$$

Step 1: Check whether the two denominators, n and k, are equal;

Step 2: If the two denominators are equal, that is, $n = k$, find $m + p$ and the sum is:

$$\frac{m}{n} + \frac{p}{n} = \frac{m + p}{n};$$

Step 3: If the denominators are not equal, that is, $n \neq k$, find the products $n \times k$, $m \times k$ and $p \times n$, then the sum is:

$$\frac{m}{n} + \frac{p}{k} = \frac{m \times k + p \times n}{n \times k}.$$

In explaining this procedure to children, one would start with numerical examples as in

$$\frac{2}{9} + \frac{4}{9} = ?, \quad \frac{3}{19} + \frac{7}{19} = ?, \quad \frac{11}{15} + \frac{4}{15} = ?$$

and so forth. After facility is achieved, one would proceed to examples that require the third step as in

$$\frac{2}{3} + \frac{4}{9} = ?, \quad \frac{3}{7} + \frac{1}{3} = ?, \quad \frac{4}{15} + \frac{1}{5} = ?$$

In doing examples requiring Step 3, one would want to remind children of the order of precedence rule that requires products to be computed before sums (see §8.2.2). It is expected that children will be able to fluently perform calculations with common fractions and mixed numbers by using the equations in Steps 2 and 3 in Grade 5.

If the procedures for addition and multiplication were limited to those above, fractions would not be the hardest topic in the elementary school curriculum. Much to the chagrin of many, there are topics yet to consider. However, if your child understands and can perform fluently the procedures described to this point, they have the knowledge to complete the task. The point is, first mastering these procedures at the level of recall is the key to future success because they apply to all quantities that are real numbers.

14.4 What Your Child Needs to Know

This chapter has been about how to compute with common fractions, the most difficult topic in elementary arithmetic. How can we best communicate this material to children?

To answer this question, I want to digress here for a minute and discuss the teaching of reading.

There are essentially two ways to teach reading: **phonics** and **word recognition**. Phonics depends on the sounds attached to each letter and enables a child to figure out the word. Word recognition is modeled on how adults read, and is based on recognizing individual words. Phonics works well for something like 90% of children and not at all for the remainder.[1] Word recognition works equally well for all children, but nowhere near as well as phonics for those who can learn by that method.[2]

If one merely thinks about the relative size of the databases required to learn by either method, it is clear that word recognition, which requires *a priori* knowledge of a word prior to reading it, quickly generates a very large data requirement. Further, there is little, or no, ability to "figure out" words not previously introduced. Phonics, on the other hand, has a small number of rules which the learner applies to "figure out" words not yet learned, hence a smaller initial data base. These rules also provide the child with a method to read on their own. In the author's view, the small initial data base has great application in that it lets children figure things out and is the reason why the phonics method works so well for most children.

The author is convinced that the same ideas apply to the teaching of arithmetic. Specifically, we can teach children a relatively small number of rules and how to apply these rules. For example, suppose a child is asked find: $23 + 48$. To the author, the simplest solution for a child is to use the standard procedure, that is:

$$
\begin{array}{r}
23 \\
+48 \\
\hline
71
\end{array}
$$

The essential point is that the actual calculation, whether mental or written, is performed by using the facts in the addition table and the standard procedure.

Modern curriculum documents are replete with the notion of *individual strategies*. Let me give an example of what is meant by an individual strategy as derived from curriculum documents and applied to the problem above. One strategy might consist of recognizing that 48 is 2 short of 50, so we can borrow 2 from 23 and rewrite the problem as: $21 + 50 = 71$. Another child's individual strategy might consist of thinking of the sum as $23 + 48 = (20 + 40) + (3 + 8)$. The point about individual strategies is they are *individual*. To the degree that strategies are a tool for teaching children about the application of the Laws, for example, the Associative Law, strategies are useful. When they become a substitute for computational fluency with the

---

[1] At the time I was teaching my own three children to read using phonics, I did a bit of research on this. My comments are based on that research.

[2] In school systems where differential learning rates for reading are a concern, the less successful word recognition method may be used to teach reading because it gives no child an advantage.

standard procedures and knowledge of the addition and multiplication tables, they are a tragic mistake in my opinion.[3]

My complaint with individual strategies as a substitute is that they have little application beyond the arithmetic problem at hand, and certainly no application in the context of symbolic algebra such as is required in a calculus course. Using the standard procedures, reinforces the learning of the general arithmetic algorithms thereby building fluency. But more than this, since these same procedures are used throughout our modern world, their repetitive use prepares the student for the requirements of later life. In other words, the standard procedure has great application. Individual strategies as a substitute have almost none.

We have given step-by-step procedures for performing calculations at key points in this Chapter. For example, all computations with fractions can be correctly performed using the eight rules in §13.10. This is a very small list.

Does your child, in Grade 4 need to know how they are all derived as shown in Chapter 12? Not then. What the child needs to be able to do is correctly and fluidly use these rules and understand why they are true based on the concept of unit fraction and the Fundamental and Notation Equations.

That said, the factual content in Chapter 12 is the foundation of all our mathematics and by Grade 7 your child will be expected to understand and apply these facts. As has been continually stressed, these facts are derived from our observations of the world. Together they comprise the smallest and most powerful fact base underlying any branch of human knowledge. This is why the CCSS-M stresses the learning and application of these facts. What every child needs to know is how to apply these facts to fluidly perform computations.

Goals for Grades 1-3 are in the last chapter and are not repeated here.

14.4.1   Goals for Grade 4

The child should be able to

1. recognize that a whole number is a multiple of each of its factors. e.g., 21 is a multiple of 3 and 7;

2. determine whether any counting number ≤ 100 is prime or composite;

3. explain computationally why $\frac{n}{m} = \frac{n \times p}{m \times p}$ (see §10.6 and R10);

4. understand that for $m > 1$, the fraction $\frac{m}{n}$ is the sum of m unit fractions $\frac{1}{n}$;

---

[3]Some Canadian curriculum documents suggests that the intent is that children should substitute strategies for knowledge at the level of recall. Quite clearly, that is contrary to the intent of the CCSS-M.

(a) understand addition and subtraction of fractions as joining and separating parts in respect to the same whole;

(b) decompose a fraction into a sum in multiple ways, e.g., $\frac{3}{8} = \frac{1}{8} + \frac{1}{8} + \frac{1}{8}$ or $\frac{3}{8} = \frac{1}{8} + \frac{2}{8}$;

(c) decompose a mixed number into a sum in multiple ways, e.g., $2\frac{3}{8} = 2\frac{1}{8} + \frac{1}{8} + \frac{1}{8}$ or $2\frac{3}{8} = 1\frac{1}{8} + 1\frac{2}{8}$, or $2\frac{3}{8} = 2 + \frac{1}{8} + \frac{2}{8}$, etc.;

(d) add and subtract mixed numbers involving like denominators by replacing the mixed numbers with equivalent fractions;

(e) solve word problems involving addition and subtraction of fractions having like denominators by using equations to represent the problem;

5. apply previous understanding of multiplication, i.e., $\frac{m}{n} = m \times \frac{1}{n}$, to multiply fractions by whole numbers;

(a) fully understand $p \times \frac{m}{n} = \frac{p \times m}{n}$, computationally;

(b) fully understand $p \times \frac{m}{n} = \frac{p \times m}{n}$, visually, e.g., $3 \times \frac{2}{5}$ is equivalent to $(3 \times 2) \times \frac{1}{5}$;

(c) solve word problems involving multiplication of fractions by whole numbers using equations and visual models. e.g., If each child at a party drinks $\frac{2}{3}$ of liter of lemonade and five children are attending the party, how much lemonade must be made?;

14.4.2 Goals for Grade 5

The child should be able to

1. use equivalent fractions as a means to add fractions with unlike denominators and be fully aware that

$$\frac{m}{n} + \frac{p}{q} = \frac{mq + pn}{nq};\, [4]$$

2. understand that a correct interpretation of a fraction is as division of the numerator by the denominator, i.e., $\frac{m}{n} = m \div n$;

(a) use this interpretation to solve problems (A 20 kg sack of sugar is to be equally shared by 25 people. Express each person's share as a fraction.);

---

[4]The reader should be aware that two letters written together as in nq is shorthand for $n \times q$ and such expressions appear in general usage and in the school curriculum. We have retained the use of $\times$ for clarity.

3. understand the equations

$$\frac{m}{n} \times q = \frac{m \times q}{n} = m \times \frac{q}{n}$$

in all the aspects that follow:

(a) interpret the product $\frac{m}{n} \times q$ as m parts of a partition of q into n equal parts by using the Notation Equation;

(b) obtain this partition as the result of a sequence of operations leading to $m \times \frac{q}{n}$;

(c) understand this is equivalent to partitioning $m \times q$ into n equal parts;

(d) for example, use a visual fraction model to show $\frac{2}{3} \times 4 = \frac{8}{3}$, and create a story context for this equation;

(e) do the same with $\frac{2}{3} \times \frac{4}{5} = \frac{8}{15}$;

(f) more generally, understand and interpret $\frac{m}{n} \times \frac{q}{p} = \frac{m \times q}{n \times p}$;

4. interpret multiplication as scaling (resizing), by:

(a) comparing the size of a product to the size of one factor on the basis of the size of the other factor, without performing the indicated multiplication;

(b) explaining why multiplying a given number by a fraction greater than 1 results in a product greater than the given number (recognizing multiplication by whole numbers greater than 1 as a familiar case); explaining why multiplying a given number by a fraction less than 1 results in a product smaller than the given number; and relating the principle of fraction equivalence $\frac{m}{k} = \frac{nm}{nk}$ to the effect of multiplying $\frac{m}{k}$ by 1;

5. find the area of a rectangle with fractional side lengths by tiling it with unit squares of the appropriate unit fraction side lengths, and show that the area is the same as would be found by multiplying the side lengths;[5]

6. use multiplication of fractions to find the area of rectangles having sides with fractional lengths and represent fraction products as rectangular areas;

7. solve real world problems involving multiplication of fractions and mixed numbers;

---

[5]What is being requested here is a visual demonstration of $A = L \times W$ when the length and width are fractions.

8. apply and extend the meaning of division to division of fractions by counting numbers;

   (a) interpret division of a unit fraction by a whole number computationally using the relationship between division and multiplication, e.g., $\frac{1}{4} \div 3 = \frac{1}{12}$, because $\frac{1}{12} \times 3 = \frac{1}{4}$;

   (b) solve real world problems involving division of unit fractions by whole numbers.

Chapter 15

Fractions for Kids: Advanced Topics

Chapter Overview. The operations of subtraction and division in respect to fractions are discussed. The computations surrounding reducing a common fraction to lowest terms are presented.

15.1   Subtraction and Division of Common Fractions

Subtraction and division of fractions were not discussed in Chapter 14 because the the computations are more difficult. In particular, subtraction involves negative numbers which are not introduced until Grade 6. That said, the basic formulae developed in Chapter 14 still apply so that the development below is merely applying previous rules.

15.1.1   Subtracting Common Fractions

There are two essential facts we need to recall from Chapter 12. These are that subtraction is defined by the equation (see §12.3.5)

$$x - y \equiv x + (-y)$$

and R6 (see §12.6.1) which asserts

$$(-1) \times x = -x.$$

Thus the operation of subtraction is replaced by addition of the additive inverse, and additive inverses are found by multiplying by -1, which is the additive inverse of 1. Consider then that we have a common fraction: $\frac{m}{n}$. For fractional forms like this, R6 tells us that

$$(-1) \times \frac{m}{n} = -\frac{m}{n}$$

where the expression on the RHS is the standard centered dash notation for the additive inverse of $\frac{m}{n}$.

Now recall our rule for multiplying a whole number times a common fraction (see §14.1.1), which tells us that

$$(-1) \times \frac{m}{n} = \frac{(-1) \times m}{n}.$$

Since $(-1) \times m = -m$, we have

$$(-1) \times \frac{m}{n} = \frac{-m}{n}$$

so that the additive inverse of the fraction $\frac{m}{n}$ is obtained by replacing the numerator by its additive inverse.

Thus, to subtract $\frac{m}{n}$ from the fraction $\frac{p}{q}$, we apply the definition to obtain

$$\frac{p}{q} - \frac{m}{n} \equiv \frac{p}{q} + \left(-\frac{(m)}{n} \right).$$

But the additive inverse inside parentheses on the RHS is $\frac{-m}{n}$, so that required sum is

$$\frac{p}{q} - \frac{m}{n} \equiv \frac{p}{q} + \frac{-m}{n}$$

which can be computed using the procedure in §14.3.2 to give

$$\frac{p}{q} - \frac{m}{n} = \frac{p \times n - q \times m}{q \times n}.$$

Methods for finding both the numerator and denominator were developed in Chapter 9. Dealing with fractions involving negative integers is a Grade 6 requirement. A simple numerical example at the Grade 6 level might be:

$$\frac{1}{2} - \frac{2}{3} = \frac{1 \times 3 - 2 \times 2}{2 \times 3} = \frac{3 - 4}{6} = \frac{-1}{6}.$$

15.1.2 Dividing Common Fractions

Recall that in §12.3.5 the operation of division was defined by:

$$x \div y = x \times y^{-1}$$

where $y \neq 0$ and y^{-1} is the multiplicative inverse of y. The reasoning supporting this definition is discussed in §12.3.5. We remind the reader that the key idea in division is that the **quotient times the divisor should be equal to the dividend**.

Let us now suppose we have the two common fractions $\frac{m}{n}$ and $\frac{p}{q}$. Using the definition above, we have

$$\frac{m}{n} \div \frac{p}{q} = \frac{m}{n} \times \left(\frac{p}{q}\right)^{-1}.$$

If we apply the Notation Equation to $\frac{p}{q}$ the RHS becomes:

$$\frac{m}{n} \div \frac{p}{q} = \frac{m}{n} \times \left(p \times \frac{1}{q}\right)^{-1}.$$

Recalling R8 of §12.6.1 that says $(x \times y)^{-1} = x^{-1} \times y^{-1}$, we have:

$$\left(p \times \frac{1}{q}\right)^{-1} = p^{-1} \times \left(\frac{1}{q}\right)^{-1}$$

so that

$$\frac{m}{n} \div \frac{p}{q} = \frac{m}{n} \times \left[p^{-1} \times \left(\frac{1}{q}\right)^{-1}\right].$$

Focus on the terms inside the square brackets on the RHS. What we need to know is how to find the multiplicative inverse of a unit fraction, $\frac{1}{q}$. The Fundamental Equation which tells us that for any counting number q, the unit fraction $\frac{1}{q}$ is its multiplicative inverse. However, since the test of whether two quantities are multiplicative inverses of one another is simply to compute their product and see if it is equal to 1, the Fundamental Equation

$$q \times \frac{1}{q} = 1$$

also tells us that q is the multiplicative inverse of $\frac{1}{q}$. In other words,

$$\left(\frac{1}{q}\right)^{-1} = q$$

which readers will recognize as an instance of R3 of §12.6.1 which asserts that

$$(y^{-1})^{-1} = y.$$

Applying Fundamental Equation again, this time to p^{-1}, we have $p^{-1} = \frac{1}{p}$, so that the product in square brackets that we needed to find is:

$$p^{-1} \times \left(\frac{1}{q}\right)^{-1} = \frac{1}{p} \times q.$$

Applying the Commutative Law and the Notation Equation to the expression on the RHS gives us:

$$p^{-1} \times \left(\frac{1}{q}\right)^{-1} = \frac{q}{p}$$

Applying this result to the RHS of

$$\frac{m}{n} \div \frac{p}{q} = \frac{m}{n} \times \left(\frac{p}{q}\right)^{-1}$$

produces the equation that governs division:

$$\frac{m}{n} \div \frac{p}{q} = \frac{m}{n} \times \frac{q}{p} = \frac{m \times q}{n \times p}$$

The division equation is often described as **invert and multiply** because the divisor, $\frac{p}{q}$, is inverted to become $\frac{q}{p}$ and then the product with $\frac{m}{n}$ is taken.

To summarize, what we have demonstrated in respect to finding the multiplicative inverse of a common fraction is that

$$\left(\frac{p}{q}\right)^{-1} = \frac{q}{p}.$$

That this is obviously so follows from the following computation:

$$\frac{q}{p} \times \frac{p}{q} = \frac{q \times p}{p \times q} = 1.$$

In the usual fractional notation, we would write:

$$\left(\frac{p}{q}\right)^{-1} = \frac{1}{\frac{q}{p}} = \frac{q}{p}.$$

A simple numerical example of division would be computing $\frac{2}{3} \div \frac{3}{4}$. Using the procedure above,

$$\frac{2}{3} \times \left(\frac{3}{4}\right)^{-1} = \frac{2}{3} \times \frac{4}{3} = \frac{2 \times 4}{3 \times 3} = \frac{8}{9}.$$

Now let us return to the key idea: **quotient times the divisor should be equal to the dividend**. In the problem above of finding

$$\frac{m}{n} \div \frac{p}{q},$$

if $\frac{m}{n}$ is the dividend and $\frac{p}{q}$ is the divisor, what is the quotient? The only thing it can be is the product

$$\frac{m \times q}{n \times p}.$$

To see that this is in fact the case, we compute the product of the proposed quotient times the divisor $\frac{p}{q}$ to obtain:

$$\frac{m \times q}{n \times p} \times \frac{p}{q} = \frac{m}{n} \times \frac{q \times p}{p \times q} = \frac{m}{n}.$$

By Grade 5, it is expected that children will come to understand the relationship between multiplication and division as expressed by the relation for integers where d is a factor of k

$$\text{if } k \div d = q \text{ then } k = q \times d$$

which is the basis of the division operation (see §12.3.5). It is expected in the context of fractions that children will understand that

$$\frac{1}{8} \div 4 = \frac{1}{32}$$

exactly because

$$4 \times \frac{1}{32} = \frac{1}{8}$$

when reduced to lowest terms (see below) And by Grade 6 it is expected that children will be fluent with division of common fractions.

15.2 R10 and Lowest Terms

The reader may well recall from their own school years that there is a process known as **reducing a fraction to lowest terms**. It is related to the process for finding a common denominator by producing equivalent fractions using R10 of §12.6.1 which for common fractions states:

$$\frac{m}{n} = \frac{m \times k}{n \times k}$$

where m, n, $k \in \mathcal{I}$ and n, $k \neq 0$. R10 is a consequence of the Fundamental Equation which asserts that $\frac{k}{k} = 1$ and is used to obtain common denominators in §14.2.2 above as follows:

$$\frac{m}{n} = \frac{m}{n} \times 1 = \frac{m}{n} \times \frac{k}{k} = \frac{m \times k}{n \times k}.$$

The CCSS-M expect that children in Grade 4 will understand these steps applied to common fractions.

15.2.1   A Visual Interpretation of R10

We want to develop a physical interpretation of R10 which can be used with children. Thus consider the RHS of R10, namely, $\frac{m \times k}{n \times k}$. The essential feature of the RHS is that the denominator is a product. So our visual interpretation begins with a unit fraction having a denominator that factors, for example, $\frac{1}{6} = \frac{1}{2 \times 3}$.

Box #1   Box # 2   Box # 3

The three boxes have the same area which we take to be 1 unit. The area of Box #1 is divided into *sixths*, the area of Box #2 is divided into *thirds*, and the area of Box #3 is divided into *halves*.

In the diagram above, Box #1 is divided into six equal parts. What we want to think about is:

What fractions can be formed from *sixths*?

One way to answer this question is simply to start adding $\frac{1}{6}$ to itself and see if any other fractions turn up. Thus for example we could form the sum

$$\frac{1}{6} + \frac{1}{6} = 2 \times \frac{1}{6}.$$

In our diagram, forming this sum corresponds to removing the horizontal line from Box #1 to form Box #2. Box #2 is divided into 3 equal parts each of which is

305

composed of 2 equal parts as shown in Box #1. Since each one of the 3 equal parts of Box #2 has area $\frac{1}{3}$, we know that

$$2 \times \frac{1}{6} = \frac{1}{3}.$$

Visually starting with the diagram of Box #2 and inserting the horizontal line produces Box #1 and illustrates why R10 tells us that $\frac{1}{3} = \frac{2}{6}$ through the computation

$$\frac{1}{3} = \frac{2}{2} \times \frac{1}{3} = \frac{2}{2 \times 3} = \frac{2}{6}.$$

Continuing in this way using Box #1 and #2, we see that

$$\frac{1}{6} + \frac{1}{6} + \frac{1}{6} + \frac{1}{6} = 4 \times \frac{1}{6}.$$

Combining the four subunits on the left side of Box #1, we see that the result must be $\frac{2}{3}$ which visually illustrates R10 for the case

$$\frac{2}{3} = \frac{2}{2} \times \frac{2}{3} = \frac{2 \times 2}{2 \times 3} = \frac{4}{6}.$$

Similarly, if we remove the vertical lines from Box #1 to form Box #3, we know the area of either the top or bottom will given by

$$\frac{1}{6} + \frac{1}{6} + \frac{1}{6} = 3 \times \frac{1}{6}.$$

Either way, we know the total area is

$$3 \times \frac{1}{6} = \frac{1}{2}.$$

The visual representations above can be reinterpreted numerically as

$$\frac{1}{2} = \frac{3}{3} \times \frac{1}{2} = \frac{3 \times 1}{3 \times 2} = \frac{3}{6}.$$

The reader might want to construct a similar diagram involving *tenths*, *fifths* and *halves* to better understand these ideas.

Visual results like those above definitely enhance understanding. But they are not a substitute for computational knowledge as the CCSS-M standards make clear.

Thus, every child is expected to know and understand that these results all follow from the application of the Associative Law and R8 as in:

$$
\begin{aligned}
4 \times \frac{1}{6} &= (2 \times 2) \times \frac{1}{2 \times 3} = (2 \times 2) \times \left(\frac{1}{2} \times \frac{1}{3} \right) \\
&= 2 \times \left(\left(2 \times \frac{1}{2} \right) \times \frac{1}{3} \right) = 2 \times \left(1 \times \frac{1}{3} \right) \\
&= 2 \times \frac{1}{3} = \frac{2}{3}.
\end{aligned}
$$

In the next section we introduce a simplified procedure referred to as **cancelling**. But we stress that these are the computations that underlie the cancelling process.

15.2.2   Lowest Terms and Common Factors

The effect of R10 is that there are an infinite number of notations that all give rise to the same number in $\mathcal{Q}$ and identify the same place on the number line. For example:

$$
\frac{3}{8}, \ \frac{6}{16}, \ \frac{9}{24}, \ \frac{15}{40}, \ \frac{-3}{-8}, \ \frac{-21}{-56}, \ \frac{30}{80}, \ \text{and} \ \frac{-2400}{-6400}
$$

are all valid representations of the same rational number, $\frac{3}{8}$. One of the things that makes fractions complicated is that there is a preferred expression for each rational number. That preferred expression is the one having the least positive denominator. In the above list, we see that this is $\frac{3}{8}$.

Let us recall how these multiple expressions arise. They all come from the application of R10:

$$
\frac{m}{n} = \frac{m \times k}{n \times k}
$$

Notice on the RHS of R10, the numerator is $m \times k$, and the denominator is $n \times k$. Thus the numerator is a product, the denominator is a product, and both contain the factor k. The factor k is referred to as a **common factor** of the numerator and denominator. In this situation, R10 is reinterpreted as follows:

$$
\frac{m \times k}{n \times k} = \frac{m \times \cancel{k}}{n \times \cancel{k}} = \frac{m}{n},
$$

where the overstrike denotes **cancelling**, and therefore eliminating, the common factor k from both denominator and numerator. This process preserves the value of the fraction, but **reduces** the fraction to a more compact form. Indeed, this process can be repeated until the numerator and denominator have no common factor, in which case, we say the fraction is in **lowest terms**.

In our visual example with *sixths*, we would reduce $\frac{4}{6}$ as follows

$$\frac{4}{6} = \frac{2 \times 2}{2 \times 3} = \frac{\cancel{2} \times 2}{\cancel{2} \times 3} = \frac{2}{3}.$$

The concept *lowest terms* applies more generally. For example, in the fractional expression

$$\frac{\sqrt{2}}{2 \times \sqrt{2}},$$

the quantity $\sqrt{2}$ is a factor of both the numerator and denominator, since

$$\frac{\sqrt{2}}{2 \times \sqrt{2}} = \frac{1 \times \sqrt{2}}{2 \times \sqrt{2}},$$

whence after cancelling the $\sqrt{2}$ we have,

$$\frac{\sqrt{2}}{2 \times \sqrt{2}} = \frac{1 \times \cancel{\sqrt{2}}}{2 \times \cancel{\sqrt{2}}} = \frac{1}{2},$$

which is clearly a simpler form.

However, our focus is on common fractions where both the numerator and denominator are integers and on determining how to ensure a common fraction is in lowest terms.

15.3 Factoring Whole Numbers

Fractions are not in lowest terms if there is a common factor in both the numerator and the denominator. Since what we are trying to determine is whether a given denominator and a given numerator have a common factor, the first thing to ask is whether the numerator and denominator factor at all. If a number, n, can be factored, we say it is **composite** which means that $n = k \times p$, where neither k nor p is 1.

Since for the most part the calculations required of children in elementary school deal with fractional forms in which both the denominator and the numerator are one or two-digit numbers, much of what children need to know can be at the level of recall.

The body of the multiplication table contains all single digit products. As such, it provides answers to the question: $n \times m = ?$. This is the initial way in which the table is used. However, it can also be used in the opposite way as follows:

given n in the body of the multiplication table, what are its factors:
$n = ? \times ??$

For example, for entries like 45, 12 and 72 in the body of the multiplication table, a child needs to immediately recognize that

$$45 = 9 \times 5,$$

$$12 = 4 \times 3 = 2 \times 6,$$

$$72 = 9 \times 8,$$

and so forth.

Notice that from the first example, using the fact that $9 = 3 \times 3$, we get

$$45 = (3 \times 3) \times 5 = 3 \times (3 \times 5) = 3 \times 15.$$

Although the last product, 3×15, isn't in the multiplication table, but the factorization quickly follows from $(3 \times 3) \times 5$.

For 72, we can obtain the following factorizations not in the table

$$72 = 18 \times 4 = 36 \times 2 = 3 \times 24 = 6 \times 12,$$

by starting with $9 = 3 \times 3$ and $8 = 2 \times 2 \times 2$ which are in the table and using the Associative and Commutative Laws.

Clearly, immediate knowledge of the multiplication table at the level of recall is an essential prerequisite for success with factoring.

15.3.1   Prime and Composite Numbers

Once you are thoroughly familiar with the multiplication table, you notice that not every counting number, n, can be factored into a product of counting numbers in which neither of the factors is 1.[1] There are some numbers which can only be expressed as a product of 1 and themselves, for example 2 and 3. Counting numbers, other than 1, having this property, are called **prime** numbers. Counting numbers having a proper factorization, that is, that can be expressed as $n = k \times p$, where k, $p \in \mathcal{N}$ and k, $p \neq 1$, are called **composite**. In the event n is composite and $n = k \times p$, we say p **divides** n, or equivalently, p **is a divisor of** n.

A list of the first several prime numbers is:

$$2,\ 3,\ 5,\ 7,\ 11,\ 13,\ 17,\ 19,\ 23,\ \ldots$$

---

[1]Expressing $n = 1 \times n$ is called the **trivial** factorization.

Integers between 2 and 23 not appearing in this list are composite, for example $22 = 2 \times 11$.

There are lots of fun facts about primes. For example, there is no largest prime. However, these are not relevant here. What is relevant is that using this short list of primes will enable you to factor any number less than $23 \times 23 = 529$.

15.3.2   Procedure for Finding Factors of a Counting Number

Given a counting number, n, that is less than 529, we want to determine whether n is prime or composite, and, if it is composite, to completely factor n as a product of primes. We use the following procedure to factor n:

Step 1: make a list of primes in increasing order up to $\sqrt{529} = 23$, as shown;

$$2, \ 3, \ 5, \ 7, \ 11, \ 13, \ 17, \ 19, \ 23;$$

Step 2: starting with 2, successively test divide n by primes in the list in increasing order until either, a divisor is found, or 23 is found not to be a divisor of n;

Step 3: if a divisor is found, so that $n = p_1 \times q$, where p_1 is the least prime divisor of n, record p_1 and q as factors of n;

Step 4 starting with p_1 (the previously found least prime divisor of n), successively test divide q by primes in the list in increasing order until either, a divisor is found, or 23 is found not to be a divisor of q;

Step 5: if a divisor is found, so that $q = p_2 \times q_1$, where p_2 is the least prime divisor of q, record p_2 and q_1 as factors of n and repeat Step 4 with q_1 in place of q and p_2 in place of p_1;

Step 6: if none of the primes up to 23 is a divisor of n, or the test factor q, stop.

There are three points to make here before considering a numerical example.

The first is, because $23 \times 23 = 529$, if n is going to factor, it will have a prime divisor ≤ 23. So the number of prime divisors that have to be tested is limited.

The second is that if we find a least prime divisor, labelled p_1 above, then the other factor, labelled q above, will be much smaller than n and the next computation will be simpler because

$$q = n \div p_1 \leq n \div 2$$

since $2 \leq p_1$.

The third point is that because p_1 is the **least** prime divisor of n, none of its predecessors in the list of primes can be a divisor of the residual factor q. This is why in Step 4 we can repeat the procedure starting at p_1 instead of 2 without having to worry we will miss a divisor.

Let's see how this procedure works in practice by factoring 375. We start with 2, which does not divide 375, because the *ones* digit in 375 is **odd**, that is, not divisible by 2. The next prime is 3 and 3 divides 375, since $375 = 3 \times 125$. Thus, 3 and 125 are factors of 375. While we know 3 is prime, we do not know about 125, so we apply the procedure to 125.

To factor 125, we can start with 3 instead of 2 because we already know 2 is not a factor of $375 = 3 \times 125$. Test dividing 3 into 125 gives $q = 41$ and $r = 2$. So 3 is not a divisor of 125. Trying 5, we find $125 = 5 \times 25$, whence $375 = 3 \times 5 \times 25$. Thus, $375 = 3 \times 5 \times 25$, where 3 and 5 are primes.

What is left to do is test 25, which we recognize as 5×5 from the multiplication table. Thus,

$$375 = 3 \times 5 \times 5 \times 5$$

where all the factors are prime.

Next, consider 243. As in the last example, since 243 ends with an odd digit, it is not divisible by 2. Thus, try 3. Division shows $243 = 3 \times 81$. Moreover, knowledge of the multiplication table gives $81 = 9 \times 9$ and $9 = 3 \times 3$, so we quickly arrive at:

$$243 = 3 \times 3 \times 3 \times 3 \times 3.$$

Lastly, consider 164. Since 164 is **even** (its *ones* digit is divisible by 2), we know 2 is a divisor, and we have $164 = 2 \times 82$, where 82 is again even. So another division by 2 gives $164 = 2 \times 2 \times 41$. Again, knowledge of the multiplication table tells us 41 is not in the body of the table, whence it does not factor and it must be prime. Thus,

$$164 = 2 \times 2 \times 41.$$

15.4 Lowest Terms Procedure

We are now in a position to give a procedure for putting any fraction in lowest terms. We assume the fraction is positive, because given a positive fraction $\frac{m}{n}$, its additive inverse can be written as $(-1) \times \frac{m}{n}$ and the factor -1 has no effect on whether the fractional part is in lowest terms.

Thus, let m, $n \in \mathcal{N}$ be any positive integers. We consider the fraction $\frac{m}{n}$. The following procedure will put $\frac{m}{n}$ in lowest terms:

Step 1: express m as a product of prime factors;

Step 2: express n as a product of prime factors;

Step 3: if no prime factor of m is a prime factor of n, then m and n have no common factor and $\frac{m}{n}$ is in lowest terms so we stop;

Step 4: if some prime factor p of m is also a factor of n, then $m = m_1 \times p$ and $n = n_1 \times p$, so $\frac{m}{n} = \frac{m_1 \times p}{n_1 \times p}$ which is not in lowest terms;

Step 5: restart the procedure to determine whether $\frac{m_1}{n_1}$ is in lowest terms.

We consider some examples.

Let's start with $\frac{18}{21}$. Using the previous procedure for determining prime factors, we find:

$$18 = 2 \times 3 \times 3 \quad \text{and} \quad 21 = 3 \times 7.$$

This takes us to Step 3, where we note 3 occurs in both factorizations, so we go to Step 4. In Step 4 we write the numerator as 6×3 and the denominator as 7×3, whence the revised fraction is

$$\frac{18}{21} = \frac{3 \times 6}{3 \times 7} = \frac{\cancel{3} \times 6}{\cancel{3} \times 7} = \frac{6}{7}.$$

Next we go to Step 5 which tells us to apply the process to the revised fraction. Repeating Steps 1 and 2, we have $6 = 2 \times 3$, and 7 is already prime. Since there is no common factor, we conclude

$$\frac{18}{21} = \frac{6}{7}$$

is in lowest terms.

As our next example, consider $\frac{63}{72}$. The prime factorizations for the numerator and denominator are:

$$63 = 3 \times 3 \times 7 \quad \text{and} \quad 72 = 2 \times 2 \times 2 \times 3 \times 3.$$

There is a common factor of 3 which can be eliminated as in:

$$\frac{63}{72} = \frac{3 \times 21}{3 \times 24} = \frac{21}{24}.$$

Applying the procedure to $\frac{21}{24}$ produces a factored numerator of 3×7 and a factored denominator of $2 \times 2 \times 2 \times 3$, which again share a common factor of 3. Eliminating the common factor produces the fraction $\frac{7}{8}$. Applying the process to this fraction shows $\frac{7}{8}$ is in lowest terms because the numerator is the prime 7 and it is not a factor in the denominator. Because of its importance, we present this calculation in complete detail:

$$\frac{63}{72} = \frac{3 \times 3 \times 7}{2 \times 2 \times 2 \times 3 \times 3}$$
$$= \frac{7 \times 3 \times 3}{2 \times 2 \times 2 \times 3 \times 3}$$
$$= \frac{7 \times \cancel{3} \times \cancel{3}}{2 \times 2 \times 2 \times \cancel{3} \times \cancel{3}}$$
$$= \frac{7}{2 \times 2 \times 2} = \frac{7}{8}.$$

The summary calculation effectively factors the numerator and denominator, and then cancels all common factors. Obviously, this is quicker than repeating the steps in the procedure. The key is that only common *factors* are cancelled. In other words, the arithmetic operation must be multiplication. As greater knowledge of the multiplication table is acquired, even shorter calculations are possible, as in:

$$\frac{63}{72} = \frac{9 \times 7}{8 \times 9} = \frac{\cancel{9} \times 7}{8 \times \cancel{9}} = \frac{7}{8}.$$

The requirement to convert arbitrary fractions into lowest terms materially increases the knowledge and skill requirements. Specifically, to master the procedures given, a child needs to be able to factor numbers and correctly implement the cancellation process. But there is more, as our next topic shows.

15.4.1 Reducing Sums to Lowest Terms

We have given a procedure for adding two fractions. This procedure, which is derived from R12 in the Summary, **works for all fractions**, including algebraic fractions, which is why it's mastery is so important. However, it is incomplete, because it does not, in general, produce a result that is in lowest terms, even when the two fractions being added are in lowest terms. The following example illustrates this failure:

$$\frac{1}{2} + \frac{1}{4} = \frac{4+2}{8} = \frac{6}{8},$$

since $\frac{6}{8}$ is not in lowest terms due to there being a common factor of 2 in the numerator and the denominator. A revised computation which includes reducing is:

$$\frac{1}{2} + \frac{1}{4} = \frac{6}{8} = \frac{3 \times \cancel{2}}{4 \times \cancel{2}} = \frac{3}{4}.$$

313

There is one pitfall in reducing fractions that the reader should be aware of because it arises regularly in respect to fractional forms having algebraic numerators and denominators. Consider the 2 sitting inside the box in the numerator in the following computation:

$$\frac{x + \boxed{2}}{6} = \frac{x + \boxed{2}}{3 \times 2}.$$

Since the denominator contains a factor of 2, many children learning algebra, and some college students, experience an irresistible urge to cancel a 2 from the denominator where it is a **factor** against the $\boxed{2}$ in the numerator where it is a **summand**, not a factor.

Think of this temptation as a **sin**. It is always incorrect. Let's see why. We know that $\frac{x}{y} = x \times \frac{1}{y}$. This means

$$\frac{x + \boxed{2}}{6} = (x + \boxed{2}) \times \frac{1}{6}.$$

The parentheses have to be there because the numerator is a sum! Thus the addition has to be carried out **before** the multiplication. This is why

> **only factors appearing in both the numerator and the denominator may be cancelled**.

The 2 in the box in the numerator is not a factor and therefore cannot be cancelled.

A complete procedure for the adding fractions

$$\frac{m}{n} + \frac{p}{k} = ?$$

would need to include a fourth step:

Step 1: Check whether the two denominators, n and k, are equal;

Step 2: If the two denominators are equal, that is, $n = k$, find $m + p$ and the sum is:

$$\frac{m}{n} + \frac{p}{n} = \frac{m + p}{n}$$

and go to Step 4;

Step 3: If the denominators are not equal, that is, $n \neq k$, find the products $n \times k$, $m \times k$ and $p \times n$, and the sum is:

$$\frac{m}{n} + \frac{p}{k} = \frac{m \times k + p \times n}{n \times k}$$

and go to Step 4;

314

Step 4: verify the sum found in Step 3 or 4 is in lowest terms; if not, reduce it to lowest terms.

Remember, when dealing with Steps 3 and 4 it is important to discuss how order of precedence affects the computations.

Consider the following:

$$\frac{13}{36} + \frac{5}{36} = \frac{18}{36} = \frac{1 \times 18}{2 \times 18} = \frac{1 \times \cancel{18}}{2 \times \cancel{18}} = \frac{1}{2}.$$

This example demonstrates that even in the simplest cases where both fractions have the same denominator and are in lowest terms, the sum will not automatically have to be in lowest terms. More importantly, it shows that when the numerator is a factor of the denominator, as in 2×18, we can represent the numerator as a product by writing it as 1×18. Seeing this spelled out completely, shows where the residual 1 in the numerator of $\frac{1}{2}$ comes from.

Even as revised, the procedure for adding fractions is not all that complicated. But there is one more twist, namely, the requirement that the addition process use the smallest denominator that will serve the purpose. Since this may well be part of you child's instruction, we turn to that now.

15.4.2 Least Common Denominator or Multiple

Suppose we want to add two fractions having the denominators 4 and 6. Using the procedure given above, the common denominator would be 4×6. The question is whether there is a smaller denominator that will do? It turns out that a number known as the **least common multiple** of the two denominators (also called the **least common denominator**) is the number we are looking for.

The procedure for finding the least common multiple of two counting numbers n and k is simply to list their multiples until the least common multiple is found. Since the least common multiple must be $\leq n \times k$, the listing process is sure to stop.

As a numerical example, consider 4 and 6. If we start listing multiples of 6 and 4, we have the following:

$$6, \ 12, \ 18, \ 24$$
$$4, \ 8, \ 12.$$

We list multiples of the larger number, 6, first, because there will always be fewer computations required to get to the product than if we start with the smaller number. To produce the entire list of multiples for 6 up to $6 \times 4 = 24$ has four numbers. The equivalent list for 4 contains six numbers, but our list for 4 stops at 12 because 12 is in both lists and is the least common multiple.

We use the least common multiple to add fractions in the following manner:

$$\frac{1}{4} + \frac{5}{6} = \frac{1 \times 3}{4 \times 3} + \frac{5 \times 2}{6 \times 2}$$
$$= \frac{3}{12} + \frac{10}{12} = \frac{13}{12}.$$

Let's compare the addition procedure using least common multiple with the computation given by the general rule set down as Step 3 in §14.2.3 above:

$$\frac{1}{4} + \frac{5}{6} = \frac{1 \times 6}{4 \times 6} + \frac{5 \times 4}{6 \times 4}$$
$$= \frac{6}{24} + \frac{20}{24} = \frac{26}{24}.$$

The answer has to be reduced since both the numerator and denominator have a common factor. We immediately have:

$$\frac{26}{24} = \frac{13 \times 2}{12 \times 2} = \frac{13 \times \cancel{2}}{12 \times \cancel{2}} = \frac{13}{12}.$$

In this case, the procedure using Step 3 appears to require the extra step of reducing to lowest terms which was not present in the first computation using the least common denominator, 12. However, reducing the final answer may still be required as the following example shows.

Find $\frac{1}{2} + \frac{1}{6}$. As before, to find the least common multiple, we begin by listing multiples of 6:

$$6,\ 12,$$

of which only two multiples are needed. Since 2 is a divisor of 6, we know that 6 is the least common multiple of 2 and 6. Thus,

$$\frac{1}{2} + \frac{1}{6} = \frac{1 \times 3}{2 \times 3} + \frac{1}{6} = \frac{3}{6} + \frac{1}{6} = \frac{4}{6}.$$

Notice that 4 and 6 are both even, so this fraction is not in lowest terms. It has to be reduced and this process gives:

$$\frac{4}{6} = \frac{2 \times 2}{3 \times 2} = \frac{2 \times \cancel{2}}{3 \times \cancel{2}} = \frac{2}{3}.$$

Again, we compare with Step 3 of the basic procedure in §14.2.3:

$$\frac{1}{2} + \frac{1}{6} = \frac{1 \times 6}{2 \times 6} + \frac{1 \times 2}{6 \times 2} = \frac{6}{12} + \frac{2}{12} = \frac{8}{12}.$$

Reducing this fraction involves two repetitions of cancelling a common factor of 2, or one step cancelling a common factor of 4, as in:

$$\frac{8}{12} = \frac{2 \times 4}{3 \times 4} = \frac{2 \times \not{4}}{3 \times \not{4}} = \frac{2}{3}.$$

The advantage of using the least common multiple of the denominators, n and k, instead of the product $n \times k$ resides totally in the fact that the least common multiple **may** be smaller than the product. But it may not be smaller, in which case the investment in making a list of multiples has been wasted.

15.5   Common Fractions: The Essentials

We have studied all the standard arithmetic computations with fractions. It is now possible to identify the essential facts that must be known to a child to succeed with fractions. The reader will note that we have dealt with the multiplication of fractions **before** their addition. In the author's view, this approach is simpler and more understandable to any child who knows what unit fractions are and understands the Fundamental and Notation Equations.

For purposes of this discussion, we assume m, n, p, $q \in \mathcal{I}$ and n, $q \neq 0$.

A child must know that the unit fraction $\frac{1}{n}$ is the result of dividing a whole into n equal parts. As such the Fundamental Equation results in:

$$n \times \frac{1}{n} = \frac{1}{n} \times n = 1.$$

A consequence of the Distributive Law, or equivalently that multiplication is repetitive addition, is the Notation Equation:

$$\frac{m}{n} = m \times \frac{1}{n}.$$

These first two items are absolutely essential to understanding the arithmetic of fractions.

To multiply fractions a child must know and be able to apply:

$$\frac{m}{n} \times \frac{p}{q} = \frac{m \times p}{n \times q}.$$

To divide fractions a child must understand that division is multiplication by the reciprocal. Using the standard notation which is introduced by Grade 6, the reciprocal of $\frac{m}{n}$ is given by:

$$\left(\frac{m}{n}\right)^{-1} \equiv \frac{1}{\frac{m}{n}} = \frac{n}{m}$$

and that to write this, we have the added requirement that $m \neq 0$. Using the reciprocal, we have

$$\frac{m}{n} \div \frac{p}{q} \equiv \frac{m}{n} \times \frac{1}{\frac{p}{q}} = \frac{m}{n} \times \frac{q}{p}.$$

where we have the added requirement that $p \neq 0$.

To succeed with addition, a child must know and be able to apply the following:

1. to find common denominators and reduce fractions:

$$\frac{k}{k} = 1 \quad \text{and} \quad \frac{m \times k}{n \times k} = \frac{m}{n};$$

2. to add fractions having the same denominator:

$$\frac{m}{n} + \frac{p}{n} = \frac{m + p}{n};$$

3. to add fractions having different denominators:

$$\frac{m}{n} + \frac{p}{q} = \frac{m \times q + n \times p}{n \times q}.$$

Finally, at the point that negative numbers are introduced, the child should understand that subtraction is a defined operation and that:

$$\frac{p}{q} - \frac{m}{n} \equiv \frac{p}{q} + \left(-\frac{m}{n} \right).$$

As we have seen, these equations can be converted to mechanical prescriptions for performing any required computation with fractions. The reasons why they are true are based on the Fundamental Equation and the Notation Equation which are explained in Chapter 13.

Most important is that these procedures apply generally. Inspection of the rules in §12.6.1 shows that these same rules are there. So anyone who knows and can apply these equations can succeed at the manipulations of algebra which involve fractions, whether these fractions are symbolic or numerical. This includes your child!

15.6 Improper Fractions and Mixed Numbers

A fraction $\frac{m}{n}$, m, $n \in \mathcal{N}$, is called **improper** if $m > n$, that is if the numerator exceeds the denominator as in $\frac{15}{4}$. We can think of $\frac{m}{n}$ in terms of division, as in

$m \div n$, and by applying the Division Algorithm we would obtain integers q and r such that

$$m = n \times q + r,$$

where $0 \leq r < n$. Thinking this way allows us to write the numerator as a sum, which after applying the Division Algorithm to $\frac{15}{4}$ gives

$$15 = 4 \times 3 + 3.$$

Using this sum we can rewrite the fraction as:

$$\frac{m}{n} = \frac{n \times q + r}{n} = \frac{n \times q}{n} + \frac{r}{n} = q + \frac{r}{n}$$

which for $\frac{15}{4}$ produces

$$\frac{15}{4} = \frac{3 \times 4 + 3}{4} = \frac{3 \times 4}{4} + \frac{3}{4} = 3 + \frac{3}{4}.$$

Note that because $m > n$, q will be at least 1. The notation for $q + \frac{r}{n}$ is referred to as a **mixed number** because it consists of a whole number and a **residual fraction** that satisfies $0 \leq \frac{r}{n} < 1$. The whole number will be the largest whole number that is less than or equal to our original fraction, $\frac{m}{n}$.

In splitting the sum into two fractions as in

$$\frac{n \times q + r}{n} = \frac{n \times q}{n} + \frac{r}{n}$$

the reader will observe that if we turn the equation around, we are simply adding two fractions having the same denominator using methods developed in §14.3.1:

$$\frac{n \times q}{n} + \frac{r}{n} = \frac{n \times q + r}{n}.$$

As such, it is simply another example of how every mathematical equation can be used in two directions.

The notation for mixed numbers can be ambiguous. For example, the last sum, $3 + \frac{3}{4}$, is often written suppressing the plus sign as:

$$3 + \frac{3}{4} = 3\frac{3}{4}$$

which makes clear why these are called mixed numbers. The notation with the plus suppressed is confusing because it could be misinterpreted as $3 \times \frac{3}{4}$ with the multiplication sign suppressed.

As another example, consider:

$$\frac{35}{8} = \frac{4 \times 8 + 3}{8} = \frac{4 \times 8}{8} + \frac{3}{8} = 4 + \frac{3}{8} = 4\frac{3}{8}.$$

15.6.1 Adding Mixed Numbers

Suppose we want to add these two mixed numbers:

$$4\frac{3}{8} + 3\frac{3}{4} = ?$$

No problems will arise provided the complete notation is used:

$$4\frac{3}{8} + 3\frac{3}{4} = \left(4 + \frac{3}{8}\right) + \left(3 + \frac{3}{4}\right) = 7 + \left(\frac{3}{8} + \frac{3}{4}\right)$$
$$= 7 + \left(\frac{3}{8} + \frac{6}{8}\right) = 7 + \frac{9}{8}.$$

The reader will notice the fraction in the last mixed number is improper, because $9 > 8$. Since $\frac{9}{8} = 1 + \frac{1}{8}$, the final answer will be

$$7 + \frac{9}{8} = 8 + \frac{1}{8}.$$

In its most compact form, this would be written as the mixed number

$$8\frac{1}{8}.$$

Here is where the ambiguity in mixed numbers and the use of juxtaposition as a substitute for $\times$ can lead to error. The temptation to multiply can be overwhelming, particularly for one with knowledge of the Fundamental Equation. However, if the plus sign is not suppressed, there can be no question of what is meant, namely, $8 + \frac{1}{8}$. Alternatively, one can take as a rule that

> **juxtaposition never substitutes for times in a purely numerical expression**.

Thus, $8\frac{1}{8}$, and the like, always means $8 + \frac{1}{8}$.

In summary, adding mixed numbers can be accomplished in a straight forward manner, as shown above:

> **add the whole numbers, add the fractions, reduce if necessary, record the result**.

Multiplying mixed numbers is a different story.

15.6.2 Multiplying Mixed Numbers

Multiplying mixed numbers is difficult because each mixed number is actually a sum. Since products of sums are common place, we illustrate the computation in general by finding the product of $a + b$ and $c + d$. Applying the Distributive Law to the sums gives:

$$\begin{aligned}(a + b) \times (c + d) &= a \times (c + d) + b \times (c + d) \\ &= (a \times c + a \times d) + (b \times c + b \times d).\end{aligned}$$

There are a total of four products to be found which then have to be summed. Let's see how this works out for the pair of mixed numbers above:

$$\begin{aligned}4\frac{3}{8} \times 3\frac{3}{4} &= \left(4 + \frac{3}{8}\right) \times \left(3 + \frac{3}{4}\right) \\ &= 4 \times \left(3 + \frac{3}{4}\right) + \frac{3}{8} \times \left(3 + \frac{3}{4}\right) \\ &= \left(4 \times 3 + 4 \times \frac{3}{4}\right) + \left(\frac{3}{8} \times 3 + \frac{3}{8} \times \frac{3}{4}\right) \\ &= \left(12 + \frac{4 \times 3}{4 \times 1}\right) + \left(\frac{3 \times 3}{8} + \frac{3 \times 3}{8 \times 4}\right) \\ &= (12 + 3) + \left(\frac{9}{8} + \frac{9}{32}\right) = 15 + \left(1 + \frac{1}{8} + \frac{9}{32}\right) \\ &= 16 + \left(\frac{4}{32} + \frac{9}{32}\right) = 16 + \frac{13}{32} = 16\frac{13}{32}.\end{aligned}$$

This computation is tedious, and there are many places in which errors can creep in. However, it is typical of how the Distributive Law is used and it is expected that children in Grade 6 will be able to perform similar computations.

In the case of multiplying mixed numbers, there is an alternative procedure which first converts the two mixed numbers to improper fractions by reversing the process of converting an improper fraction to a mixed number as described at the start of this section. Thus,

$$4\frac{3}{8} = \frac{4 \times 8}{8} + \frac{3}{8} = \frac{32}{8} + \frac{3}{8} = \frac{35}{8}$$

and

$$3\frac{3}{4} = \frac{3 \times 4}{4} + \frac{3}{4} = \frac{12}{4} + \frac{3}{4} = \frac{15}{4}.$$

We then find the product of the two improper fractions in the usual way:

$$4\frac{3}{8} \times 3\frac{3}{4} = \frac{35}{8} \times \frac{15}{4} = \frac{35 \times 15}{8 \times 4}$$

$$= \frac{525}{32} = \frac{16 \times 32 + 13}{32}$$
$$= \frac{16 \times 32}{32} + \frac{13}{32} = 16 + \frac{13}{32} = 16\frac{13}{32}.$$

The second calculation appears substantially easier to the author, with much less to keep track of for someone learning the procedure.

15.7 What Your Child Needs to Know

Goals for Grades 1-3 are in Chapter 13 and are not repeated here.

15.7.1 Goals for Grade 4

The child should be able to

1. recognize that a whole number is a multiple of each of its factors. e.g., 21 is a multiple of 3 and 7;

2. determine whether any counting number ≤ 100 is prime or composite;

3. explain computationally why $\frac{n}{m} = \frac{n \times p}{m \times p}$ (see §12.6.1 and R10);

4. understand that for $m > 1$, the fraction $\frac{m}{n}$ is the sum of m fractions $\frac{1}{n}$;

    (a) understand addition and subtraction of fractions as joining and separating parts in respect to the same whole;

    (b) decompose a fraction into a sum in multiple ways, e.g., $\frac{3}{8} = \frac{1}{8} + \frac{1}{8} + \frac{1}{8}$ or $\frac{3}{8} = \frac{1}{8} + \frac{2}{8}$;

    (c) decompose a mixed number into a sum in multiple ways, e.g., $2\frac{3}{8} = 2\frac{1}{8} + \frac{1}{8} + \frac{1}{8}$ or $2\frac{3}{8} = 1\frac{1}{8} + 1\frac{2}{8}$, or $2\frac{3}{8} = 2 + \frac{1}{8} + \frac{2}{8}$, etc.;

    (d) add and subtract mixed numbers involving like denominators by replacing the mixed numbers with equivalent fractions;

    (e) solve word problems involving addition and subtraction of fractions having like denominators by using equations to represent the problem;

5. apply previous understanding of multiplication, i.e., $\frac{m}{n} = m \times \frac{1}{n}$, to multiply fractions by whole numbers;

    (a) fully understand $p \times \frac{m}{n} = \frac{p \times m}{n}$, computationally;

322

(b) fully understand $p \times \frac{m}{n} = \frac{p \times m}{n}$, visually, e.g., $3 \times \frac{2}{5}$ is equivalent to $(3 \times 2) \times \frac{1}{5}$;

(c) solve word problems involving multiplication of fractions by whole numbers using equations and visual models. e.g., If each child at a party drinks $\frac{2}{3}$ of liter of lemonade and five children are attending the party, how much lemonade must be made?;

15.7.2   Goals for Grade 5

The child should be able to

1. use equivalent fractions as a means to add common fractions with unlike denominators and be fully aware that

$$\frac{m}{n} + \frac{p}{q} = \frac{mq + pn}{nq}; {}^{2}$$

2. understand that a correct interpretation of a fraction is as division of the numerator by the denominator, i.e., $\frac{m}{n} = m \div n$;

(a) use this interpretation to solve problems (A 20 kg sack of sugar is to be equally shared by 25 people. Express each person's share as a fraction.);

3. understand the equations

$$\frac{m}{n} \times q = \frac{m \times q}{n} = m \times \frac{q}{n}$$

in all the aspects that follow:

(a) interpret the product $\frac{m}{n} \times q$ as m parts of a partition of q into n equal parts;

(b) obtain this partition as the result of a sequence of operations leading to $m \times \frac{q}{n}$;

(c) understand this is equivalent to partitioning $m \times q$ into n equal parts;

(d) for example, use a visual fraction model to show $\frac{2}{3} \times 4 = \frac{8}{3}$, and create a story context for this equation;

---

[2]The reader should be aware that two letters written together as in nq is shorthand for $n \times q$ and such expressions appear in general usage and in the school curriculum. We have retained the use of $\times$ for clarity.

(e) do the same with $\frac{2}{3} \times \frac{4}{5} = \frac{8}{15}$;

(f) more generally, understand and interpret $\frac{m}{n} \times \frac{q}{p} = \frac{m \times q}{n \times p}$;

4. interpret multiplication as scaling (resizing), by:

   (a) comparing the size of a product to the size of one factor on the basis of the size of the other factor, without performing the indicated multiplication;

   (b) explaining why multiplying a given number by a fraction greater than 1 results in a product greater than the given number (recognizing multiplication by whole numbers greater than 1 as a familiar case); explaining why multiplying a given number by a fraction less than 1 results in a product smaller than the given number; and relating the principle of fraction equivalence $\frac{m}{k} = \frac{nm}{nk}$ to the effect of multiplying $\frac{m}{k}$ by 1;

5. find the area of a rectangle with fractional side lengths by tiling it with unit squares of the appropriate unit fraction side lengths, and show that the area is the same as would be found by multiplying the side lengths;[3]

6. use multiplication of fractions to find the area of rectangles having sides with fractional lengths and represent fraction products as rectangular areas;

7. solve real world problems involving multiplication of fractions and mixed numbers;

8. apply and extend the meaning of division to division of fractions by counting numbers;

   (a) interpret division of a unit fraction by a whole number computationally using the relationship between division and multiplication, e.g., $\frac{1}{4} \div 3 = \frac{1}{12}$, because $\frac{1}{12} \times 3 = \frac{1}{4}$;

   (b) solve real world problems involving division of unit fractions by whole numbers.

15.7.3   Goals for Grade 6

The child should be able to

1. interpret and compute quotients of fractions, e.g., $\frac{3}{4} \div \frac{2}{3}$;

---

[3]What is being requested here is a visual demonstration of $A = L \times W$ when the length and width are fractions.

2. solve real world problems involving quotients of fractions, e.g., What is the width of a piece of land having length $\frac{3}{4}$ mi and area $\frac{2}{3}$ sq mi?, How many $\frac{3}{4}$ cup servings can be made from 3 and $\frac{1}{2}$ cups of yogurt?, etc.;

3. write, read, evaluate and interpret algebraic expressions in which letters stand for numbers;

   (a) write expressions that record numerical operations. e.g., subtract 5 from x as $x - 5$;

   (b) identify parts of a mathematical expression using descriptors like sum, product, term, factor, quotient. e.g., in the expression

   $$4(7 + 3),$$

   4 is a factor, $7 + 3$ is a sum composed of two terms, the entire expression is a product;[4]

   (c) evaluate expressions for specific values of the variables, e.g., find the value of A given $A = (s + 2) \times t$ when $s = 3$ and $t = 7$;

4. apply the essential properties to generate equivalent algebraic expressions, e.g., $5(2 + 3x)$ becomes $10 + 15x$ by applying the Distributive Law, $2y + 4$ becomes $(y + 2)2$ again using the Distributive Law;

5. identify when two expressions are equivalent, e.g., $x + 2y$ and $(x + y) + y$ can be identified as equivalent using Associative and Distributive Laws;

6. understand that solving an equation for an unknown is a process of determining whether a particular value from a specified set substituted for the unknown will make the equation true;

7. solve equations of the form $x + p = q$ and $px = q$ for cases in which p, q and x are non-negative rational numbers;

8. understand ratio and proportion in relation to fractions, e.g. The ratio of chairs to tables is four to one. We have 80 chairs. What fraction should we multiply by 80 to determine the number of tables?;

---

[4]This type of material introduces your child to algebra which is not really covered in this text. However, to help your child with this type of material, using the grounding provided by this book, all you have to do is to go through the descriptors in your child's book and you should have no trouble figuring out what is required.

9. understand rates as a result of a division process, e.g., speed is distance divided by time;

10. use two variables to represent quantities in real world problems, e.g., d and t to represent time and distance in $d = 4t$ where 4 is speed in appropriate units.

15.7.4   Goals for Grade 7

The child should be able to

1. understand that for any pair of rational numbers p and q, $p - q = p + (-q)$ where $-q$ is the additive inverse of q;

2. understand that the addition and multiplication properties of the rationals arise by extending the Commutative, Associative and Distributive Laws to all numbers;

3. know and be able to use the fact that $(-1)(-1) = (-1) \times (-1) = 1$;

4. know that $(-p) \times q = p \times -q$ and more generally

$$-\left(\frac{p}{q}\right) = \frac{-p}{q} = \frac{p}{-q};$$

5. use the rules of arithmetic to generate equivalent linear algebraic expressions $\frac{2x}{3} = \frac{2}{3} \times x$;

   (a) add linear algebraic expressions with rational coefficients, e.g., $2x + 5x = 7x$;

   (b) subtract algebraic expressions with rational coefficients;

   (c) factor algebraic expressions with rational coefficients, e.g., $\frac{2x}{3} + 3x = (\frac{2}{3} + 3)x$;

   (d) multiply algebraic expressions with rational coefficients;

The are many additional expectations in the CCSS. But these become far more algebraic in nature. What you should realize is that by Grade 7 there is an increasing expectation that students will understand and be able to apply the complete set of rules identified in this chapter.

Chapter 16

Order Properties of $\mathcal{R}$

Chapter Overview. The order properties of the real numbers are developed using four properties of **positive** numbers. These are used to generate ten general order properties that apply to all real numbers. Methods for ordering common fractions are presented. These are used to examine the placement of common fractions and mixed numbers on the real line. Grade level goals are discussed.

16.1   Ordering Integers: A Brief Review

In Chapter 10 we we began our discussions of ordering. At the heart of the order idea is the concept of **more** applied to collections via the pairing process (see Chapter 2). Pairing led directly to counting and to the counting numbers which are primary in that their properties are derived directly from the behavior of collections in the real world. With the counting numbers came addition, zero and the additive inverses of counting numbers. As we know, these three types of numbers comprise the integers.

These distinctions between types of numbers are algebraic; that is, they are based on the behavior of a particular number in relation to the counting numbers. Thus, zero was defined as being the **additive identity** and defined by its behavioral property, $0 + n = n$. Similarly, negative numbers were identified as the **additive inverses** of counting numbers; that is, given $n \in \mathcal{N}$, the negative integer $-n$ satisfies $n + (-n) = 0$.

We turned these ideas into an order relation applying to all integers in §10.2 by first identifying the counting numbers as **positive** integers and then by asserting that given two integers m and n:

$$m < n \text{ if and only if } m + p = n$$

for some counting number p. It was also shown that this relation was completely equivalent to:

$$m < n \text{ if and only if } n + (-m) = p$$

is a counting number.

Using the above equivalences, in §10.5-6 we developed the properties of order as they interacted with the operations of addition and multiplication. As well, the important concept of the **number line** was developed as a means of illustrating the ordering of the whole numbers.

We shall follow the same approach to ordering the real numbers as we did with the integers. For integers, we started with the idea that the counting numbers are positive. With reals, we start with the idea that some reals are positive and list some behavioral axioms. These axioms are then used to identify which real numbers are positive and from there to defining the order relations in exactly the manner laid out above.

16.2 Axioms Governing Positive Numbers

The following four statements are axioms:

P1: some real numbers are **positive**;

P2: given $x \in \mathcal{R}$, exactly one of the following three assertions is true:

$$x = 0; \ x \text{ is positive}; \ -x \text{ is positive};$$

P3: if x and $y \in \mathcal{R}$ are positive, then $x + y$ is positive;

P4: if x and $y \in \mathcal{R}$ are positive, then $x \times y$ is positive.

P1 simply asserts some reals have the special property of being positive. P2 classifies each real number with respect to the property of being positive and is analogous to the three types of integers as either being a counting number, the additive inverse of a counting number, or zero. P3 and P4 simply say that the positive reals are closed under addition and multiplication. In this respect, positive reals are exactly like counting numbers.

Recall that the integers are a subset of the real numbers, and further that every counting number can be obtained by adding 1 to itself. So P3 guarantees that if 1 is positive, then every counting number is positive, and hence our axioms are identifying the correct subset of the integers as being positive. For this reason, it is important to verify that 1 is positive.

To accomplish this, we apply P2 and eliminate the possibility that 1 is either zero or the additive inverse of a positive real number.

We know that $1 \neq 0$, since this is a requirement that our arithmetic reflect the real world. The only remaining choices permitted by P2 are that 1 is positive, or -1 is positive, but both cannot be true. We consider the latter. If -1 is positive, applying R7 of §12.6.1 gives

$$(-1) \times (-1) = 1.$$

Since P4 tells us the product of positives is positive, 1 must also be positive as well as -1. But this is forbidden by P2. So 1 must be positive instead of -1, the only choice left under P2.

As noted, since 1 is positive, all the counting numbers are positive in just the way we expect. Further, if n is a counting number, $-n$ is identified as being in the remaining group of real numbers, since $-(-n) = n$ and the classification of integers into positive, negative and zero is exactly replicated.

16.3   The Nomenclature of Order

We shall use the following nomenclature:

1. $x \in \mathcal{R}$ is **negative** provided $-x$ is positive;

2. for x and $y \in \mathcal{R}$, we write $x < y$, and say x **is less than** y, provided there is a positive $z \in \mathcal{R}$ such that $x + z = y$;

3. for x and $y \in \mathcal{R}$, we write $x \leq y$, and say x **is less than or equal to** y, provided $x < y$ or $x = y$;

4. for x and $y \in \mathcal{R}$, we write $y > x$, and say y **is greater than** x, provided $x < y$;

5. for x and $y \in \mathcal{R}$, we write $y \geq x$, and say y **is greater than or equal to** x, provided $x < y$ or $x = y$.

This usage is consistent with previous usage specified in Chapter 10 and at the beginning of this chapter. We will appeal to these definitions in what follows.

16.4   Laws Respecting $<$ on $\mathcal{R}$

Let x, y and $z \in \mathcal{R}$ stand for arbitrary reals. Then the following four laws are satisfied by $<$:

1. the **Transitive Law**:

$$\text{if} \quad x < y \quad \text{and} \quad y < z, \quad \text{then} \quad x < z;$$

2. the **Trichotomy Law**, given any x and y, exactly one of the following holds:

$$x < y \quad \text{or}, \quad y < x \quad \text{or} \quad x = y;$$

3. the **Addition Law**:

$$\text{if} \quad x < y \quad \text{then} \quad x + z < y + z;$$

4. the **Multiplication Law**:

$$\text{if} \quad x < y \quad \text{and} \quad 0 < z, \quad \text{then} \quad x \times z < y \times z.$$

These four laws are identical to the laws in §10.5, except that they are made about real numbers and not integers. In §10.5 they were derived based on the definition

$$m < n \quad \text{if and only if} \quad m + p = n$$

for some counting number p. This definition has been replaced by

$$x < y \quad \text{if and only if} \quad x + z = y$$

for some positive real number z, which is the second definition in the section on nomenclature. The arguments and reasons underlying the truth of these four laws are the same as those given in §10.5.

16.5   Rules for Order and $<$

The following **Order Rules** have many useful applications. Again, let x, y, $z \in \mathcal{R}$ stand for arbitrary reals:

OR 1 $x < y$ if and only if[1] $y - x$ is positive.

OR 2 $0 < 1$, and if $n \in \mathcal{N}$, i.e., n is a counting number, then $1 \le n$;

---

[1]We remind the reader that the phrase *if and only if* between two assertions means that either assertion is a conclusion of the other. In the present case this means, given $x < y$, we can conclude $y - x$ is positive. Alternatively, given $y - x$ is positive, we can conclude $x < y$. This means the statements are **logically equivalent**. See §9.4.2 for further discussion.

OR 3 x is negative if and only if $x < 0$;

OR 4 the product of two negative numbers is positive;

OR 5 the product of a negative number and a positive number is a negative number;

OR 6 if $x \neq 0$, then $0 < x \times x$;

OR 7 if $0 < x$, then $0 < x^{-1}$;

OR 8 if $x < 0$, then $x^{-1} < 0$;

OR 9 if $0 < x < y$, then $0 < y^{-1} < x^{-1}$, where now the order is **reversed**;

OR 10 if x, y, w, and z are positive, then

$$\frac{x}{y} < \frac{w}{z} \quad \text{if and only if} \quad x \times z < w \times y;$$

OR 11 if $x < y$, then $x < \frac{x+y}{2} < y$.

We verify these rules in order.

Let's see why OR 1, which asserts that $x < y$ is equivalent to $y - x$ is positive is valid. First we recall that $y - x \equiv y + (-x)$. From the definition of $<$, we know that $x < y$, means there is a positive z, such that

$$x + z = y.$$

Adding $-x$ to left on both sides of this equation gives:

$$(-x) + (x + z) = (-x) + y.$$

The LHS reduces to z via

$$(-x) + (x + z) = (-x + x) + z = 0 + z = z.$$

The RHS to $y - x$ via

$$(-x) + y = y + (-x) \equiv y - x.$$

So $y - x$ is positive, since $z = y - x$, and z is positive. Conversely, from $y - x$ is positive, we set $z = y - x$, and add x to both sides producing $x + z = x + (y - x)$. Reversing the steps leading to $y - x$ above, gives $x + z = y$, whence since z is positive, we have $x < y$.

331

The first statement in OR 2 asserts that $0 < 1$. Since 1 is positive and 0 is the additive identity, we have

$$0 + 1 = 1,$$

whence by definition of $<$, $0 < 1$. The second statement is $1 \leq n$. Since every counting number, n, can be obtained by adding 1 to itself, and 1 is positive, $1 \leq n$ also follows.

OR 3 asserts x being negative is equivalent to $x < 0$. To see this, for x negative, $-x$ is positive, and $x + (-x) = 0$, so $x < 0$ by the definition of $<$. Conversely, given $x < 0$, we know there exists positive z so that $x + z = 0$. But uniqueness of additive inverses tells us that $z = -x$, so $-x$ is positive and x is negative as claimed.

OR 4 asserts the product of two negative numbers is a positive number. Recall Rule 7 of 12.6.1 that $x \times y = (-x) \times (-y)$. Now if x and y are negative, then $-x$ and $-y$ are positive, whence their product is positive and by Rule 7, so is $x \times y$. Further, if $x \neq 0$, then either x is positive, or $-x$ is positive. In either case, $x \times x$ has to be positive, and this is the result of OR 6.

OR 5 asserts the product of a negative x and a positive y is negative. To see this, we know from OR 3 that x negative is equivalent to $x < 0$. The Multiplication Law (§15.4) tells us that inequalities are preserved when we multiply through an inequality by positive numbers like y. Thus,

$$x \times y < 0 \times y = 0,$$

so that $x \times y$ is negative by OR 3, as claimed.

OR 7 and 8 assert x and its multiplicative inverse, $x^{-1} = \frac{1}{x}$, are both be positive, or both negative. The key fact is that the multiplicative inverse is defined by the equation that asserts the product, $x \times \frac{1}{x} = 1$ and this product is positive, irrespective of whether x is positive or x is negative. Since OR 5 tells us the product of a positive number and a negative number is negative, it is clear both x and $\frac{1}{x}$ must both be positive or both negative.

This brings us to OR 9 asserting that $0 < x < y$ implies $0 < \frac{1}{y} < \frac{1}{x}$. Thus assume $0 < x < y$. From OR 7, the multiplicative inverses of x and y must also be positive. We also know the product of positive numbers is positive (P4), so $\frac{1}{x} \times \frac{1}{y}$ is positive, whence multiplying through $0 < x < y$ by this product preserves the inequalities by the Multiplication Law so that

$$0 \times \left(\frac{1}{x} \times \frac{1}{y} \right) < x \times \left(\frac{1}{x} \times \frac{1}{y} \right) < y \times \left(\frac{1}{x} \times \frac{1}{y} \right).$$

Performing the indicated operations using the Associative and Commutative Laws leaves:

$$0 < \frac{1}{y} < \frac{1}{x},$$

which is what we wanted. Restated in the notation of common fractions and counting numbers we have

$$0 < n < m \quad \text{if and only if} \quad 0 < \frac{1}{m} < \frac{1}{n}.$$

This result must be applied with care because the **inequalities reverse**. One good way to think about this result is to remember

> **when you make the denominator of a fraction larger, you make the value of the fraction smaller**.

OR 10 asserts that if $0 < x,\ y,\ z,\ w$ ($x,\ y,\ z,\ w$ are all **positive**),

$$\frac{x}{y} < \frac{w}{z} \quad \text{if and only if} \quad x \times z < w \times y.$$

OR 6 tells us $\frac{1}{y}$ and $\frac{1}{z}$ are both positive. Since the product of positives is positive, we have that the products $y \times z$ and $\frac{1}{y} \times \frac{1}{z}$ are both positive. Thus, we know from the Multiplication Law that multiplying through any inequality by either one of these products will preserve the inequality. If we multiply the inequality on the LHS of OR 10 by $y \times z$ we have

$$\frac{x}{y} \times (y \times z) < \frac{w}{z} \times (y \times z).$$

Since the y's cancel on the LHS and the z's on the RHS, we see this is just

$$x \times z < w \times y$$

which is the RHS of OR 10. Conversely, if we multiply through this last inequality by the positive product $\frac{1}{y} \times \frac{1}{z}$, we obtain

$$(x \times z) \times \left(\frac{1}{y} \times \frac{1}{z} \right) < (w \times y) \times \left(\frac{1}{y} \times \frac{1}{z} \right).$$

Cancelling the z's on the LHS and the y's on the RHS and applying the Notation Equation to what is left produces the LHS of OR 10 and completes the equivalence. Restating OR 10 for positive common fractions, we have

$$\frac{q}{m} < \frac{p}{n} \quad \text{if and only if} \quad q \times n < p \times m.$$

The reader will notice that the products being formed are the **numerator of one fraction times the denominator of the other**. The process of forming these products is referred to as **cross multiplication**.

Lastly, consider OR 11. Given $x < y$, it follows from the Addition Law that $x + x < x + y$ and $x + y < y + y$. Since $x + x = x \times 2$ and $y + y = y \times 2$, we have

$$x \times 2 < x + y < y \times 2.$$

Since $0 < 2$ implies $0 < 2^{-1}$, we can multiply through the inequality by the multiplicative inverse of 2, to obtain

$$(x \times 2) \times 2^{-1} < (x + y) \times 2^{-1} < (y \times 2) \times 2^{-1}$$

which, after computation, is exactly:

$$x < \frac{x + y}{2} < y.$$

We note that OR 11 merely asserts that

> **the arithmetic average of two unequal numbers lies strictly between them**.

16.6 Ordering Positive Common Fractions

The properties and rules developed above apply to all real numbers. Knowledge of these properties can be used to order algebraic fractions of the type studied in higher grades. For this reason the CCSS-M places heavy emphasis that children be able to correctly apply these rules to common fractions, so we turn our attention there. In the primary and elementary grades, negative integers are not treated until Grade 6, so we will focus on fractions which are ratios of counting numbers, hence positive. Our focus will be on developing easily used computational rules for comparing numerical fractions.

16.6.1 Comparing Unit Fractions

In elementary school, studying the order properties of fractions begins with ordering **unit fractions**. These have been studied extensively in Chapter 13. Unit fractions can be thought of as being the result of dividing a unit whole into some number of equal parts. As part of their initial study of fractions, every child should arrive at the conclusion that the more equal parts a given whole is divided into, the smaller each part must be. This fact about unit fractions is summarized as:

$$1 > \frac{1}{2} > \frac{1}{3} > \frac{1}{4} > \frac{1}{5} > \frac{1}{6} > \frac{1}{7} > \ldots > 0.$$

Each unit fraction is positive because it is the multiplicative inverse of a positive number. The remaining relations are a consequence of OR 9 which when stated for unit fractions asserts: for any pair of counting numbers, n, m, if $n < m$, then

$$0 < \frac{1}{m} < \frac{1}{n}.$$

In summary, comparing unit fractions is straight forward and an easily remembered rule is:

Given two unit fractions, the one having the larger denominator is the smaller of the two.

16.6.2 Comparing Positive Fractions With the Same Numerator

Next we consider the case of comparing two fractions having the same numerator as in $\frac{q}{m}$ and $\frac{q}{n}$. The Notation Equation tells us that

$$\frac{q}{m} = q \times \frac{1}{m} \quad \text{and} \quad \frac{q}{n} = q \times \frac{1}{n}.$$

By using the Multiplication Law and the fact that q is positive

$$q \times \frac{1}{m} < q \times \frac{1}{n} \quad \text{if and only if} \quad \frac{1}{m} < \frac{1}{n}.$$

Since the fractions on the RHS are unit fractions, this relation holds only in the case that $m > n$. This case simply extends our previous rule to arbitrary numerators.

Once again, this fact should be consistent with a child's initial learning experience:

given any quantity whatsoever, the more equal parts it is divided into, the smaller each part must be.

16.6.3 Comparing Fractions With The Same Denominator

Let's fix a counting number n as the single denominator and note that $0 < \frac{1}{n}$. Given any other counting numbers p and q, we know from the Multiplication Law that

$$q < p \quad \text{if and only if} \quad q \times \frac{1}{n} < p \times \frac{1}{n}.$$

The RHS is $\frac{q}{n} < \frac{p}{n}$ which is what we want and gives us the following rule:

Given two fractions having the same denominator, the one having the larger numerator will be the larger.

16.6.4 Comparing Arbitrary Fractions

Comparing arbitrary fractions is difficult because, as the computations in §16.6.2 and 16.6.3 show, increasing the numerator of a fraction has the opposite effect of increasing the denominator of the fraction. For example, consider a fixed fraction, $\frac{q}{m}$ and that we want to know whether it is more or less than another fraction of the form

$$\frac{q + k}{m + j}$$

where k, j are counting numbers. If we consider the effects on the denominator and numerator separately, we know

$$\frac{q}{m + j} < \frac{q}{m} \quad \text{but} \quad \frac{q}{m} < \frac{q + k}{m}.$$

The first effect makes the denominator larger and the fraction smaller. The second makes the numerator larger and the fraction larger. But which will dominate? There is no way to tell without further computation.

Cross multiplication using OR 10 provides a simple procedure to find the answer. Namely, for any two positive common fractions $\frac{q}{m}$ and $\frac{p}{n}$

$$\frac{q}{m} < \frac{p}{n} \quad \text{if and only if} \quad q \times n < p \times m.$$

In words, cross multiply and compare the resulting products (numerator of first times denominator of second with numerator of second times denominator of first) which will be positive integers.

The equivalence expressed in OR 10 is also valid for the other order relations as listed below:

$$\frac{q}{m} < \frac{p}{n} \quad \text{if and only if} \quad q \times n < p \times m$$

$$\frac{q}{m} = \frac{p}{n} \quad \text{if and only if} \quad q \times n = p \times m$$

$$\frac{q}{m} > \frac{p}{n} \quad \text{if and only if} \quad q \times n > p \times m,$$

where in each case the numbers being compared on the RHS are positive integers!

In the next section we give a variety of numerical examples. But let's examine one example to illustrate how these facts are used. Suppose we want to know the relationship between $\frac{25}{26}$ and $\frac{26}{27}$. What we know is

$$\frac{25}{26} \; \boxed{?} \; \frac{26}{27} \quad \text{if and only if} \quad 25 \times 27 \; \boxed{?} \; 26 \times 26$$

where what goes in the box is either $<$, $=$, or $>$. Since

$$25 \times 27 = 675 < 676 = 26 \times 26,$$

we know what goes in the box is $<$.

Summarizing for the general case,

$$\frac{q}{m} \; \boxed{?} \; \frac{p}{n} \quad \text{if and only if} \quad n \times q \; \boxed{?} \; m \times p$$

where again what goes in the box is either $<$, $=$, or $>$.

16.6.5 Some Numerical Examples

Given counting numbers m, n, p, q, the following three steps provide a procedure for determining the order of the two common fractions, $\frac{q}{m}$ and $\frac{p}{n}$.

Step 1 if $q = p$ (**Equal Numerators**), the fraction having the smaller denominator is larger;

Step 2 if $m = n$ (**Equal Denominators**), the fraction having the larger numerator is larger;

Step 3 if $q \neq p$ and $m \neq n$ (**Unequal Numerators and Unequal Denominators**), cross multiply to obtain $q \times n$ and $p \times m$ and apply the summary line to the cross products to conclude;

337

(a) if the cross products are equal, the fractions are equal;

(b) if $q \times n < p \times m$, then $\frac{p}{n}$ is the larger, that is, $\frac{q}{m} < \frac{p}{n}$;

(c) if $q \times n > p \times m$, then $\frac{q}{m}$ is the larger, that is, $\frac{q}{m} > \frac{p}{n}$;

(c') if $p \times m < q \times n$, then $\frac{q}{m}$ is the larger, that is, $\frac{p}{n} < \frac{q}{m}$.

(We have included (c') because it uses **less than**.)

We apply this procedure to the following list of fractions:

$$\frac{1}{2}, \frac{1}{3}, \frac{3}{5}, \frac{3}{7}, \frac{3}{8}, \frac{6}{13}, \frac{7}{14}, \frac{11}{23}.$$

For $\frac{1}{2}$ and $\frac{1}{3}$, we have $\frac{1}{3} < \frac{1}{2}$ by Step 1.

To compare $\frac{1}{2}$ with the remaining fractions which have the form $\frac{p}{n}$, we must determine the order relation that goes in the box as shown:

$$\frac{1}{2} \quad \boxed{?} \quad \frac{p}{n}$$

for each fraction in the list. We find the relation by using cross multiplication as described in Step 3. Thus, we compare, $1 \times n$ with $p \times 2$, in other words, the denominator of the second fraction with the numerator of the second fraction multiplied by 2. The relationship between n and $p \times 2$ will be the same as the relationship between $\frac{1}{2}$ and $\frac{p}{n}$. Thus,

$$3 \times 2 < 7, \quad \text{so} \quad \frac{3}{7} < \frac{1}{2},$$
$$3 \times 2 < 8, \quad \text{so} \quad \frac{3}{8} < \frac{1}{2},$$
$$6 \times 2 < 13, \quad \text{so} \quad \frac{6}{13} < \frac{1}{2},$$
$$11 \times 2 < 23, \quad \text{so} \quad \frac{11}{23} < \frac{1}{2}.$$

Ordering $\frac{1}{3}$ with any of the remaining fractions, $\frac{p}{n}$, again uses cross multiplication by comparing the order relationship between $n = 1 \times n$ and $p \times 3$. Thus,

$$5 < 3 \times 3, \quad \text{so} \quad \frac{1}{3} < \frac{3}{5},$$
$$7 < 3 \times 3, \quad \text{so} \quad \frac{1}{3} < \frac{3}{7},$$
$$8 < 3 \times 3, \quad \text{so} \quad \frac{1}{3} < \frac{3}{8},$$

$$13 < 6 \times 3, \quad \text{so} \quad \frac{1}{3} < \frac{6}{13},$$
$$14 < 7 \times 3, \quad \text{so} \quad \frac{1}{3} < \frac{7}{14},$$
$$23 < 11 \times 3, \quad \text{so} \quad \frac{1}{3} < \frac{11}{23}.$$

To compare $\frac{3}{5}$ with $\frac{3}{7}$ or $\frac{3}{8}$, we use Step 1 which reveals

$$\frac{3}{8} < \frac{3}{7} < \frac{3}{5} \quad \text{since} \quad 5 < 7 < 8.$$

To compare $\frac{3}{5}$ with the remaining fractions requires Step 3. In this case the cross multiples are $3 \times q$ and $5 \times p$. Thus, $\frac{6}{13} < \frac{3}{5}$ since

$$6 \times 5 < 13 \times 3, \quad \text{so} \quad \frac{6}{13} < \frac{3}{5}.$$

We can also use previous relationships as in

$$\frac{11}{23} < \frac{7}{14} = \frac{1}{2} < \frac{3}{5}.$$

We have $\frac{3}{7} < \frac{6}{13}$, since $3 \times 13 = 39 < 42 = 6 \times 7$ and $\frac{6}{13}$ with $\frac{11}{23}$ since

$$6 \times 23 = 138 < 143 = 13 \times 11,$$

so $\frac{6}{13} < \frac{11}{23}$.

Combining all these facts and applying the Transitive Law gives:

$$\frac{1}{3} < \frac{3}{8} < \frac{3}{7} < \frac{6}{13} < \frac{11}{23} < \frac{1}{2} = \frac{7}{14} < \frac{3}{5}.$$

The procedures are computational and always work. For this reason, use of Steps 1-3 in the author's view it is the simplest way of determining the relative order of common fractions.

16.7   Positioning Fractions on the Real Line

The CCSS-M demands that children understand the relationship of numbers to the line so for this reason we return to it and focus on the placement of fractions.

In §10.4 we developed a graphical description of the integers called the real line and described its construction. For purposes of discussion and illustration, we reproduce our previous diagram:

A graphical description of the integers. The integers are placed on the line in such a way that there is one unit of distance between each pair of successive integers.

A review of the construction process shows that a key step is the positioning of 0 and 1 which determines the physical length of the **unit interval** in the diagram being constructed.

Depending on the purpose of the diagram of the real line, we might choose the physical unit of length to be one inch, or one centimeter, or one foot. Either of the first two choices would be appropriate in making a diagram to fit in a book, but the last would not because it is too long. So it is clear exactly how the choice of unit affects the diagram, we redraw the diagram using a larger unit of length.

A second graphical description of the integers. The physical length of the unit interval from 0 to 1 in this diagram measures twice that in the diagram above. This physical length is the same between any integer n and its successor, $n + 1$. That this length changes between diagrams is why measurement units are essential when it comes to working in the world.

16.7.1   Placing Fractions in the Unit Interval

A discussion of placing fractions on the line generally begins with proper fractions in which the numerator is less than the denominator, whence they satisfy

$$0 < \frac{q}{m} < 1.$$

Recalling the discussion in §13.2, we know the unit fraction $\frac{1}{m}$ (m a counting number) subdivides the unit interval into m equal parts (see diagrams in §13.2). This fact together with the Notation and Fundamental Equations determine the placement of each proper fraction with denominator m. For example, for $m = 4$, there are three proper fractions,

$$\frac{1}{4}, \quad \frac{2}{4} = \frac{1}{2}, \quad \text{and} \quad \frac{3}{4}$$

which are positioned in the unit interval as shown.

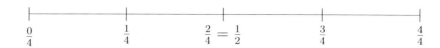

A graph showing the placement of proper fractions having denominator 4. The fractions divide the unit interval into four equal parts. Labelling starts at the left with $\frac{1}{4}$.

The graph below shows the position of the six proper fractions having denominator 7 in the unit interval.

A graph showing the placement of fractions in the unit interval having denominator 7. These fractions divide the interval into seven equal parts. The naming starts from the left and uses the principles developed in §13.2.

Suppose we want to construct a single diagram showing both sets of fractions. The essential fact that governs everything is that fractions with denominator 4 divide the unit interval into four equal parts and those with denominator 7 divide the same interval into seven equal parts. Constructing a diagram for *sevenths* can be done by setting 7 unit lengths end-to-end to form a unit interval and marking each subdivision with the appropriate fraction. This will produce the last diagram. The required divisions for *fourths* can be added using the fact that $\frac{1}{2}$ must divide the whole interval into 2 equal parts, then $\frac{1}{4}$ and $\frac{3}{4}$ subdivide each of the intervals created by the placement of $\frac{1}{2}$ into 2 equal parts. This diagram is shown next.

A graph showing the placement of fractions in the unit interval having denominator 7 followed by those with denominator 4.

To further illustrate the complex order relationships between fractions, we re-draw this diagram, this time including all the fractions with denominator 9. You might want to think about where fractions with denominator 8 would be placed, and whether knowing this would be helpful in respect to placing the fractions with denominator 9.

341

A graph showing the placement of fractions having denominator 4, 7 and 9 that are in the unit interval. The relative placement of fractions having different denominators requires calculation. Note the symmetry about $\frac{1}{2}$.

The fraction $\frac{1}{2}$ is referred to as a **benchmark**. This is because it divides the unit interval into two equal parts. If you consider the other fractions in the interval having the same denominator, for example fractions with denominator 7, they are distributed symmetrically about $\frac{1}{2}$. This means we can use the information from placing fractions less than $\frac{1}{2}$ on the line to tell us how to place fractions on the line that are greater than $\frac{1}{2}$.

While the idea of benchmarks is useful, it cannot substitute for knowledge of the principles and rules respecting the ordering of fractions, and the ability to correctly perform computations required to order common fractions. We see this when we try to place the list of fractions from our first example into a graph of the unit interval. While knowledge that most of the fractions have to be to the left of $\frac{1}{2}$ in the diagram, their exact placement requires calculations as discussed above.

A graphical description of the fractions from the initial list of fractions to be ordered. As the reader can see, $\frac{1}{2}$ sits half way between 0 and 1. The remaining fractions are placed according to their fraction of the unit length, the determination of which requires calculation.

Review of all the diagrams in this section leads to the conclusion that deciding the order relationships among arbitrary fractions is difficult. That is why children need to have Step 3 to fall back on. It always works.

We will return to the problem of placing numbers on the line when we discuss decimals.

16.7.2   Placing Mixed Numbers on the Real Line

The reader will recall our discussion of positive mixed numbers in Section 15.6. There we asserted that the numerator of every improper fraction, $\frac{m}{n}$, can be written in the form: $q + \frac{r}{n}$, where q is an integer, and r is an integer satisfying, $0 \leq r < n$. This fact is a result of applying the Division Algorithm to $m \div n$.

Suppose we consider a rational number, $\frac{m}{n}$, in the interval from 25 to 26 so that:

$$25 \leq \frac{m}{n} < 26.$$

By applying the Division Algorithm to $m \div n$, we obtain $m = q \times n + r$ where $0 \leq r < n$. Substituting into $\frac{m}{n}$ gives

$$\frac{m}{n} = \frac{q \times n + r}{n}$$
$$= \frac{q \times \cancel{n}}{\cancel{n}} + \frac{r}{n} = q + \frac{r}{n}$$

where q, r, $n \in \mathcal{N}$ and $0 < r < n$ so that $\frac{r}{n}$ is a proper fraction and hence is in the unit interval. Since $25 \leq q + \frac{r}{n} < 26$ and $0 \leq \frac{r}{n} < 1$, we have

$$25 - \frac{r}{n} \leq q < 26 - \frac{r}{n}.$$

Since there is only one integer in this interval, namely 25, we conclude that $q = 25$ and

$$\frac{m}{n} = 25 + \frac{r}{n}$$

and

$$m = 25 \times n + r.$$

Thus, to place the improper fraction, $\frac{m}{n}$, on the line, we would write the numerator as $m = q \times n + r$ using the Division Algorithm and understand that this means that $\frac{m}{n}$ is the sum of the quotient q and the residual $\frac{r}{n}$ which is a **proper fraction**.

Given any other fraction in the unit interval, that is, $\frac{p}{k}$ where p, $k \in \mathcal{N}$ and $p < k$, the Addition Law tells us that

$$\frac{r}{n} < \frac{p}{k} \quad \text{if and only if} \quad 25 + \frac{r}{n} < 25 + \frac{p}{k}.$$

Thus, the relative position on the line of the improper fraction $\frac{m}{n} = 25 + \frac{r}{n}$, in respect to the position of any **other** fraction $25 + \frac{p}{k}$ in the interval between 25 and 26, has to be the same as the relative position of the residual, $\frac{r}{m}$ in respect to the residual $\frac{p}{k}$ when these two fractions are considered as members of the unit interval between 0 and 1.

Of course there is nothing special about 25, so that every interval defined by consecutive integers merely repeats the unit interval in respect to the relative positions of its constituent numbers.

A graph showing the placement of fractions having denominator 4 in the interval from 25 to 26. They are equally spaced and divide the interval into four equal parts. The last member coincides with 26. Written as improper fractions, the numbers are

$$\frac{100}{4},\ \frac{101}{4},\ \frac{102}{4},\ \frac{103}{4},\ \text{and}\ \frac{104}{4}.$$

The reader may wonder about negative numbers. We illustrate this situation by presenting the analogous graph for the interval from -8 to -7.

A graph showing the placement of fractions having denominator 4 in the interval from -8 to -7. See text.

Notice that the governing equation for mixed numbers with denominator 4 in the interval is:

$$-\frac{31}{4} = \frac{(-8) \times 4 + 1}{4} = \frac{(-8) \times 4}{4} + \frac{1}{4}$$
$$= \frac{(-8) \times \cancel{4}}{\cancel{4}} + \frac{1}{4} = -8 + \frac{1}{4}.$$

In summary, each interval between successive integers has the same order structure as the unit interval. For this reason, the focus on the relative ordering of fractions can be confined to the unit interval. We will see further application of these ideas when we take up decimals, the topic of the next chapter.

16.8 Intervals and Length

We have defined the **unit interval** to be the set of all real numbers x that satisfy

$0 \le x$ and $x \le 1$, that is, all real numbers that are at least 0 and no more than 1.

Further, we know that this interval defines the **unit length** (see §10.4 and 12.1.2). In particular, we know from the previous discussion that the position of any counting number, n, on the real line is exactly n units of distance to the right of 0. Further, as discussed in §12.1.2, the position of any proper fraction $\frac{m}{n}$ is exactly $m \times \frac{1}{n}$ units of distance to the right of 0. From these two facts it follows that for any positive real number x, the position of x on the real line is x units of distance to the right of 0. Negative real numbers y are identified by the property that $-y$ is **positive** and positioned a distance of $-y$ units to the left of 0.

Since each real number is either positive, negative or 0, each has a position on the real line which corresponds to its distance from 0. We want to generalize the these ideas.

Given two real numbers a and b that satisfy $a \leq b$, the set of real numbers x that satisfy

$a \leq x$ and $x \leq b$, that is, x is at least as large as a and x is no more than b

is called the **interval from** a **to** b. In set notation, the interval from a to b is written as:

$$\{x : a \leq x \leq b \quad \text{and} \quad x \in \mathcal{R}\}.$$

There is nothing very difficult going on here. The only thing you need to remember is the left endpoint, a, is less than or equal to the right endpoint b. This requirement ensures that an interval contains at least one real number, namely, a.

In terms of this notation the unit interval would be described as the interval from 0 to 1 and written in set notation as:

$$\{x : 0 \leq x \leq 1 \quad \text{and} \quad x \in \mathcal{R}\}.$$

With any interval from a to b, we associate a **length**, given by $b - a$. Here is another place where the requirement that $a \leq b$ becomes important. Specifically, we have

$$a \leq b \quad \text{if and only if} \quad a + z = b$$

for some $z \geq 0$. From our previous calculations in the chapter, we know that in fact $z = b - a$. Thus, the **length of an interval is a non-negative number**, which is what we would expect from the real world. The unit interval has length $1 = 1 - 0$ which is also as it should be.

16.9 Size of a Real Number

There are lots of quantities in the real world that represent the **size** of something. Let's consider a few.

With respect to geometry, length provides a number that tells us the size of the distance between two points. Area provides a number that tells us the size of plots of land. Volume tells us the size of a cube, or a ball, or a quantity of liquid. In finance we have numbers that tell us the amount of money that a given bill represents. In physics we have mass as a numerical measure of amount of resistance to motion, and so on.

The one thing we can take from all these examples is that a numerical measure of size should be a non-negative number.

Mathematicians considered this question in respect to real numbers. Specifically, they asked

What is the size of a real number?

What is being looked for is an exact numerical measure that answers the question: How big is x?

For example, we might ask: How big is 50? Now if your immediate response is that the best answer to this question is 50, you'd be absolutely right! In respect to positive real numbers, to say the size of x is x exactly preserves the order relationships between positive real numbers developed in this chapter. To be clear, using this measure, that the size of x is x, for positive real numbers satisfies the following:

the size of x is less than the size of y if and only if $x < y$.

So we might consider saying the size of x was less than the size of y provided $x < y$ as a way of extending the notion of size to all real numbers.

But consider,

$$-75 < 50.$$

Do we want to say the size of -75 is less than the size of 50?

Or, consider

$$-75 < -10.$$

Do we want to say the size of -75 is less than the size of -10?

Or worse yet,

$$-75 < 0.$$

Surely we want zero to have a smaller size than any other real number.

The solution to these questions is that we should take the size of a number, x, to be the length of the interval from 0 to x when x is non-negative, and the length of the interval from x to 0 when x is negative. In short, the distance from 0 to x.

This leads to the following rule for calculating the size of a real number, x:

if $0 \leq x$, the size of x is x; if $x < 0$, the size of x is $-x$.

The reader can verify that this rule gives exactly the length of the interval from 0 to x if x is positive, and the length of the interval from x to 0 if x is negative.

For the examples above, the size of 50 is 50; the size of -75 is $-(-75) = 75$ and the size of -10 is $-(-10) = 10$. So -10 has a smaller size than 50 which again has a smaller size than -75. As well, using this rule we have the size of 0 is less than or equal to the size of every other real number.

16.9.1 Absolute Value

If you think about it, you'd probably guess there is a special name for the size of a number. In this you'd be right. The size of a number is called its **absolute value**. Needless to say, there is a special notation too. The absolute value of x is denoted by $|x|$. Using this notation, we rewrite the rule for its calculation as:

$$\text{if } 0 \leq x, \text{ then } |x| = x, \text{ and if } x < 0, \text{ then } |x| = -x.$$

The CCSS-M identifies Grade 6 as the appropriate time to introduce absolute value to children.

It is also important to remember that the absolute value of any number is its distance from 0. Thus, if x is positive, the number x will be found x units to the right of 0 on the line. If x is negative, then x will be found x units to the left of 0. This is illustrated below.

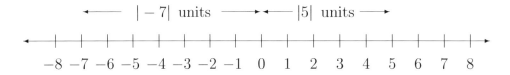

A graphical illustration of the positions of 5 and -7 in terms of their absolute values $|5|$ and $|-7|$. Geometrically, $|x|$ is the length of the interval from 0 to x.

16.10   Distance and Addition

In §10.4.2 we discussed the effect of adding an integer to a fixed integer m in respect to the real line. The CCSS-M demand children have a thorough understanding of these ideas, so we review them here for real numbers.

Consider the problem of finding the quantity $3+x$ on the real line. To accomplish this we might think of starting at 3 on the line and then moving to $3+x$. Using the ideas concerning intervals, we know the distance between any two real numbers , a and b, is the length of the interval from a to b where a is taken to name the smaller number. So we need to determine the distance between 3 and $3+x$ and then, by starting at 3 and moving the required distance, either to the right or to the left, we will end up at $3+x$.

How do we find the distance we need to move? We calculate the length of the interval from $3+x$ to x. There are two possibilities concerning x, either it is positive, or it is negative. If the former, then $3 \le 3+x$ and the length of the interval from 3 to $3+x$ is $(3+x)-3 = x$. If the latter, then $3+x \le 3$ and the length of the interval from 3 to $3+x$ is $3-(3+x) = -x$. In other words, length is given by

$$|(3+x)-3| = |x|.$$

This is one of the really useful properties of the absolute value. The absolute value of the difference between two numbers gives the distance between them **irrespective** of which is larger.

In which direction do we move? The answer is straight forward. If $0 < x$, we move to the right because we know $3 < 3+x$. If $x < 0$ we move to the left because we know $3+x < 3$. These ideas are illustrated below.

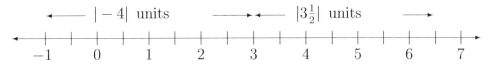

A geometric illustration of the result of adding $3\frac{1}{2}$ to 3 and -4 to 3. The first moves a distance equal to $|3\frac{1}{2}|$ to the right of 3 and the second moves a distance $|-4|$ to the left of 3.

The reader should be clear: there is nothing special about 3. If we fix a real number y on the line, then adding x simply moves one x units to the right or left of y depending on whether x is positive or negative as discussed above.

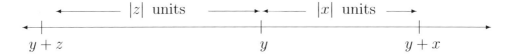

A geometric illustration of the result of adding $x > 0$ and $z < 0$ to a fixed y. The addition of x moves to a position $|x|$ units to the right from y. The addition of z moves to a position $|z|$ units to the left from y.

16.11 What Your Child Needs to Know

16.11.1 Goals for Grade 2

It is expected that your child will be able to:

1. measure the length of an object by selecting and using appropriate tools such as rulers, yardsticks, meter sticks, and measuring tapes;

2. measure the length of an object twice, using length units of different lengths for the two measurements;

3. describe how two measurements using different units relate to the size of the unit chosen;

4. estimate lengths using units of inches, feet, centimeters, and meters;

5. measure to determine how much longer one object is than another, expressing the length difference in terms of a standard length unit.

16.11.2 Goals for Grade 3

It is expected that your child will be able to:

1. compare two fractions with the same numerator or the same denominator by using the methods and principles discussed;

2. recognize that numerical comparisons of real-world fractions are valid only when the two fractions refer to the same whole, i.e., we can not compare $\frac{1}{2}$ a cantaloupe with $\frac{1}{3}$ of a watermelon;

3. record the results of numerical comparisons with the symbols $>$, $=$, or $<$, and justify the conclusions, e.g., by making and/or using a visual fraction model.

16.11.3   Goals for Grade 4

It is expected that your child will be able to:

1. compare two fractions with different numerators and different denominators, e.g., by creating common denominators or numerators as in $\frac{2}{9} < \frac{3}{8}$ because $\frac{2}{9} < \frac{3}{9} < \frac{3}{8}$, or by comparing to a benchmark fraction such as $\frac{1}{2}$;

2. recognize that numerical comparisons are valid only when the two fractions refer to the same whole, and numbers are in the same units, e.g. to compare $5°$ and $25°$ we must know both numbers refer to the same temperature scale;

3. record the results of comparisons with symbols $>$, $=$, or $<$, and justify the conclusions, e.g., by using a visual fraction model.

16.11.4   Goals for Grade 5

It is expected that your child will be able to:

1. use a pair of perpendicular number lines, called axes, to define a coordinate system, with the intersection of the lines (the origin) arranged to coincide with the 0 on each line and a given point in the plane located by using an ordered pair of numbers, called its coordinates;

2. understand that the first number in an ordered pair indicates how far to travel from the origin in the direction of one axis, and the second number indicates how far to travel in the direction of the second axis, with the convention that the names of the two axes and the coordinates correspond (e.g., x-axis and x-coordinate, y-axis and y-coordinate);

3. represent real world and mathematical problems by graphing points in the first quadrant of the coordinate plane, and interpret coordinate values of points in the context of the situation.

The reference to the first quadrant means that both x and y will be non-negative.

16.11.5   Goals for Grade 6

It is expected that your child will be able to:

1. understand that positive and negative numbers are used together to describe quantities having opposite directions or values (e.g., temperature above/below zero, elevation above/below sea level, credits/debits, positive/negative electric charge);

2. use positive and negative numbers to represent quantities in real-world contexts, explaining the meaning of 0 in each situation;

3. understand a rational number as a point on the number line;

4. extend number line diagrams and coordinate axes familiar from previous grades to represent points on the line and in the plane with negative number coordinates;

   (a) recognize opposite signs of numbers as indicating locations on opposite sides of 0 on the number line; recognize that the opposite of the opposite of a number is the number itself, e.g., $-(-3) = 3$, and that 0 is its own opposite;

   (b) understand signs of numbers in ordered pairs as indicating locations in quadrants of the coordinate plane;

   (c) recognize that when two ordered pairs differ only by signs, the locations of the points are related by reflections across one or both axes;

   (d) find and position integers and other rational numbers on a horizontal or vertical number line diagram;

   (e) find and position pairs of integers and other rational numbers on a coordinate plane.

5. understand ordering and absolute value of rational numbers;

   (a) interpret statements of inequality as statements about the relative position of two numbers on a number line diagram. For example, interpret the assertion, $-3 > -7$, as having the effect that -3 is located to the right of -7 on a number line oriented from left to right.

   (b) write, interpret, and explain statements of order for rational numbers in real-world contexts; for example, write $-3°C > -7°$ C to express the fact that $-3°$ C is warmer than $-7°$ C;

   (c) understand the absolute value of a rational number as its distance from 0 on the number line; interpret absolute value as magnitude for a positive or negative quantity in a real-world situation, for example, for an account balance of -30 dollars, write $|-30| = 30$ to describe the size of the debt in dollars;

   (d) distinguish comparisons of absolute value of two numbers from statements about their order; for example, recognize that an account balance less than -30 dollars represents a debt having a size of greater than 30 dollars;

6. solve real-world and mathematical problems by graphing points in all four quadrants of the coordinate plane.

Chapter 17

Exponentiation and Decimals

Chapter Overview. The notation 10^n is defined for any integer n. This notation is applied to Arabic Notation to generate decimals. The representation of rational numbers in this notation is discussed. Exponential notation in general is defined and rules for its behavior are presented.

We studied the Arabic System of notation for integers at length. Among its great features was that it supported the computations of arithmetic. In this chapter we shall see that this support extends to all real numbers when they are represented as decimals. Since the representation of reals as decimals is best understood in the context of powers of 10, our study begins with exponents and exponentiation.

17.1   What is Exponentiation?

We have already seen the exponent -1 and exponential notation used in a very limited way to specify **multiplicative inverses** as in §12.3.5:

$$4^{-1}, \quad \text{and} \quad x^{-1}.$$

The computations discussed in this chapter will extend this notation to include arbitrary integers as exponents, not just -1.

The reader will recall our emphasis on the idea that multiplication is repetitive addition. That this is so was particularly clear when one factor in a product is an integer. The intention with respect to exponentiation is that when the exponent is an integer, the result should be **repetitive multiplication of the base** times 1. Keeping this intuitive idea in mind should be helpful in understanding this material.

As discussed in §10.3, there are exactly three types of integers, positive, negative and 0 and all can occur as exponents. Thus, we must be able to interpret exponentiation as repetitive multiplication for each type of integer. Central to our implementation of exponentiation will be the facts from from the summary list in §12.6.1:

$$x = 1 \times x,$$

$$-x = (-1) \times x = x \times (-1)$$

and that an integer k is negative exactly if it is the additive inverse of a positive integer, in other words exactly if for some positive integer n,

$$k = -n = n \times (-1)$$

which is the form we will use below. Lastly, in providing meaning to quantities like 10^{-4} and 10^{-6}, we will have to ensure that we retain the existing meaning of the negative exponent -1 as specifying the **multiplicative inverse**.

17.2   Non-negative Integer Powers of 10

Our initial focus will be on exponents used in powers of 10, as in

$$10^n$$

because of their essential role in the Arabic Notation system.

To start, we define what we mean by 10^n where $0 \leq n \in \mathcal{I}$. We again note that the integer n is referred to as an **exponent** and the number 10 is called the **base**. The result, 10^n, is referred to as a **power of** 10 and we say

10^n is: 10 **raised to the** n^{th} **power**.

To give an entirely numerical example, we would say

10^4 is: 10 **raised to the** 4th **power**.

For $n \geq 0$, the rules E1 and E2 define the calculation of 10^n:

E1: $10^0 = 1$;

E2: for $n > 0$, $10^n = 10 \times 10^{n-1}$.

To find any actual value of 10^n, we need to remember that every non-negative integer can be obtained by starting with 0 and successively adding 1s. The way we actually use this result is to notice that if we start with any non-negative integer and repetitively subtract 1s, eventually we get to 0. To solidify the idea, let's do some calculations.

Suppose we want to know the value of 10^1. Using E2, we see that

$$
\begin{aligned}
10^1 &= 10 \times 10^{(1-1)} = 10 \times 10^0 \\
&= 10 \times 1 = 10,
\end{aligned}
$$

where to get the second line, we use E1.

To find a value for 10^2:

$$
\begin{aligned}
10^2 &= 10 \times 10^{(2-1)} = 10 \times 10^1 \\
&= 10 \times 10 = 100.
\end{aligned}
$$

Notice that in calculating 10^2 we can use the fact that we have already found $10^1 = 10$ to get the second line.

Similarly, we have

$$
10^3 = 10 \times 10^{(3-1)} = 10 \times 100 = 1000,
$$

and

$$
10^4 = 10 \times 10^3 = 10 \times 1000 = 10000,
$$

and so forth. So for positive n, the above simply translates into:

10^n denotes the number written as a 1 followed by n zeros.

A little thought should convince you that these rules for calculating non-negative powers of 10 implement our intention that exponentiation should be nothing more than **repetitive multiplication**. Thus, 10^7 is merely 7 copies of 10 multiplied together with 1. Further, 10^{12} should be twelve copies of 10 multiplied together with 1, or in our usual Arabic notation, a 1 followed by 12 zeros.

17.3   Negative Integer Powers of 10

To complete the task of defining all powers of 10 for integer exponents, we have to ask what 10^k is for k a negative integer, in other words for $k = -n$ where n is a positive integer.

Recall that an essential requirement on any specification for how to compute 10^{-n} is that we must preserve the notation $10^{-1} \equiv \frac{1}{10}$. In other words, however we compute 10^{-1}, we must end up with the **multiplicative inverse** of 10. Keeping this in mind leads to the following rule:

E3: for $n \geq 1$, $10^{-n} \equiv (10^n)^{-1} = \frac{1}{10^n}$.

The effective part of E3 is in the relation

$$10^{-n} \equiv (10^n)^{-1}$$

because on the RHS we have now have an exact instruction for how to do a calculation. Since $n > 0$, we can use E2 to compute 10^n. We then use the fact that $(10^n)^{-1}$ is the multiplicative inverse of 10^n, and this leads directly to

$$(10^n)^{-1} = \frac{1}{10^n}.$$

Further, since $-n = (-1) \times n = n \times (-1)$, we conclude

$$10^{(-1) \times n} = 10^{n \times (-1)} = (10^n)^{-1} = \frac{1}{10^n}.$$

If we apply E3 to 10^{-1} we have,

$$\begin{aligned}
10^{-1} &= (10^1)^{-1} \\
&= 10^{-1} = \frac{1}{10},
\end{aligned}$$

which is exactly what we needed. The actual use of E3 in the first line takes 10^{-1} and turns it into $(10^1)^{-1}$. Since $10^1 = 10$ by E2, we get what we want.

For higher powers, we have

$$\begin{aligned}
10^{-2} &= (10^2)^{-1} \\
&= 100^{-1} = \frac{1}{100},
\end{aligned}$$

$$\begin{aligned}
10^{-3} &= (10^3)^{-1} \\
&= 1000^{-1} = \frac{1}{1000}
\end{aligned}$$

and

$$\begin{aligned}
10^{-6} &= (10^6)^{-1} \\
&= 1000000^{-1} = \frac{1}{1000000}.
\end{aligned}$$

All the calculations work because we already know how to find 10^n for non-negative integers and m^{-1} for any non-zero integer.

There is one other extremely useful fact:

$$10^{-n} = \left(\frac{1}{10}\right)^n.$$

The reason why this is true is due to the fact that $\left(\frac{1}{10}\right)^n$ is also a multiplicative inverse for 10^n. Since E3 specifies that 10^{-n} is a multiplicative inverse for 10^n, we know these two quantities must be the same as shown in §12.3.3.

The key idea to remember for negative integer powers of 10 is:

if $-n$ is a negative integer, 10^{-n} is the unit fraction whose denominator is a 1 followed by n zeros, that is, $\frac{1}{10^n}$.

17.4 Powers of 10 and Arabic Notation

Let us recall the role of place in the Arabic Notation System for non-negative integers. Specifically, if we had a four digit number, the place of each digit had a different meaning, starting at the right and working left. Thus, in 8547, the 7 is in the *ones* place, the 4 in the *tens* place, the 5 in the *hundreds* place and the 8 is in the *thousands* place. The number represented by 8547 is the same as:

$$8547 = 8 \times 1000 + 5 \times 100 + 4 \times 10 + 7 \times 1,$$

as we discussed in Chapter 5. In that discussion, we used the names for each place, rather than powers of 10 as developed above.

There were two reasons for this. First, as taught to children, place is identified by its name, *ones, tens, thousands, ten thousands,* and so forth. Second, in order to discuss powers of 10 properly, we need a minimal knowledge of arithmetic which children in primary and elementary do not have, and which we did not have in Chapter 5. But now we have that knowledge, and so we can rewrite the equation above using exponents as:

$$8547 = 8 \times 10^3 + 5 \times 10^2 + 4 \times 10^1 + 7 \times 10^0.$$

Indeed, to express a much larger number, for example, we have:

$$\begin{aligned} 7653429 &= 7 \times 10^6 + 6 \times 10^5 + 5 \times 10^4 + 3 \times 10^3 \\ &\quad + 4 \times 10^2 + 2 \times 10^1 + 9 \times 10^0. \end{aligned}$$

The ultimate simplicity of this system for expressing a positive integer is now apparent. First each digit is multiplied by a power of 10 that is determined by its place. What is that power of ten? Well, starting with the right most digit, the power is

0, since $1 = 10^0$. With each step (place) to the left, we increment the power of 10 by 1. So, in our example, 7653429, the 5 is **four steps** to the left of the 9, so the power of 10 associated with that place is $0 + 4 = 4$, and the value of 5 in 7653429 is 5×10^4.

As we already know, the Arabic System provides a notation for each non-negative integer, and this notation extends to all integers via the centered dash. But we know there are many numbers that are not integers, for example, $\frac{1}{2}$. The question is:

Can we extend the Arabic System to apply to all real numbers?

17.5   Representing Rational Numbers

The Arabic System as discussed provides a notation for every integer. More importantly, the Arabic System supports arithmetic computations. The problem we face is that the real numbers include numbers that are not integers. Thus, we need to extend our system of notation to include all real numbers, and to do so in a manner that supports numerical computations.

Recall that in §15.6 we used the Division Algorithm and the Fundamental Equation to show how every rational number could be expressed as the sum of an integer and a non-negative proper common fraction. Expressing this in mathematical form, we see that any rational number, x, can be written as

$$x = m + \frac{q}{n}$$

where m is an integer and q, n are counting numbers with $q < n$. Since the RHS of this equation involves only integers, the reader may well ask:

Isn't this a perfectly good notation for x?

The answer is yes, but as every reader who has ever dealt with fractions knows, computing with fractions is not easy in the sense that adding or multiplying integers is easy. Moreover, once we get to mixed numbers, things get really messy (see §15.6).

The problem of extending the Arabic System to the rationals can be solved provided we are willing to make a sacrifice. We can use the notation for fractions as ratios of integers and give up on *supports computations*, or we can support computations and give up on **having an exact notation for each rational number**. In other words to solve the problem, we have to make a trade-off.

The hope that we might achieve a system that supports computations for rational numbers comes from the fact that when two fractions have the same denominator, their sum is obtained simply by adding the numerators as in:

$$\frac{m}{n} + \frac{q}{n} = \frac{m+q}{n}.$$

Since the numerators are integers, and we have a good system for integers that supports computations, by sticking with the one denominator idea we can find a way to make things work.

17.6 Extending the Arabic System to $\mathcal{Q}$

If we think about extending the Arabic System of notation to all of $\mathcal{Q}$, we see that an essential difficulty is how to extend the notion of place in respect to providing notations for positive mixed numbers. The difficulty arises because in any expression for an integer, the right-most place is the *ones* place and all places to the left are used by positive powers of 10.

To be clear, we need a numerical example, say 25. When I say all places to the left of the two are taken, what I mean is that if we add another digit, say 3, as in

$$325,$$

the 3 has a fixed, predetermined meaning, namely, 3×10^2 in this expression. Further, as we have seen, the predetermined meanings extend as many places to the left as we might try.

On the other hand, if we put the 3 on the right, as in

$$253,$$

the previous meanings attached to the 2 and 5 change, becoming 2×10^2 and 5×10^1, instead of 2×10^1 and 5×10^0, respectively.

More thought suggests that what is needed is a marker, that is, some notational device that marks the place of the *ones* digit.

The device that was chosen is the **decimal point** which is positioned immediately to the right of the digit intended to be the *ones* digit in a numeral.[1] The only function of the decimal point in Arabic numerals for numbers like 25.3 which are not integers is to locate the *ones* digit.

---

[1]The decimal point was introduced to the western world by the Persian mathematician al-Khwāizmhrī in the early 800s (see Wikipedia) based on Indian mathematics.

17.6.1 Interpreting Digits to the Right of the Decimal Point

Once we have figured out how to mark the place of the *ones* digit, we can add as many digits on the right as we choose. But we have to say how those digits will be interpreted.

The rule is simple:

> if a digit k in a numeral is n places to the right of the decimal point, its value is $k \times 10^{-n}$.

Thus, every digit to the right of the decimal point can be thought of as a fraction having a denominator that is a power of 10. Fractions having denominators that are powers of 10 are called **decimal fractions**. Numbers written in this form will be referred to as **decimal numbers**.

Let's see how this idea works out in practice. In the example just used, 25.3, we have

$$25.3 = 2 \times 10^1 + 5 \times 10^0 + 3 \times 10^{-1} = 25 + \frac{3}{10}$$

where the expression on the far right emphasizes that

> **every decimal number is the sum of an integer and a residual decimal fraction found in the unit interval**.

We will use this fact repetitively in what follows.

The integer part of a decimal number is determined by the digits to the left of the decimal point interpreted in their usual manner. Since we already know how to work with integers expressed in Arabic notation, we will concentrate on calculations with the residual, that is, the number represented by a decimal fraction that comes from the unit interval.

For example, consider:

$$
\begin{aligned}
0.205 &= 0 \times 10^0 + 2 \times 10^{-1} + 0 \times 10^{-2} + 5 \times 10^{-3} \\
&= \frac{2}{10} + \frac{0}{100} + \frac{5}{1000} \\
&= \frac{200}{1000} + \frac{0}{1000} + \frac{5}{1000} \\
&= \frac{205}{1000}.
\end{aligned}
$$

The last line of this expression illustrates how the residual ends up being a single decimal fraction even though each place to the right involves a different negative power of 10.

The names of the places in which digits to the right of the decimal point occur are simple; we use the name of the unit fraction multiplier. Thus, the first place to the right of the decimal point is the *tenths* place, the second digit to the right is the *hundredths* place, the third place to the right is the *thousandths* place, and so forth.

In the example 25.3, the 3 is in the *tenths* place. In 0.205, the 2 is in the *tenths* place, the 0 is in the *hundredths* place, and the 5 is in the *thousandths* place.

Let's summarize what we know so far. Every rational number expressed by a decimal numeral is the sum of an integer and a positive residual that can be found in the unit interval, that is, between 0 and 1. For the residual to be positive, the integer must be the largest integer that is **less than** the number in question. The residual has an exact representation as a proper decimal fraction, that is, a fraction having a power of 10 in the denominator and a numerator that is less than the denominator. For clarity, three numerical examples are

$$25.47 = 25 + \frac{47}{100}, \quad 1.9863 = 1 + \frac{9863}{10000}, \quad \text{and} \quad -4.24 = -5 + \frac{76}{100}.$$

The first two are straight forward because for positive numbers the largest integer **less than** the number is found by ignoring the decimal part which results in 25 and 1, respectively. The third example -4.24 is a negative number. Since $-4 \not< -4.24$, the largest integer less than -4.24 will be -5. Finding the **positive** residual can be a bit tricky because it will not simply be the decimal part as in the first two examples. So let's go through the computation in detail.

We know every real number that is not an integer lies in an interval between two consecutive integers. In this case the integers are -5 and -4, since adding the negative number $-.24$ moves one to the left on the line as shown in §10.4.2 and 16.10

$$-5 < -4.24 = -4 + (-.24) < -4.$$

Adding 5 through the inequality puts the negative decimal residual $1 + (-.24)$ in the interval from 0 to 1 as we see from

$$0 = -5 + 5 < -4.24 + 5 = 1 + (-.24) < -4 + 5 = 1.$$

If we call the positive residual we are trying to find r, then $r = 1 + (-.24)$ so that

$$r = \frac{100}{100} - \frac{24}{100} = \frac{76}{100}.$$

17.6.2 The Problem of $\frac{1}{3}$

In extending the Arabic System of notation to the rationals, we indicated that a trade-off was being made: the simple fact is that there are some rational numbers that cannot be expressed as an exact decimal number. What do we mean by this? Any number for which we can find a decimal expression involving a finite number of digits is said to have be an **exact decimal number**. For example,

$$\frac{1}{5} = \frac{2}{10} = 0.2,$$

so we would say $\frac{1}{5}$ has an exact decimal representation. There is a simple test for whether a given rational $\frac{m}{n}$ has an exact decimal representation. If it does, it means we can find a decimal number which we call d such that

$$d - \frac{m}{n} = 0.$$

Note that the expression for d can only involve a finite number of digits and it is not allowed to use any type of code to indicate additional digits. Clearly, the decimal number above representing $\frac{1}{5}$ has this property.

But the rational number $\frac{1}{3}$ does not have such an exact representation. To see why, suppose we consider 0.3 as a decimal number candidate to represent $\frac{1}{3}$. We apply the test to find

$$0.3 - \frac{1}{3} = \frac{3}{10} - \frac{1}{3} = \frac{9}{30} - \frac{10}{30} = -\frac{1}{30}.$$

The result is not zero. We might try $.33$. Performing the required calculation gives

$$0.33 - \frac{1}{3} = \frac{33}{100} - \frac{1}{3} = \frac{99}{300} - \frac{100}{300} = -\frac{1}{300}$$

which is closer to zero, but still not zero. Indeed, whatever decimal number we try, we will not get zero. The essential reason is that 3 is not a divisor of any power of 10, and hence no power of 10 will serve as a common denominator with 3 (see §15.4.2).

17.7 Placing Decimals on the Real Line

In what follows, it will be convenient to have a notational scheme for discussing arbitrary decimal fractions. Similar to the notational scheme set up in §8.3.1, we shall denote digits to the right of the decimal place by n_t for *tenths*, n_h for *hundredths* and n_{th} for *thousandths*. Thus,

$$0.n_t n_h n_{th} = \frac{n_t}{10} + \frac{n_h}{100} + \frac{n_{th}}{1000},$$

where the numerator in each case is one of the digits 0 - 9. Now let us return to the problem of finding decimals on the real line.

We have already discussed (Section 16.7) the placement of fractions on the real line. So, in a sense, we already know how to do this. Nevertheless, a careful description of how decimal numbers are placed on the real line will be helpful from several perspectives.

Since every decimal number, e.g., 27.6, is the sum of an integer and a decimal fraction, we really need only explore the position and relationship of decimal fractions found in the unit interval. So consider

$$0.736 = 0 + \frac{7}{10} + \frac{3}{100} + \frac{6}{1000} = \frac{736}{1000}.$$

In terms of our notational scheme, $n_t = 7$, $n_h = 3$ and $n_{th} = 6$.

To interpret the portion of decimal notation to the right of the decimal point, we work left to right, the opposite of the way we work with integers. So the first place to the right of the decimal point is the *tenths* place and the digit in this place is 7. The contribution to the sum from this place is:

$$.7 = \frac{7}{10}.$$

It is the largest individual contribution from any digit to the right of the decimal point and its position is shown in the illustration below showing the unit interval subdivided into 10 equal parts.

A graph showing the placement of fractions having denominator 10 in the unit interval. Each subinterval must have the same length, $\frac{1}{10}$, making the total length 1. The position of the decimal $.7$ is also shown.

Next consider the *hundredths* place. Because the digit n_h must be one of the digits from 0 to 9, the decimal $.7 n_h$ satisfies

$$.7 \le .7 n_h = \frac{7}{10} + \frac{n_h}{100} < \frac{7}{10} + \frac{10}{100} = \frac{8}{10} = .8.$$

For $n_h = 3$, we have

$$.7 \leq .73 = \frac{7}{10} + \frac{3}{100} < \frac{8}{10} = .8.$$

To be completely clear as to why this is true, we put all fractions over 100:

$$.7 = \frac{70}{100} < \frac{73}{100} < \frac{80}{100} = .8$$

where we are applying the standard procedure for ordering fractions having the same denominator (§16.6.3).

Now the really critical point that needs to be recognized is that the fractions $\frac{71}{100}$, $\frac{72}{100}$, $\ldots$, $\frac{79}{100}$, subdivide the interval from 0.7 to 0.8 into ten equal parts in exactly the same way that the fractions $\frac{1}{10}$, $\frac{2}{10}$, $\ldots$, $\frac{9}{10}$, subdivide the interval from 0 to 1 into ten equal parts. We show this, together with the placement of $.73$, graphically as follows:

A graph showing the placement of fractions having denominator 100 in the interval between $.7$ and $.8$. The decimal $.73$ is also shown. The reader should also notice that the fractions with denominator 100 divide the interval into ten equal parts each having length $\frac{1}{100}$.

Lastly, Consider the *thousandths* place in 0.736. Analogous to what we have already observed for *hundredths*, we have:

$$.73 \leq .73\,n_{th} \leq \frac{73}{100} + \frac{n_{th}}{1000} < \frac{73}{100} + \frac{10}{1000} = \frac{74}{100} = .74,$$

because n_{th} must be one of the digits from 0 to 9, so that in the example where $n_{th} = 6$:

$$.73 \leq .736 = \frac{73}{100} + \frac{6}{1000} < \frac{74}{100} = .74.$$

For clarity, we put all fractions over 1000 and apply the rules for ordering fractions:

$$.73 = \frac{730}{1000} < \frac{736}{1000} < \frac{740}{1000} = .74.$$

Again, we make the critical point, namely, that that the fractions $\frac{731}{1000}$, $\frac{732}{1000}$, $\ldots$, $\frac{739}{1000}$, subdivide the interval from 0.73 to 0.74 into ten equal parts in exactly the same way that the fractions $\frac{71}{100}$, $\frac{72}{100}$, $\ldots$, $\frac{79}{100}$, subdivide the interval from 0.7 to 0.8 into ten equal parts. We show this, together with the placement of $.736$, graphically as follows:

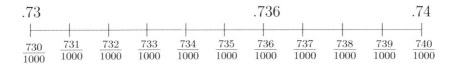

A graph showing the placement of fractions having denominator 1000 in the interval between .73 and .74. The placement of 0.736 is also shown.

The series of graphs above show ever finer division by concentrating on smaller intervals. Each interval being represented has a length that is one tenth the length of the preceding interval. Thus the interval from .73 to .74 has a length that is one tenth the length of the interval from .7 to .8, and so on. It is clear that this process of subdividing each interval into ten equal parts can continue for as long as we want. Consistent with the CCSS standards, we stop at *thousanths*.

The following diagram shows the placement of 0.736 in the original unit interval:

A graph showing the placement of 0.736 in the unit interval. Fractional forms have been replaced by their decimal equivalents.

Let's review what we've learned. Consider a decimal numeral $0.n_t n_h n_{th}$ where n_t, n_h and n_{th} are any of the digits from $0 - 9$. We know this numeral identifies a fraction in the unit interval. For clarity of exposition, we will take $n_t = 2$, $n_h = 9$ and $n_{th} = 5$, so we may think of our number as 0.295.

1. We know the unit interval is divided into ten equal parts by

$$.1, .2, .3, .4, .5, .6, .7, .8 \text{ and } .9;$$

the interval from .2 to .3 is divided into ten equal parts by

$$.21, .22, .23, .24, .25, .26, .27, .28, .29,$$

and the interval from .29 to .30 is divided into ten equal parts by

$$.291, .292, .293, .294, .295, .296, .297, .298, .299.$$

2. Given the *tenths* digit is 2, we know that 0.295 must lie in the interval between 0.2 and 0.3.

3. Given the *hundredths* digit 9, we know 0.295 must lie in the interval between 0.29 and 0.30.

To summarize, in terms of meaning, we have

$$0.n_t n_h n_{th} = \frac{n_t n_h n_{th}}{1000},$$

so that each such decimal in the unit interval is equivalent to a fraction having denominator 1000. More generally, each decimal in the unit interval is equivalent to a proper fraction having the denominator be a power of 10.

The discussion above tells us how to interpret any decimal number of the form: $0.n_t n_h n_{th}$. But what about numbers of the form $205.n_t n_h n_{th}$, or any other decimal form having non-zero digits to the left of the decimal point? Here, we simply use the fact that:

$$205.n_t n_h n_{th} = 205 + 0.n_t n_h n_{th} = 205 + \frac{n_t n_h n_{th}}{1000}.$$

In other words simply find 205 on the real line and treat the interval from 205 to 206 as though it were the unit interval to place 0.295. The position found in this interval will be the position of $205.n_t n_h n_{th}$. The essential fact which needs to be stressed here is that the interval between any two successive integers, k and $k+1$, looks exactly like the unit interval. The only difference is its position on the real line.

The above tells us how to interpret decimal notation. It does not tell us how to represent particular numbers in decimal notation. For example, we know we have a number which is represented by the fraction, $\frac{1}{4}$. Does this number have a decimal notation? If so, is there a procedure for finding it? The answer to both questions are: yes. But to provide that answer, we have to discuss the arithmetic of decimals.

17.8 Exponents Again

In the next chapter we discuss calculations with decimals. Since these calculations involve exponents, we develop the key rules governing integer exponents. The first step is to define x^n for x non-zero.

17.8.1 Non-negative Integer Exponents

Let x be any real number other than 0. For n a non-negative integer, the rules E1 and E2 define the calculation of x^n:

E1: $x^0 = 1$;

E2: for $n > 0$, $x^n = x \times x^{n-1}$.

As the reader can see, when $x = 10$ these rules are exactly the rules given previously in §17.2. Using the computations there as a guide, we have

$$x^1 = x \times 1, \quad x^2 = x \times x \times 1, \quad x^3 = x \times x \times x \times 1,$$

$$x^4 = x \times x \times x \times x \times 1, \quad x^5 = x \times x \times x \times x \times x \times 1$$

and so forth. If you count the number of factors of x in each product, it is the same as the exponent. Notice that we have not used parentheses in these calculations because so long as the operations entirely consist of multiplication, there is no need.

E2 can be restated as:

$$x^{n+1} = x \times x^n = x^n \times x.$$

17.8.2   Inverses, Exponents and Order of Precedence

Additive Inverses and Exponents

Every non-zero real number is the additive inverse of some **other** non-zero number. The rules E1 and E2 do not ask anything about x other than whether it is zero. If a given number is not zero, then E1 and E2 apply. Any other property of the particular number in question has no effect on the discussion. However, as pointed out in Chapter 12, ambiguities can arise when other operations are involved, for example, the centered dash which is used to denote the additive inverse. An order of precedence rule is used to eliminate this ambiguity.

Consider the expression -2^4. The centered dash instructs: *find the additive inverse of* $\boxed{?}$. The exponent 4, instructs: *apply E2 to* $\boxed{?}$. There are two questions. The first question is: Which operation is performed first? The second question is: To what does the operation apply, in other words, What's in the box?

The precedence rule tells us that

raising to a power always happens first.

The answer to the second question is that the operation of exponentiation is applied to the smallest expression possible. Let's see how this works for -2^4.

The first rule tells us that we have to compute the power first, so that the computation will look like:

$$- \left(\boxed{?}^4 \right) = - \left(\boxed{?} \times \boxed{?} \times \boxed{?} \times \boxed{?} \right)$$

where we have used parentheses to enforce computing the fourth power first. The answer to the second question is that the smallest expression that can go in the box is the 2 sitting to the left of the power in -2^4. Thus, these two rules force the following computation

$$-2^4 = -(2 \times 2 \times 2 \times 2) = -16.$$

If we want to raise -2 to the power of 4, then we have to use parentheses as in

$$(-2)^4 = (-2) \times (-2) \times (-2) \times (-2) = 16.$$

Here, the parentheses force the exponent to apply to -2, instead of simply 2.

Another example is:

$$x \times y^2 = x \times y \times y,$$

whereas

$$(x \times y)^2 = (x \times y) \times (x \times y) = x^2 \times y^2$$

and so forth.

Multiplicative Inverses and Exponents

As the reader knows, every non-zero real number has a multiplicative inverse. Indeed, since the criterion for determining whether two numbers are multiplicative inverses of one another is simply to check whether the product of those two numbers is 1, every non-zero real number can be viewed as a multiplicative inverse. This fact forces E1 and E2 to apply to multiplicative inverses in the same way as any other real number.

For example, consider, $x = \frac{1}{5}$. Then

$$\left(\frac{1}{5}\right)^2 = \frac{1}{5} \times \frac{1}{5} = \frac{1 \times 1}{5 \times 5} = \frac{1}{25} = \frac{1}{5^2}.$$

The fact that $\frac{1}{5}$ is a multiplicative inverse does not change the application of E1 and E2. Moreover, we already know how to do the calculation shown above (see §14.2.1). Further, **no matter how we choose to represent the numbers involved, we must get the same answer**. Thus, for example, all of the following must have the value $\frac{1}{25}$

$$5^{-1} \times 5^{-1} = (5^{-1})^2 = (5^2)^{-1} = 25^{-1}.$$

These considerations tell us how we should interpret negative integer exponents. We should simply take E3 to be a definition which asserts:

E3: If $m = -n$ for some $n \in \mathcal{N}$, $x^m = (x^n)^{-1} = \frac{1}{x^n}$.

Defining the effect of negative exponents in this way preserves all the notation developed for multiplicative inverses.

In the next chapter we will use these rules applied to powers of 10. For the moment we ask the reader to recognize the following notational equalities

$$\frac{1}{10} = 10^{-1}, \quad \frac{1}{100} = 10^{-2}, \quad \text{and} \quad \frac{1}{1000} = 10^{-3}$$

in understanding decimal arithmetic.

17.8.3   General Rules

We want to develop the key rules for dealing with integer exponents. To accomplish this, we will consistently use our ideas regarding x^n as a guide. Specifically, every calculation with integer exponents must come down to x^n for some positive integer n and some non-zero x.

The Addition Law for Exponents

The first question we consider is what happens when two non-negative powers of x are multiplied together? Thus, we ask:

$$\text{What is } x^n \times x^m?$$

The first thing to note is that the entire computation involves multiplication of factors of x. There are n factors of x associated with x^n and m factors of x associated with x^m. Determining the total number of factors comes down to counting, or using the power of addition, simply computing $n + m$. The second thing is to recall that because the only operation is multiplication, the Commutative and Associative Laws say we can compute the product in any order we want as long as we don't change the total number of factors of x which is fixed at $n + m$. Since the number of factors of x in x^{n+m} is exactly $n + m$, these facts convince us that

$$x^n \times x^m = x^{n+m}.$$

This equation is known as the **Addition Law for Exponents**.

It is essential to remember that in applying these results, there is a common base, as in

$$z^4 \times z^7, \quad 3^6 \times 3^5, \quad \text{or} \quad (x-5)^3 \times (x-5)^2.$$

The Addition Law does not apply in a situation of $a^3 \times b^2$ because a and b are different bases.

369

Using the Addition Law for Exponents in the cases above we get

$$z^{11}, \quad 3^{11}, \quad \text{and} \quad (x-5)^5,$$

respectively.

As the following computation shows, E3 is completely consistent with the Addition Law for Exponents:

$$7^{-2} = 7^{((-1)+(-1))} = 7^{-1} \times 7^{-1}.$$

We stress there is no other way to assign a meaning to 7^{-2} that is consistent with E1 and E2 and their consequences.

Product Law for Exponents

The second law concerns products of exponents. Let m be a fixed positive integer. Observe:

$$(x^m)^2 = x^m \times x^m = x^{m+m} = x^{m \times 2}$$

and similarly,

$$(x^m)^3 = x^m \times x^m \times x^m = x^{m+m+m} = x^{m \times 3}.$$

We could repeat the above sequence to obtain

$$(x^m)^4 = x^{m \times 4} \quad \text{or} \quad (x^m)^5 = x^{m \times 5}$$

or indeed

$$(x^m)^n = x^{m \times n}$$

where n is any other positive integer. This equation in known as the **Product Law for Exponents**.

Another way to think about this is to ask: What is the base in $(x^n)^m$? Clearly, the base is x^n because a the smallest thing the exponent m can apply to is x^n. So the computation has to be simply multiplying m copies of x^n together. The Addition Law for Exponents now tells us to add the exponents. Since all the exponents are the same, namely n, and there are m of them, the sum must be $m \times n$, which is exactly what the Product Law for Exponents asserts.

Product of Bases Rule for Exponents

There is one last rule, namely the **Product of Bases Rule for Exponents**:

$$(a \times b)^n = a^n \times b^n.$$

To see why this is true, consider $(a \times b)^3$. Using E2 we quickly obtain

$$(a \times b)^3 = (a \times b) \times (a \times b) \times (a \times b)$$

an expression that only involves multiplication. So the Associative and Commutative Laws guarantee we can rewrite it however we want, specifically as

$$(a \times b)^3 = (a \times a \times a) \times (b \times b \times b) = a^3 \times b^3.$$

A similar computation can be performed for any integral value of n. The point is that in $(a \times b)^n$, there are n factors of a and n factors of b and multiplying these factors together generates a^n and b^n, respectively.

The reader should be warned: this rule for products raised to a power applies **only to products and never to sums**.

We need one further rule for applying exponents. It was first discussed in respect to the exponent -1. It states:

in any expression an exponent applies to the smallest part of the expression possible.

Thus, in we have:
$$x \times y^3 = x \times y \times y \times y$$

so the exponent applies only to y and not to the product $x \times y$. If we want the latter, we must use parentheses as in: $(x \times y)^3$.

17.8.4   Summary of Rules for Exponents

Let x, y be fixed non-zero real numbers and n, m denote integers. The following two equations define exponential notation for non-negative integers:

E1: $x^0 = 1$;

E2: if $n > 0$, then $x^n = x \times x^{n-1}$.

The following rules for calculation apply.

E3: If $m = -n$ for some $n \in \mathcal{N}$, $x^m = (x^n)^{-1}$.

The Addition Law for Exponents:

$$x^n \times x^m = x^{n+m}.$$

The Product Law for Exponents:

$$(x^n)^m = x^{n \times m}.$$

The Product of Bases Law for Exponents:

$$(x \times y)^n = x^n \times y^n.$$

Finally, there is one order of precedence rule that asserts:

in any expression an exponent applies to the smallest part of the expression possible.

17.9   What Your Child Needs to Know

17.9.1   Goals for Grade 5

It is expected your child will be able to

1. incorporate decimal fractions into the place value system;

2. recognize that in a multi-digit number, a digit in one place represents 10 times as much as it represents in the place to its right and $\frac{1}{10}$ of what it represents in the place to its left;

3. explain patterns in the number of zeros of the product when multiplying a number by powers of 10;

4. explain patterns in the placement of the decimal point when a decimal is multiplied or divided by a power of 10;

5. use whole-number exponents to denote powers of 10;

6. read, write, and compare decimals to thousandths;

   (a) read and write decimals to thousandths using base-ten numerals, number names, and expanded form, e.g.,

   $$347.392 = 3 \times 100 + 4 \times 10 + 7 \times 1 + 3 \times \frac{1}{10} + 9 \times \frac{1}{100} + 2 \times \frac{1}{1000};$$

   (b) compare two decimals to thousandths based on meanings of the digits in each place, using $>$, $=$, and $<$ symbols to record the results of comparisons;

7. fluently multiply multi-digit whole numbers using the standard algorithm;

8. find whole-number quotients of whole numbers with up to four-digit dividends and two-digit divisors, using strategies based on place value, the properties of operations, and/or the relationship between multiplication and division;

9. illustrate and explain the calculations by using equations, rectangular arrays, and/or area models;

10. add, subtract, multiply, and divide decimals to hundredths, using concrete models or drawings and strategies based on place value, properties of operations, and/or the relationship between addition and subtraction;

11. relate the strategy to a written method and explain the reasoning used.

There are no goals related to decimals in higher grades. Thus, it is expected that on completion of Grade 5 your child has a complete understanding of the decimal system of arithmetic and can fluently perform the operations of addition, subtraction, multiplication and division using the standard algorithms which we cover in the next chapter.

Chapter 18

Arithmetic Operations with Decimals

Chapter Overview. The Arabic System of notation is reviewed using the full power of exponentials. The theory and practice of performing arithmetic operations using decimal numerals are discussed. Approximate notations for irrational numbers are discussed.

The true beauty of decimals is that there is almost nothing more to learn when it comes to doing calculations. For addition and subtraction, all we have to do is line up the decimal points and use the previously developed algorithms. For multiplication, we apply the previous algorithm as though the decimal numbers were integers and then count the decimal places in the two factors to determine the position of the decimal point in the product. Division of decimal numbers is only slightly more complicated.

To see why these procedures work, we will carefully explain two ways of looking at decimal numerals. In the first way, we will consider each numeral as a sum. Alternatively, we can consider each decimal numeral as the product of an integer and a negative power of 10. We will use these two representations to explain why the standard algorithms work.

18.1   Decimal Representations Again: Theory

Because a thorough knowledge of decimal representations is a prerequisite to understanding computations with decimals, we again review decimal notation making full use of exponential notation.

Recall that an arbitrary decimal numeral having three places on the right of the decimal point is written as:

$$n_{1000}n_{100}n_{10}n_1.n_t n_h n_{th}$$

where each entry is one of the single digits from the list of ten Arabic numerals multiplied by an integer power of 10. Thus in the numerical example 785.316,[1]

$$n_{100} = 7 \quad \text{and} \quad n_h = 1.$$

In this form, the integer power of 10 is not explicitly shown but is determined by the position of each digit in relation to the others and the decimal point. Making this information explicit, we have

$$n_{100} = 7 \times 100 = 7 \times 10^2, \quad \text{and} \quad n_h = \frac{1}{100} = 1 \times 10^{-2}.$$

Thus, if we write out our numerical example in complete detail, we have

$$
\begin{aligned}
785.316 &= 7 \times 100 + 8 \times 10 + 5 \times 1 + \frac{3}{10} + \frac{1}{100} + \frac{6}{1000} \\
&= 7 \times 10^2 + 8 \times 10^1 + 5 \times 10^0 + 3 \times 10^{-1} + 1 \times 10^{-2} + 6 \times 10^{-3}
\end{aligned}
$$

where the second line now makes full use of the exponential notation. Clearly, the decimal numeral on the LHS is more compact. That a decimal numeral, such as 785.316, conveys the same information as the expression on the RHS is due entirely to the fact that there is a common agreement as to what integer power of 10 is associated with each place. Thus, anywhere in the world if one were to ask: What power of 10 multiplies the second digit to the right of the decimal point?, the answer will always be the same: 10^{-2}. This fact is crucial to performing addition.

There is an alternate way to think about decimal numerals which is also useful and was mentioned above. Consider the following manipulation of our arbitrary decimal numeral:

$$n_{1000}n_{100}n_{10}n_1.n_t n_h n_{th} = n_{1000}n_{100}n_{10}n_1 + \frac{n_t n_h n_{th}}{1000}.$$

The sum on the RHS puts the fractional parts over the common denominator which is 1000. Because the Addition Law for Exponents tells us that

$$10^3 \times 10^{-3} = 10^0 = 1,$$

---

[1]Restricting the discussion to *thousandths* is consistent with CCSS-M requirements for Grade 5 but is not necessary in general.

we can recast the RHS of the last decimal numeral equation as:

$$n_{1000}n_{100}n_{10}n_1 + \frac{n_t n_h n_{th}}{1000} = n_{1000}n_{100}n_{10}n_1 + (n_t n_h n_{th}) \times 10^{-3}$$
$$= n_{1000}n_{100}n_{10}n_1 \times (10^3 \times 10^{-3}) + (n_t n_h n_{th}) \times 10^{-3}$$
$$= (n_{1000}n_{100}n_{10}n_1 000 + n_t n_h n_{th}) \times 10^{-3}$$
$$= (n_{1000}n_{100}n_{10}n_1 n_t n_h n_{th}) \times 10^{-3}.$$

The last expression is the product of an integer and a negative power of 10. The point is that

> **every decimal number can be expressed as the product of an integer and a negative power of 10 that records the number of places to the right of the decimal point in the decimal numeral for that number**

(in our case three), as in:

$$n_{1000}n_{100}n_{10}n_1.n_t n_h n_{th} = (n_{1000}n_{100}n_{10}n_1 n_t n_h n_{th}) \times 10^{-3}.$$

A numerical example will be helpful, so consider 8136.207. Applying the sequence above gives

$$8136.207 = 8136 + \frac{207}{1000}$$
$$= 8136 \times (10^3 \times 10^{-3}) + 207 \times 10^{-3}$$
$$= (8136 \times 10^3) \times 10^{-3} + 207 \times 10^{-3}$$
$$= (8136000 + 207) \times 10^{-3}$$
$$= 8136207 \times 10^{-3}.$$

All we have done is to create a common factor, 10^{-3}, which we can then pull out using the Distributive Law to obtain a single integer times a negative power of 10:

$$8136.207 = 8136207 \times 10^{-3}.$$

A simple rule is being applied here, namely,

> count the places to the right of the decimal point, in this case 3, remove the decimal point to obtain an integer and then multiply that integer by 10 raised to -1 times the integer count, in this case, 10^{-3}.

The next three examples illustrate this rule:

$$31.4 = 314 \times 10^{-1}, \ 1.57 = 157 \times 10^{-2}, \ \text{and} \ 963.882 = 963882 \times 10^{-3}.$$

Finally, the reader will note that this process can be applied in both directions. Given an integer and a negative of a power of 10, we can immediately convert their product to a decimal number, for example:

$$2651 \times 10^{-2} = 26.51, \quad 87946 \times 10^{-4} = 8.7946,$$

and

$$3248 \times 10^{-5} = .03448.$$

The value of the exponent tells us how many places we need in every case. Then count the places starting at the right and put in the decimal point.

18.2 Theory of Decimal Arithmetic: Multiplication

The theory underlying multiplication of decimal numbers is easy to explain and understand if we use the alternate representation of a decimal number as a product of an integer and a negative power of 10 as discussed in the last section. The key theoretical reason is that products in general are subject to the Associative and Commutative Laws and products of powers of 10 use the Addition Law for Exponents. Let's see how.

Consider multiplying two decimal numbers, p and n.

Step 1: Express the two numbers in standard decimal notation;

$$p_{1000}p_{100}p_{10}p_1.p_t p_h p_{th} \quad \text{and} \quad n_{1000}n_{100}n_{10}n_1.n_t n_h n_{th};$$

Step 2: rewrite each numeral as an integer times a negative power of 10 using the methods of the last section as in

$$p_{1000}p_{100}p_{10}p_1 p_t p_h p_{th} \times 10^{-3} \quad \text{and} \quad n_{1000}n_{100}n_{10}n_1 n_t n_h n_{th} \times 10^{-3};$$

Step 3: compute the product, m, of the two integers using the standard algorithm;

Step 4: compute the sum, k, of the two negative exponents in the powers of 10;

Step 5: introduce a decimal point into m by counting k places to the left of the rightmost digit in m.

Step 4 is an application of the Addition Law for Exponents (see §17.8.3). The introduction of the decimal point in Step 5 uses the procedure in §18.1.

This process is easily mastered by any child who knows how to multiply integers using the standard algorithm. All that has to be recognized is that the position of the decimal point in the answer is obtained by counting the total number of places to the right of the decimal points in the two factors comprising the product, since the result when multiplied by -1 will be the sum of the negative powers of 10 in the original decimal numbers.

18.2.1 Multiplication of Decimals: Numerical Examples

Example 1.

In this example, we apply the procedure in detail. Typically, a decimal multiplication problem would appear as:

$$\begin{array}{r} 5.76 \\ \times\ 4.8 \\ \hline \end{array}$$

Following Step 1, we write the two numbers as 5.76 and 4.8. Step 2 converts these to

$$576 \times 10^{-2} \quad \text{and} \quad 48 \times 10^{-1}.$$

Finding the product of the two integers amounts to performing:

$$\begin{array}{r} 576 \\ \times\ 48 \\ \hline \end{array}$$

which we already know how to do using exactly the procedure developed in Chapter 8.

Applying the Addition Law for Exponents we have

$$10^{-2} \times 10^{-1} = 10^{-3}$$

since $(-2) + (-1) = -3$. Applying these procedures gives

$$27648 \times 10^{-3} = 27.648$$

which completes Step 5.

Example 2.

Let's do another example, this time using the simple process that can easily be taught to children fluent with the standard algorithm for multiplication.

$$
\begin{array}{r}
2.6 \\
\times\ .5 \\
\hline
\end{array}
$$

Instead of going through all the steps of writing out the numbers as integers and powers of 10, let's just do the multiplication ignoring the decimal points as we would have in Chapter 8. Then we would have:

$$
\begin{array}{r}
2.6 \\
\times\ .5 \\
\hline
130
\end{array}
$$

The original problem had a total of two decimal places. What we know is that each digit to the right of the decimal point counts for one power of 10^{-1}. There are a total of two digits to the right of the decimal points in the two numbers, so there will be two places to the right of the decimal point in the answer as determined by Step 5. Thus the complete solution to the problem requires us to insert the decimal point in the answer as shown:

$$
\begin{array}{r}
2.6 \\
\times\ .5 \\
\hline
1.30
\end{array}
$$

Example 3.

We find using the simplified procedure:

$$
\begin{array}{r}
.46 \\
\times\ .13 \\
\hline
\end{array}
$$

Now, simply ignore the decimal points and perform the multiplication as integers. After multiplying by the digit 3, we have the intermediate result

$$
\begin{array}{r}
.46 \\
\times\ .13 \\
\hline
138
\end{array}
$$

The next step calls for multiplying by 1 and carefully placing the result in the *tens* column to obtain

$$
\begin{array}{r}
.46 \\
\times\ .13 \\
\hline
138 \\
+\ 46 \\
\hline
\end{array}
$$

Performing the indicated addition gives

$$
\begin{array}{r}
.46 \\
\times\ .13 \\
\hline
138 \\
+\ 46 \\
\hline
598
\end{array}
$$

At this point, the integer multiplication is complete. To place the decimal point, we count the total number of places to the right of the decimal points in the two factors. There are 4 places, so the complete solution is:

$$
\begin{array}{r}
.46 \\
\times\ .13 \\
\hline
138 \\
+\ 46 \\
\hline
.0598
\end{array}
$$

In summary, for children we can revise the 5 steps to a two-step procedure for computing the product of two decimal numbers, m and n:

Step 1: Compute $m \times n$ using the methods in Chapter 8 and ignoring the decimal points.

Step 2: Count the number of places to the right of the decimal point in m, and the number of places to the right of the decimal point in n. The total number of places in both factors is the number of places to the right of the decimal point in $m \times n$.

While these two steps are easily learned by children fluent with the standard multiplication algorithm, it should be remembered that the CCSS-M want children to understand why it works.

18.3   Theory of Decimal Arithmetic: Addition

Recall the first representation of a general decimal number as a sum given in the §18.1:

$$
n_{1000}n_{100}n_{10}n_1.n_t n_h n_{th} = n_{100} \times 100 + n_{10} \times 10 + n_1 \times 1 + n_t \times \frac{1}{10} + n_h \times \frac{1}{100} + n_{th} \times \frac{1}{1000}.
$$

Suppose we want to add two decimal numbers written in this form, for example:

$$785.316 = 7 \times 100 + 8 \times 10 + 5 \times 1 + 3 \times \frac{1}{10} + 1 \times \frac{1}{100} + 6 \times \frac{1}{1000}$$

and

$$147.375 = 1 \times 100 + 4 \times 10 + 7 \times 1 + 3 \times \frac{1}{10} + 7 \times \frac{1}{100} + 5 \times \frac{1}{1000}$$

The critical thing to notice is that in every decimal numeral

> **digits in the same place in respect to the decimal point have the same power of 10 as a multiplier**.

Thus, in our example starting at the right, the 6 and the 5 are both multiplied by $\frac{1}{1000} = 10^{-3}$. Moving one place to the left, the 1 and the 7 are both multiplied by $\frac{1}{100} = 10^{-2}$, and so forth. The fact that the multiplier is the same is the key that makes the addition algorithm work. The underlying reason is R11 of §12.6.1 which tells us how to add fractions with the **same** denominator. Thus, by carefully adding the digits in the same place starting at the right, we are always adding **apples to apples**. So for example,

$$6 \times \frac{1}{1000} + 5 \times \frac{1}{1000} = (6+5) \times \frac{1}{1000} = 11 \times \frac{1}{1000}.$$

Of course the sum of $6 + 5 = 11$ is not a single digit, but the Distributive Law and Rule 14 tell us that

$$\begin{aligned} 11 \times \frac{1}{1000} &= (10+1) \times \frac{1}{1000} = 10 \times \frac{1}{1000} + 1 \times \frac{1}{1000} \\ &= 1 \times \frac{1}{100} + 1 \times \frac{1}{1000} \end{aligned}$$

which simply means we have an additional 1 to be added in the *hundredths* place. Thus, the revised sum in the *hundredths* place is

$$1 \times \frac{1}{100} + 1 \times \frac{1}{100} + 7 \times \frac{1}{100} = (1+1+7) \times \frac{1}{100} = 9 \times \frac{1}{100}.$$

For *tenths* we have

$$3 \times \frac{1}{10} + 3 \times \frac{1}{10} = (3+3) \times \frac{1}{10} = 6 \times \frac{1}{10},$$

which completes the addition to the right of the decimal place. Now for *ones* we have

$$5 \times 1 + 7 \times 1 = (5+7) \times 1 = 12 \times 1 = 1 \times 10 + 2 \times 1,$$

for *tens*

$$1 \times 10 + 8 \times 10 + 4 \times 10 = (1 + 8 + 4) \times 10 = 13 \times 1 = 1 \times 100 + 3 \times 10$$

and finally,

$$1 \times 100 + 7 \times 100 + 1 \times 100 = (1 + 7 + 1) \times 100 = 9 \times 100.$$

Expressing this sum in the usual setup would be:

$$
\begin{array}{r}
785.316 \\
+ \quad 147.375 \\
\hline
932.691
\end{array}
$$

As the reader can see, simply by lining up the two numbers, one under the other so that the decimal points are in the same column, and then adding the columns starting at the right ensures we are repeating the processes described above. The utility of aligning the decimal points will be evident as we work some numerical examples because aligning the decimal points causes all the other places to be aligned, one above the other, for example *hundredths* above *hundredths*, and so forth.

18.3.1 Addition of Decimals: Numerical Examples

Example 4.

Add 132.165 and 25.204.

In this first example we will carefully explain each step so you understand how the theory is applied and can explain it to your child.

The standard setup for addition starts by lining up the right-most place in columns. Because both numbers have exactly three places to the right of the decimal point, the decimal points will be automatically aligned as the following shows:

$$
\begin{array}{r}
132.165 \\
+25.204 \\
\hline
\end{array}
$$

The two numbers can be written as:

$$132.165 = 1 \times 100 + 3 \times 10 + 2 \times 1 + 1 \times \frac{1}{10} + 6 \times \frac{1}{100} + 5 \times \frac{1}{1000}$$

and

$$25.204 = 2 \times 10 + 5 \times 1 + 2 \times \frac{1}{10} + 0 \times \frac{1}{100} + 4 \times \frac{1}{1000}.$$

Performing the addition by adding like terms to like terms gives:

$$132.165 + 25.204 = (1+0) \times 100 + (3+2) \times 10 + (2+5) \times 1 +$$
$$(1+2) \times \frac{1}{10} + (6+0) \times \frac{1}{100} + (5+4) \times \frac{1}{1000}$$
$$= 1 \times 100 + 5 \times 10 + 7 \times 1 + 3 \times \frac{1}{10} + 6 \times \frac{1}{100} + 9 \times \frac{1}{1000}$$
$$= 157.369.$$

The last sequence illustrates again that the algorithm works because the digits in each place have the same power of 10 as a multiplier.

Now let's repeat the process in the fluid manner we expect children to master in Grade 5. The first step is to position the two numerals one above the other so the decimal points are aligned in a column as shown:

$$\begin{array}{r} 132.165 \\ +25.204 \\ \hline \end{array}$$

We emphasize that aligning the decimal points assures that the digits from the two numerals are in the **correct place**, that is, *tenths* above *tenths*, and *hundreds* above *hundreds*, and so forth. The addition begins by applying the standard procedure to the right-most column and continuing from there. For the three columns on the right, this produces:

$$\begin{array}{r} 132.165 \\ +25.204 \\ \hline 369 \end{array}$$

where the 3 is in the *tenths* column, so it must be marked on the left with a decimal point. This is accomplished by bringing down the decimal point as shown below.

$$\begin{array}{r} 132.165 \\ +25.204 \\ \hline .369 \end{array}$$

The remainder of the process continues using the standard procedure to give:

$$\begin{array}{r} 132.165 \\ +25.204 \\ \hline 157.369 \end{array}$$

The central issue in using the revised procedure is making sure **the decimal points are aligned in a single column at the point of setup**.

The problem above does not require carrying. The next problem requires carrying and illustrates that the same procedure applies.

Example 5.

$$2.8$$
$$+5.6$$

Starting on the right, we sum the *tenths* column. The result is 14, so we record the 4 below the line and carry the 1. The 1 is, in fact, 10 *tenths*, and clearly belongs in the *ones* column. So this is where we put it. In doing this, we carry the 1 across the decimal point to the next column consisting of digits exactly as if the decimal point were not there. But, **the decimal point must be recorded below the line in the same column**, as shown in the intermediate result.

$$1$$
$$2.8$$
$$+5.6$$
$$.4$$

The last step is to sum the *ones* column with the final result shown below.

$$1$$
$$2.8$$
$$5.6$$
$$8.4$$

The examples above involve adding numbers having the same number of places to the right of the decimal point. We address the added complexity of a different number of places in our last example.

Example 6.

Add 374.9 and 8.235. The setup is

$$374.9$$
$$+\ 8.234$$

The important thing to remember here is an empty place is essentially filled with a 0. Thus,

$$374.9 = 374.900$$

as explained in the theory section. Rewriting the above using 374.900 and finding the sums in the first two columns working right to left gives:

$$374.900$$
$$+\ 8.234$$
$$34$$

384

The sum of the *tenths* column is 11, so put a 1 below the line, bring down the decimal point, and carry a 1 to the top of the *ones* column, as shown below:

$$
\begin{array}{r}
1 \\
374.900 \\
+\ 8.234 \\
\hline
.134
\end{array}
$$

The sum of the *ones* column is 13, so we write the 3 below the line in the *ones* place and carry a 1 to the top of the *tens* column, as shown:

$$
\begin{array}{r}
11 \\
374.900 \\
+\ 8.234 \\
\hline
3.134
\end{array}
$$

Summing the remaining columns gives:

$$
\begin{array}{r}
11 \\
374.900 \\
+\ 8.234 \\
\hline
383.134
\end{array}
$$

18.4 Theory of Decimal Arithmetic: Subtraction

Let's recall that in developing the full arithmetic of the integers and later the real numbers, we came to understand that subtraction was really addition of additive inverses. Thus, the theory developed for addition of decimals applies to subtraction of decimals and the reasons why things work for addition must also apply to subtraction. However, in Grade 5 negative numbers are not available and subtraction is still taught as take-away. Thus, the only problems we need consider are $p - n$ where p and n are decimal numbers satisfying

$$0 < n < p.$$

The procedures are the same as for addition, namely, line up the decimal points and do the subtraction in the usual way.

18.4.1 Subtraction of Decimals: Numerical Examples

We do two examples illustrating how the theory is applied.

Example 7.

Suppose we want to compute

$$\begin{array}{r} .854 \\ -.623 \\ \hline \end{array}$$

The subtraction procedure is exactly the one discussed in Chapter 7 with the added fact that we must keep track of the decimal point. Thus,

$$\begin{array}{r} .854 \\ - .623 \\ \hline .231 \end{array}$$

The key to the computation is that by lining the decimal points up in a single column, we ensure that all the various places, *tenths*, *hundredths*, etc., are aligned.

Example 8.

Suppose we want to find:

$$\begin{array}{r} 7 \\ - \quad 6.23 \\ \hline \end{array}$$

Since $7 > 6.23$, the calculation is feasible as take-away, but the fact that 7 is an integer appears to be a problem. However, as shown in §17.3, $7 = 7.00$ so the computation is rewritten as:

$$\begin{array}{r} 7.00 \\ - 6.23 \\ \hline \end{array}$$

Since the decimal points are aligned, we perform the subtraction using the standard procedure, borrowing across two columns and the decimal point as shown:

$$\begin{array}{ccccc} & 6 & 9 & & \\ & 7. & \cancel{0} & 10 & \\ - & 6. & 2 & 3 & \\ \hline \end{array}$$

Notice that the first borrow of 1 put an extra 10 *tenths* in the *tenths* column, and the second borrow of $\frac{1}{10}$ put an extra 10 *hundredths* in the *hundredths* column.
   At this point it is possible to complete the subtraction:

$$\begin{array}{ccccc} & 6 & 9 & & \\ & 7. & \cancel{0} & 10 & \\ - & 6. & 2 & 3 & \\ \hline & 0. & 7 & 7 & \end{array}$$

Again, we remind the reader that the decimal point is written below the line in the same column.

18.5　Theory of Decimal Arithmetic: Division

In Chapter 11 we studied division for counting numbers. Thus, given a **dividend**, n, and a **divisor**, d, the procedure developed in Chapter 11 found a **quotient**, q, and a **remainder**, r, such that

$$n = d \times q + r,$$

where $0 \le r < n$. In the case where $r = 0$, we wrote

$$n \div d = q.$$

The procedure used to find q and r was called the Division Algorithm (see §11.4). For division involving decimal numbers, the reader may remember the process learned in school which begins by moving the decimal point in the divisor to the right so that it becomes an integer. Simultaneously, the decimal point in the dividend must also be moved the same number of places to the right. For example, in $1.1 \div .02$, the initial setup is:

$$.\underline{02} \mid \overline{1.1}$$

and becomes

$$\underline{02}. \mid \overline{1\underline{10}.}$$

after moving the decimal point in the divisor. Since $55 \times 2 = 110$, the quotient is 55. Why this process produces correct results is explained in this section. Examples for children are given in the following sections.

　　In Chapter 11, the process of division was defined for integers. There we developed the Division Algorithm which solved the problem $n \div d$ for integers n and $d \ne 0$ by finding integers q and r such that

$$n = q \times d + r, \quad \text{where} \quad 0 \le r < d.$$

In Chapter 12, the process of division was extended to all real numbers x and $y \ne 0$, by defining the operation of division in terms of multiplication via the relations:

$$x \div y \equiv x \times y^{-1} = \frac{x}{y}$$

where y^{-1} is the multiplicative inverse of y. In making this definition, it was carefully explained why we should think of $x \times y^{-1}$ as a quotient (see §12.5). Our task here is to extend the Division Algorithm from integers to arbitrary decimal numbers.

Consider then the two decimal numbers $x = 33.74$ and $y = 2.1$. As shown in §17.1 each decimal number has a representation as a product of an integer and a negative power of 10:

$$x = n \times 10^{-k} \quad \text{and} \quad y = d \times 10^{-p}$$

where we may assume $n, d, k, p \in \mathcal{N}$. So for x and y as given, we have:

$$x = 3374 \times 10^{-2} \quad \text{and} \quad y = 21 \times 10^{-1}.$$

Combining the general representations with the definition of division and applying the rules governing exponents and multiplication of fractions, we have

$$
\begin{aligned}
x \div y &= (n \times 10^{-k}) \times \left(d \times 10^{-p}\right)^{-1} \\
&= \frac{n}{10^k} \times \frac{10^p}{d} \\
&= \frac{n}{d} \times \frac{10^p}{10^k} \\
&= n \div d \times 10^{p-k}.
\end{aligned}
$$

The key fact that results from this manipulation is that n and d are integers so that the Division Algorithm as developed in §11.4 can be applied. Applying the last computations to $33.74 \div 2.1$ gives:

$$
\begin{aligned}
33.74 \div 2.1 &= (3374 \times 10^{-2}) \times \left(21 \times 10^{-1}\right)^{-1} \\
&= \frac{3374}{10^2} \times \frac{10^1}{21} \\
&= \frac{3374}{21} \times \frac{10^1}{10^2} \\
&= (3374 \div 21) \times 10^{-1}.
\end{aligned}
$$

The Division Algorithm applied to $3374 \div 21$ enables us to find q and r such that

$$n = d \times q + r \quad \text{where} \quad 0 \le r < d.$$

which for the values given yields:

$$3374 = 160 \times 21 + 14.$$

The reader will notice that we still have a power of 10 to deal with because

$$33.74 \div 2.1 = (3374 \div 21) \times 10^{-1} = \frac{3374}{21} \times 10^{-1}.$$

We simply include the factor 10^{-1} with the numerator, $3374 \times 10^{-1} = 337.4$ which leaves us with a revised setup for the original division problem, namely:

$$21 \mid \overline{337.4}$$

The reason this is a correct setup is that

$$\frac{3374}{21} \times 10^{-1} = \frac{337.4}{21} = \frac{33.74}{2.1}.$$

Further, for $q = 160$ and $r = 14$, as found above, we have

$$
\begin{aligned}
337.4 &= 3374 \times 10^{-1} = (21 \times 160 + 14) \times 10^{-1} \\
&= (21 \times 160) \times \frac{1}{10} + 14 \times \frac{1}{10} \\
&= 21 \times 16 + 1.4
\end{aligned}
$$

where the second line uses the Distributive Law and the third uses the Associative Law. So when we include the factor of 10^{-1} in the numerator, the quotient is 16 instead of 160 and the remainder is 1.4 instead of 14. Since 1.4 is not an integer, we split up the remainder into an integer and a residual as follows:

$$r = 1.4 = 1 + 0.4.$$

The Division Algorithm applied to the integers $337 \div 21$ gives a quotient of 16 and a remainder of 1 as shown below:

$$337 = 21 \times 16 + 1.$$

Putting all the pieces together in respect to the original division problem, $337.4 \div 21$, we have

$$337.4 = (21 \times 16 + 1) + .4$$

and we know where each piece comes from.

18.5.1 Setup of Division of Decimals: Examples

The following examples deal with setting up division problems in the context of decimals using the theory developed above.

Example 9.

Suppose we want to find $5.1 \div 3.02$. Following the above scheme, we know

$$
\begin{aligned}
5.1 \div 3.\underline{02} &\equiv (51 \times 10^{-1}) \times (302 \times 10^{-2})^{-1} \\
&= \frac{51}{302} \times 10^{2-1} = \frac{510}{302}.
\end{aligned}
$$

The procedure results in a ratio of integers and a setup that looks like:

$$302 \mid \overline{5\underline{10}}$$

It is important to realize that the power of 10 which we will multiply into the numerator may be positive or negative. We have underlined the two places we moved the decimal point to the right in the divisor and the corresponding two places we must move the decimal point in the numerator. In this case, we are forced to add a zero creating a new *ones* place.

Example 10.

As another example, consider $4.3 \div .05$. Again, following the scheme we have

$$
\begin{aligned}
4.3 \div 0.\underline{05} \quad &\equiv \quad (43 \times 10^{-1}) \times (5 \times 10^{-2})^{-1} \\
&= \quad \frac{43}{5} \times 10^{2-1} = \frac{430}{5}.
\end{aligned}
$$

Thus, the computational setup is:

$$0\underline{5} \mid \overline{4\underline{30}}$$

Example 11.

Next, consider $0.04 \div 0.3$. Applying the scheme gives:

$$
\begin{aligned}
.04 \div 0.\underline{3} \quad &\equiv \quad (4 \times 10^{-2}) \times (3 \times 10^{-1})^{-1} \\
&= \quad \frac{4}{3} \times 10^{-2+1} = \frac{.4}{3}.
\end{aligned}
$$

Thus, the computational setup is:

$$\underline{3} \mid \overline{0.\underline{4}}$$

and as the reader can see, the dividend is a decimal number in the unit interval.

These examples all follow a simple rule for the setup. Do the setup in the usual way. Move the decimal point in the divisor as many places to the right as needed to obtain an integer. Then move the decimal point the same number of places to the right in the dividend, adding zeros if and when required. The underlines show where the decimal points have been moved in the examples above.

18.5.2 The Division Algorithm Revised: Numerical Examples

Example 11 shows that we need to revise the Division Algorithm to accommodate numerators that are decimal numerals, not merely integers. As we will see, a revised computational version of the Division Algorithm that allows us to divide positive integers into decimal numbers is nothing more than the old algorithm with a properly placed decimal point. Because there is no substantial addition to the underlying theory, we will proceed by looking at some sample computations.

Example 12.

We begin with the simplest possible example: $1 \div 2$. We put this in the usual format to apply the Division Algorithm, except we write 1 as 1.0 and we place a decimal point on the quotient line **directly above the decimal point in the dividend** as a marker. Note that it is **exactly aligned** with the decimal point in 1.0.

$$2 \mid \overline{\begin{array}{cc} & . \\ 1 & .0 \end{array}}$$

Let us ask ourselves how we would proceed if the problem we were required to solve was:

$$2 \mid \overline{\begin{array}{cc} 1 & 0 \end{array}}$$

In such a case we would say 2 does not divide 1, but it does divide 10, the quotient being 5. So we place the five directly above the 0 in 10, as shown:

$$2 \mid \overline{\begin{array}{cc} & 5 \\ 1 & 0 \end{array}}$$

Given the presence of the decimal point, we simply do the same thing, except that the reasoning is marginally different. In this case we say 2 does not divide 1, but it does divide 1.0, since

$$2 \times .5 = .5 + .5 = 1.0.$$

The point is the quotient is no longer an integer but a decimal number. We record this by writing the 5 above the line in the place to the right of the decimal:

$$2 \mid \overline{\begin{array}{cc} & .5 \\ 1 & .0 \end{array}}$$

The next step in the procedure is to multiply $2 \times .5 = 1.0$ and record the result. The 0 must be in the same column as the 0 in 1.0. This is the same as simply requiring the decimal points to be aligned, so we have:

$$
\begin{array}{r}
.5 \\
\hline
2 \mid \quad 1 \quad .0 \\
1 \quad .0
\end{array}
$$

The last step in the procedure is to subtract, as shown and this produces 0, so we stop., as shown.

$$
\begin{array}{r}
.5 \\
\hline
2 \mid \quad 1 \quad .0 \\
- \quad 1 \quad .0 \\
\hline
0 \quad .0
\end{array}
$$

So the Division Algorithm tells us that $1.0 \div 2 = .5$, which is a well known result. More precisely, we see that $\frac{1}{2}$ has an exact representation as a decimal number.

The reader should notice that the procedure for computing $10 \div 2$ to obtain 5 and the procedure for computing $1.0 \div 2$ to obtain $.5$ are the **same, except for the decimal point**. So in essence we can proceed as though the decimal point were not there, as long as we position it properly during the setup.

Example 13.

Consider $.4 \div 3$. Since the divisor is already an integer, the setup is shown below:

$$
\begin{array}{r}
. \\
\hline
3 \mid \quad .4 \; 0
\end{array}
$$

Now 3 divides 4 once, so we put a 1 above the 4 on the quotient line. Since the 4 was in the *tenths* place, the 1 is also in the *tenths* place. Now compute $.1 \times 3$ and put the product below, as shown with the 3 under the 4:

$$
\begin{array}{r}
.1 \\
\hline
3 \mid \quad .4 \; 0 \\
- \quad .3 \; 0
\end{array}
$$

This process automatically lines up the decimals points in the same column. After performing the indicated subtraction, we have:

$$
\begin{array}{r}
.1 \\
\hline
3 \mid \quad .4 \; 0 \\
- \quad .3 \; 0 \\
\hline
.1 \; 0
\end{array}
$$

At this point, we know that

$$\frac{.4}{3} = .1 + \frac{.1}{3}.$$

The numerator of the fraction, 0.1 is a remainder term, exactly as we had when dividing counting numbers. And exactly analogous to that situation, we have

$$.4 = .1 \times 3 + .1.$$

To continue the computational process, we divide 3 into .10 to obtain .03. We find .03 by the same process of test multiplications analogous as was described in §11.4. The difference is that in this case we test multiply by .01, .02, .03, until we find the largest multiplier whose product with 3 is less than .1, as the following shows:

$$.03 \times 3 = .09 \le .1 < .04 \times 3 = .12.$$

The product $3 \times .03 = .09$ is recorded as shown:

$$
\begin{array}{r}
.1\ 3 \\
\hline
3\,|\quad .4\ 0 \\
-\quad .3\ 0 \\
\hline
.1\ 0 \\
-\quad .0\ 9 \\
\hline
\end{array}
$$

Alternatively, we can obtain the same result by ignoring the decimal point and observing that

$$3 \times 3 = 9 \le 10 < 4 \times 3 = 12.$$

We record the 3 above the 0 to give exactly the result obtained from the previous, more complicated calculation. Simply keeping the decimal points aligned in a column takes care of everything!

Performing the indicated subtraction gives:

$$
\begin{array}{r}
.1\ 3 \\
\hline
3\,|\quad .4\ 0 \\
-\quad .3\ 0 \\
\hline
.1\ 0 \\
-\quad .0\ 9 \\
\hline
.0\ 1 \\
\end{array}
$$

The continued computation revises our previous expression for $.4 \div 3$ to:

$$\frac{.4}{3} = .13 + \frac{.01}{3}.$$

So our new expression for .4 is

$$.4 = .13 \times 3 + .01 = .39 + .01$$

and the remainder term, .01, is now smaller by a factor of $\frac{1}{10}$.

We may make the remainder term even smaller still by continuing the division process. To do this, we merely add another 0 to the dividend, and continue as shown:

$$
\begin{array}{r}
.1\ 3 \\
\hline
3\ \big|\quad .4\ 0\ 0 \\
-\quad .3\ 0 \\
\hline
.1\ 0 \\
-\quad .0\ 9 \\
\hline
.0\ 1\ 0
\end{array}
$$

This time we want to divide 3 into .010. We find that

$$.003 \times 3 = .009 \leq .01 < .004 \times 3 = .012,$$

so we record the .003 in the quotient and the .009, below as shown:

$$
\begin{array}{r}
.1\ 3\ 3 \\
\hline
3\ \big|\quad .4\ 0\ 0 \\
-\quad .3\ 0 \\
\hline
.1\ 0 \\
-\quad .0\ 9 \\
\hline
.0\ 1\ 0 \\
-\quad .0\ 0\ 9 \\
\hline
\end{array}
$$

Performing the last subtraction gives:

$$
\begin{array}{r}
.1\ 3\ 3 \\
\hline
3\ \big|\quad .4\ 0\ 0 \\
-\quad .3\ 0 \\
\hline
.1\ 0 \\
-\quad .0\ 9 \\
\hline
.0\ 1\ 0 \\
-\quad .0\ 0\ 9 \\
\hline
.0\ 0\ 1
\end{array}
$$

where we now have:

$$\frac{.4}{3} = .133 + \frac{.001}{3}$$

394

and our revised expression for $.4$ is

$$.4 = .133 \times 3 + .001 = .399 + .001.$$

At this stage, the remainder, $.001$, has been reduced in size by another factor of $\frac{1}{10}$. But the fraction, $\frac{.001}{3}$, is unchanged, that is, it continues to be $\frac{1}{3}$ times a negative power of 10. Only the negative power of 10 multiplier changes with each additional division. And we can continue this process indefinitely by adding zeros to the dividend. However, since we are always, in essence, dividing 3 into 10 and getting a remainder of 1, the result will simply add another 3 to the quotient and leave a remainder of 1 times an increased power of $\frac{1}{10}$.

The situation above, in which the computation repeats itself forever, produces what are referred to as **repeating decimals**. The notation for repeating decimals is to place an over-line above the digits that repeat. We give some examples:

$$\frac{.4}{3} = .1\bar{3}$$

$$\frac{1.4}{9} = .1\bar{5}$$

$$\frac{1}{12} = .08\bar{3}$$

$$\frac{5}{7} = .\overline{714285}$$

In each case, the pattern of digits under the over-line repeats endlessly. The number of places in the repeating pattern can be of any length. It is a fact that every common fraction will generate a decimal expression that either terminates, as in, $\frac{1}{2} = .5$, or repeats, as in the cases above. So the reader is clear how a multi-place pattern arises, we show the computation for $\frac{5}{7}$.

Example 14.

Find the repeating decimal expression for $5 \div 7$. We apply the procedure discussed above, showing only the final step.

$$
\begin{array}{r}
.7\ 1\ 4\ 2\ 8\ 5 \\
\hline
7\mid\quad 5\ \ .0\ 0\ 0\ 0\ 0\ 0\ 0 \\
-\quad 4\ \ .9\ 0 \\
\hline
.1\ 0 \\
-\quad .0\ 7 \\
\hline
.0\ 3\ 0 \\
-\quad .0\ 2\ 8 \\
\hline
.0\ 0\ 2\ 0 \\
-\quad .0\ 0\ 1\ 4 \\
\hline
.0\ 0\ 0\ 6\ 0 \\
-\quad .0\ 0\ 0\ 5\ 6 \\
\hline
.0\ 0\ 0\ 0\ 4\ 0 \\
-\quad .0\ 0\ 0\ 0\ 3\ 5 \\
\hline
.0\ 0\ 0\ 0\ 5\ 0
\end{array}
$$

$$
\frac{5}{7} = .71428 + \frac{.000005}{7}
$$

so that

$$
5 = .71428 \times 7 + .000005.
$$

To see why the calculation repeats, consider what happens when the computation is extended. Continuing the computation means we need to find $.000005 \div 7$ which we know is equivalent $5 \div 7$ and multiplying the result by 10^{-6}. We have already computed $\frac{5}{7}$, and the result is shown. As we can see, the same result will be achieved if we repeat the cycle, only the place of the digits will change. Alternatively, we see that to continue the computations shown, when we bring down the next 0, we will be dividing 7 into $.0000050$, which means we will put a 7 in the quotient one place to the right of the 5, and end up subtracting $.0000049$ from $.0000050$, which is essentially where we started.

18.6 Notations for Arbitrary Real Numbers

Consider again the types of real numbers we know exist. We have integers, we have fractions that are not integers, but can be expressed as ratios of integers, and we have still other numbers like $\sqrt{2}$, or π, that arise in geometry, or elsewhere, and which have been shown to be neither integers, nor fractions. If we are to do the things required in the modern world, we need to have notations for all of these numbers.

Consider the rational numbers. As we have seen, they comprise the integers and ratios of integers. The decimal system provides a notation for each rational number.

In some cases, the division procedure terminates in a finite number of steps[2] leaving a remainder of 0. In such a case, we know the decimal notation is exact. All integers and common fractions having terminating decimals have an exact decimal representation. But repeating decimals, such as $1.\overline{3}$, and irrational numbers, such as $\sqrt{2}$ and π do not. How do we know this?

In the case of a repeating decimal, we know it represents a fraction $\frac{m}{n}$. Indeed, there is a procedure for finding the fraction from the repeating decimal, although we will not give it here. We also know that every exact decimal number is also representable by a fraction having a denominator that is a power of ten. If a fraction is equal to an exact decimal number, as in the case of

$$\frac{1}{2} = 0.5,$$

all we have to do is subtract one from the other and we must get 0. For a fraction whose representation is a repeating decimal, for example $\frac{1}{3}$, this test can never be satisfied because whatever exact decimal number we choose, when we subtract it from a fraction having a repeating decimal, the difference will not be 0. We can make the result as small as we please, but we can never make it zero.

Consider irrational numbers. Such numbers are defined by some property of their behavior. For example, $\sqrt{2}$ is defined by the equation:

$$(\sqrt{2})^2 = 2.$$

So the question is: Is there a decimal number whose square is 2? Since every decimal number is also a fraction, the question would be answered by showing that there either is, or is not, a fraction whose square is 2. As indicated earlier, it was known to the Greeks that no fraction had this property. Thus if we take any fraction $\frac{n}{m}$ whatsoever, we find

$$\left(\frac{n}{m}\right)^2 - 2 \neq 0.$$

Again, we can make the residual as small as we please. But we cannot make it zero! Thus, 1.414213562 may be what your calculator asserts is the square root of 2, but it is only an approximation. Although your calculator may tell you that the square of this number is 2, it is not and you can check this yourself by doing the multiplication long-hand using the standard procedure. The reason your calculator may "think" the square is 2 is because your calculator is limited in accuracy.

But there is one other important point about irrationals: their decimal notation is an infinite non-repeating decimal. Because of this, to identify an irrational properly,

---

[2]The definition of **finite** is that we can, in principle, count the number of digits.

we have to do so with a property. For example, by saying that when it is squared, we get 3, or that when we multiply the length of the diameter of a circle by this number, we would get the circumference provides a property that allows us to identify $\sqrt{3}$ and π, respectively. We have to do this because we do not know what digits are very far out in the decimal expansion. Indeed, it is only since the advent of modern computers that the millionth digit in the decimal expansion for π became known. Since all the digits in the decimal expansion for π, and the like, will never be known, the only way to identify these numbers is by a property of their behavior. And the only way we can give them a name is by assigning a symbol like π, or an operational form like $\sqrt{3}$.

Chapter 19

A Last Word

If you have obtained the CCSS-M from the website, you will have noticed that not every topic has been covered in our book. What has been covered are the essential topics of arithmetic that will enable you to help your child through any rough spots with the new curriculum, both in elementary and higher grades.

For topics we have not covered, to help your child, start by reading the material and examples on the topic in the child's textbook. It is supposed to be written at a level your child can understand, so you should have no difficulty coming to terms with it using the knowledge gained from this book.

Our intention is to provide help. To this end, we have set up an email address and we will deal with questions to the degree that we are able. The address is:

gaskillmath@gmail.com

We know that the requirements of the new CCSS-M are substantial. But we also know that the material is what your child needs to succeed in the modern world.

Index

25090313R00233

Made in the USA
Middletown, DE
17 October 2015